“十四五”国家重点出版物出版规划项目·重大出版工程

中国学科及前沿领域2035发展战略丛书

学术引领系列

国家科学思想库

中国分子细胞科学与技术2035发展战略

“中国学科及前沿领域发展战略研究（2021—2035）”项目组

科学出版社

北　京

内 容 简 介

分子细胞科学的发展有助于探索生命的本质，为生命的改造提供更加优化的技术策略，为破解生命科学各学科中的重大科学问题提供更全面、更深入的支撑。《中国分子细胞科学与技术 2035 发展战略》面向世界科技前沿，聚焦重大科学问题，系统梳理了分子细胞科学与技术的科学意义、战略价值与发展规律和发展现状，并从 9 个方面研究了其关键科学问题、关键技术问题和重要发展方向，包括：基因组的结构与演化，生物分子的结构、功能与设计，生物分子模块的组装与功能性机器，亚细胞结构的形成与相互作用，细胞类型的区分与确定，细胞命运的决定及其可塑性，细胞间通信及细胞与微环境的互作与功能，分子生物网络与“数字化”细胞，细胞的人工改造。同时，对相关领域的发展提出了相关政策建议。

本书为相关领域战略与管理专家、科技工作者、企业研发人员及高校师生提供了研究指引，为科研管理部门提供了决策参考，也是社会公众了解分子细胞科学与技术发展现状及趋势的重要读本。

图书在版编目（CIP）数据

中国分子细胞科学与技术 2035 发展战略 / “中国学科及前沿领域发展战略研究（2021—2035）”项目组编．—北京：科学出版社，2023.5
（中国学科及前沿领域 2035 发展战略丛书）
ISBN 978-7-03-075246-8

Ⅰ. ①中…　Ⅱ. ①中…　Ⅲ. ①细胞生物学－分子生物学－发展战略－研究－中国　Ⅳ. ① Q7

中国国家版本馆 CIP 数据核字（2023）第 047114 号

丛书策划：侯俊琳　朱萍萍
责任编辑：牛　玲　田明霞 / 责任校对：何艳萍
责任印制：赵　博 / 封面设计：有道文化

科学出版社 出版
北京东黄城根北街 16 号
邮政编码：100717
http://www.sciencep.com
北京市金木堂数码科技有限公司印刷
科学出版社发行　各地新华书店经销
*
2023年5月第　一　版　开本：720×1000　1/16
2025年3月第三次印刷　印张：28 1/2
字数：450 000

定价：198.00元

（如有印装质量问题，我社负责调换）

“中国学科及前沿领域发展战略研究（2021—2035）”

联合领导小组

组　长　常　进　李静海

副组长　包信和　韩　宇

成　员　高鸿钧　张　涛　裴　钢　朱日祥　郭　雷
杨　卫　王笃金　杨永峰　王　岩　姚玉鹏
董国轩　杨俊林　徐岩英　于　晟　王岐东
刘　克　刘作仪　孙瑞娟　陈拥军

联合工作组

组　长　杨永峰　姚玉鹏

成　员　范英杰　孙　粒　刘益宏　王佳佳　马　强
马新勇　王　勇　缪　航　彭晴晴

《中国分子细胞科学与技术 2035 发展战略》

项 目 组

组 长 李 林 张学敏

成 员（以姓氏笔画为序）

丁建平 王 玥 朱学良 刘小龙 许琛琦
苏 燕 李劲松 李党生 吴家睿 吴琼琼
张 钰 陈剑峰 陈洛南 周金秋 孟飞龙
胡光晶 施慧琳 徐 萍 葛高翔 程 红

总　序

党的二十大胜利召开，吹响了以中国式现代化全面推进中华民族伟大复兴的前进号角。习近平总书记强调“教育、科技、人才是全面建设社会主义现代化国家的基础性、战略性支撑”[①]，明确要求到 2035 年要建成教育强国、科技强国、人才强国。新时代新征程对科技界提出了更高的要求。当前，世界科学技术发展日新月异，不断开辟新的认知疆域，并成为带动经济社会发展的核心变量，新一轮科技革命和产业变革正处于蓄势跃迁、快速迭代的关键阶段。开展面向 2035 年的中国学科及前沿领域发展战略研究，紧扣国家战略需求，研判科技发展大势，擘画战略、锚定方向，找准学科发展路径与方向，找准科技创新的主攻方向和突破口，对于实现全面建成社会主义现代化“两步走”战略目标具有重要意义。

当前，应对全球性重大挑战和转变科学研究范式是当代科学的时代特征之一。为此，各国政府不断调整和完善科技创新战略与政策，强化战略科技力量部署，支持科技前沿态势研判，加强重点领域研发投入，并积极培育战略新兴产业，从而保证国际竞争实力。

擘画战略、锚定方向是抢抓科技革命先机的必然之策。当前，新一轮科技革命蓬勃兴起，科学发展呈现相互渗透和重新会聚的趋

① 习近平. 高举中国特色社会主义伟大旗帜 为全面建设社会主义现代化国家而团结奋斗——在中国共产党第二十次全国代表大会上的报告. 北京：人民出版社，2022：33.

势，在科学逐渐分化与系统持续整合的反复过程中，新的学科增长点不断产生，并且衍生出一系列新兴交叉学科和前沿领域。随着知识生产的不断积累和新兴交叉学科的相继涌现，学科体系和布局也在动态调整，构建符合知识体系逻辑结构并促进知识与应用融通的协调可持续发展的学科体系尤为重要。

擘画战略、锚定方向是我国科技事业不断取得历史性成就的成功经验。科技创新一直是党和国家治国理政的核心内容。特别是党的十八大以来，以习近平同志为核心的党中央明确了我国建成世界科技强国的“三步走”路线图，实施了《国家创新驱动发展战略纲要》，持续加强原始创新，并将着力点放在解决关键核心技术背后的科学问题上。习近平总书记深刻指出：“基础研究是整个科学体系的源头。要瞄准世界科技前沿，抓住大趋势，下好‘先手棋’，打好基础、储备长远，甘于坐冷板凳，勇于做栽树人、挖井人，实现前瞻性基础研究、引领性原创成果重大突破，夯实世界科技强国建设的根基。”②

作为国家在科学技术方面最高咨询机构的中国科学院（简称中科院）和国家支持基础研究主渠道的国家自然科学基金委员会（简称自然科学基金委），在夯实学科基础、加强学科建设、引领科学研究发展方面担负着重要的责任。早在新中国成立初期，中科院学部即组织全国有关专家研究编制了《1956—1967年科学技术发展远景规划》。该规划的实施，实现了“两弹一星”研制等一系列重大突破，为新中国逐步形成科学技术研究体系奠定了基础。自然科学基金委自成立以来，通过学科发展战略研究，服务于科学基金的资助与管理，不断夯实国家知识基础，增进基础研究面向国家需求的能力。2009年，自然科学基金委和中科院联合启动了“2011—2020年中国学科发展

② 习近平. 努力成为世界主要科学中心和创新高地 [EB/OL]. (2021-03-15). http://www.qstheory.cn/dukan/qs/2021-03/15/c_1127209130.htm[2022-03-22].

战略研究”。2012 年，双方形成联合开展学科发展战略研究的常态化机制，持续研判科技发展态势，为我国科技创新领域的方向选择提供科学思想、路径选择和跨越的蓝图。

联合开展“中国学科及前沿领域发展战略研究（2021—2035）”，是中科院和自然科学基金委落实新时代“两步走”战略的具体实践。我们面向 2035 年国家发展目标，结合科技发展新特征，进行了系统设计，从三个方面组织研究工作：一是总论研究，对面向 2035 年的中国学科及前沿领域发展进行了概括和论述，内容包括学科的历史演进及其发展的驱动力、前沿领域的发展特征及其与社会的关联、学科与前沿领域的区别和联系、世界科学发展的整体态势，并汇总了各个学科及前沿领域的发展趋势、关键科学问题和重点方向；二是自然科学基础学科研究，主要针对科学基金资助体系中的重点学科开展战略研究，内容包括学科的科学意义与战略价值、发展规律与研究特点、发展现状与发展态势、发展思路与发展方向、资助机制与政策建议等；三是前沿领域研究，针对尚未形成学科规模、不具备明确学科属性的前沿交叉、新兴和关键核心技术领域开展战略研究，内容包括相关领域的战略价值、关键科学问题与核心技术问题、我国在相关领域的研究基础与条件、我国在相关领域的发展思路与政策建议等。

三年多来，400 多位院士、3000 多位专家，围绕总论、数学等 18 个学科和量子物质与应用等 19 个前沿领域问题，坚持突出前瞻布局、补齐发展短板、坚定创新自信、统筹分工协作的原则，开展了深入全面的战略研究工作，取得了一批重要成果，也形成了共识性结论。一是国家战略需求和技术要素成为当前学科及前沿领域发展的主要驱动力之一。有组织的科学研究及源于技术的广泛带动效应，实质化地推动了学科前沿的演进，夯实了科技发展的基础，促进了人才的培养，并衍生出更多新的学科生长点。二是学科及前沿领域的发展促进深层次交叉融通。学科及前沿领域的发展越来越呈现出多学科相互渗透的发展态势。某一类学科领域采用的研究策略和技术体系所产生的基础理论与方法论成果，可以作为共同的知识基础适用于不同学科领域的多个研究方向。三是科研范式正在经历深刻变革。解决系统性复杂问题成为当前科学发展的主要目标，导致相应的研究内容、方法和范畴等的改变，形成科学研究的多层次、多尺度、动态化的基本特征。数据驱动的科研模式有力地推动了新时代科研范式的变革。四是科学与社会的互动更加密切。发展学科及前沿领域愈加重要，与此同时，“互联网 +”正在改变科学交流生态，并且重塑了科学的边界，开放获取、开放科学、公众科学等都使得越来越多的非专业人士有机会参与到科学活动中来。

“中国学科及前沿领域发展战略研究（2021—2035）”系列成果以“中国学科及前沿领域 2035 发展战略丛书”的形式出版，纳入“国家科学思想库 - 学术引领系列”陆续出版。希望本丛书的出版，能够为科技界、产业界的专家学者和技术人员提供研究指引，为科研管理部门提供决策参考，为科学基金深化改革、“十四五”发展规划实施、国家科学政策制定提供有力支撑。

在本丛书即将付梓之际，我们衷心感谢为学科及前沿领域发展战略研究付出心血的院士专家，感谢在咨询、审读和管理支撑服务方面付出辛劳的同志，感谢参与项目组织和管理工作的中科院学部的丁仲礼、秦大河、王恩哥、朱道本、陈宜瑜、傅伯杰、李树深、李婷、苏荣辉、石兵、李鹏飞、钱莹洁、薛淮、冯霞，自然科学基金委的王长锐、韩智勇、邹立尧、冯雪莲、黎明、张兆田、杨列勋、高阵雨。学科及前沿领域发展战略研究是一项长期、系统的工作，对学科及前沿领域发展趋势的研判，对关键科学问题的凝练，对发展思路及方向的把握，对战略布局的谋划等，都需要一个不断深化、积累、完善的过程。我们由衷地希望更多院士专家参与到未来的学科及前沿领域发展战略研究中来，汇聚专家智慧，不断提升凝练科学问题的能力，为推动科研范式变革，促进基础研究高质量发展，把科技的命脉牢牢掌握在自己手中，服务支撑我国高水平科技自立自强和建设世界科技强国夯实根基做出更大贡献。

“中国学科及前沿领域发展战略研究（2021—2035）”

联合领导小组

2023年3月

前　言

根据国家自然科学基金委员会-中国科学院学科发展战略研究工作联合领导小组第八次会议的决定，“分子细胞科学与技术前沿领域发展战略研究”入选“中国学科及前沿领域发展战略研究（2021—2035）”项目，于2020年1月正式启动。

项目启动后，按照国家自然科学基金委员会与中国科学院对学科发展战略研究项目要求，由项目负责人李林和张学敏领导，汇聚了国内分子细胞科学研究领域的一批优秀科学家，特别是中青年科学家组成项目组。在研究过程中，项目组召开了若干次小规模、高层次研讨会，梳理与回顾了发展战略研究的背景和要求，重点讨论了分子细胞科学与技术前沿领域的科学内涵和研究框架，研究提出了以本前沿领域科学问题为主线并值得关注的多个方向，分享交流了不同方向研究的学术思想和意见建议，同时也依托文献情报专家开展了基于文献情报的分析与支撑工作。

2020年11月初，项目组克服新冠疫情带来的诸多不利影响，形成战略研究报告初稿，同时组织召开以“细胞可塑性调控与细胞工程应用”为主题的香山科学会议第685次学术讨论会，围绕相关内容进行深入讨论。2020年11月27日，国家自然科学基金委员会-中国科学院学科发展战略研究联合工作组组织举办了学科发展战略项目进展情况交流会，项目组在此次会议后结合专家意见建议在已有基础上进行了修订。2021年5月完成战略研究报告第二稿，9月完成第三稿。2022年3月项目通过总体结题验收。

按照国家自然科学基金委员会与中国科学院的整体部署，本发展战略报告第一篇（第一章、第二章）包含分子细胞科学与技术前沿领域的科学意义与战略价值、发展历程、发展现状和竞争力分析等内容；第二篇（第三章至第十一章）重点包含分子细胞科学与技术前沿领域的关键科学问题、关键技术问题、重要研究方向和我国发展方向建议等部分。具体而言，第一篇主要依托文献情报专家王玥、徐萍及其领导的研究团队完成，第二篇研究框架则分解为相对独立但逻辑上具有相互联系的9个主题，由不同的项目组成员撰写完成，具体分工如下：①基因组的结构与演化——孟飞龙、石建涛、刘行、刘珈泉、柳欣、姜雨、童明汉等；②生物分子的结构、功能与设计——程红、吴立刚、吴兴中、汪胜、宋保亮、张少庆、陆豪杰、屈良鹄等；③生物分子模块的组装与功能性机器——丁建平、丛尧、杨荟、张少庆、陈勇等；④亚细胞结构的形成与相互作用——李劲松、王晓晨、朱学良、孙丽明、李丕龙、李栋、吴聪颖、邱小波、张宏、陈佺、陈玲玲、胡荣贵、胡俊杰、俞立、酒亚明、温文玉、鲍岚等；⑤细胞类型的区分与确定——吴家睿；⑥细胞命运的决定及其可塑性——陈剑峰、王红艳、李劲松、李振斐、张雷、苟兰涛、周波、姜海、高栋、惠利健、童明汉、曾艺、曾安等；⑦细胞间通信及细胞与微环境的互作与功能——葛高翔、王红艳、孙丽明、李振斐、杨巍维、邹卫国、沈义栋、宋保亮、欧阳波、周斌、胡苹、隋鹏飞等；⑧分子生物网络与“数字化”细胞——陈洛南、石建涛、胡荣贵等；⑨细胞的人工改造——许琛琦、王广川、王锋、王皞鹏、李天晴、黄行许、韩为东、逯冬冬等。

项目组希望通过此项研究，不但能对分子细胞科学这一新兴学科的定义和内涵做出诠释，而且能在已有分子生物学与细胞生物学基础上展现更为前沿与交叉的学科发展态势——利用学科交叉新技术新手段，对生物分子、功能元件和细胞进行定性到定量、延滞到实时、静态到动态、离体到原位、单一到网络、局部到整体的研究，揭示其数量、形态、结构和功能的因果关系，为实现对细胞的操控与人工改造提供理论基础和技术手段。

限于篇幅，参与撰写、咨询和研讨的专家无法一一列出姓名，在此一并表示感谢。此外，本项目得到了依托单位中国科学院分子细胞科学卓越创新中心的大力支持，特表感谢。

由于认识局限，本报告难免存在缺陷与不足，敬请广大读者批评指正。

李 林 张学敏
《中国分子细胞科学与技术2035发展战略》
项目组组长
2022年12月

摘　要

细胞是生命体结构和功能的基本单位。生物的生长、发育、繁殖与进化等一切生命活动都以细胞为基础。多细胞生物包含多种形态与功能各异的细胞，所有这些细胞都是由单个的受精卵分裂、增殖、分化而来的。为什么受精卵可以产生所有类型的细胞？为什么在生命个体发育过程中细胞的形态和功能会发生转变？为什么某种组织的细胞（如神经细胞、肌肉细胞）一旦形成就会稳定维持它的特性？为什么有机体的正常细胞会转变为不正常的癌细胞或其他病变细胞？关于细胞的这些待解之谜是当代生命科学基础研究的热点前沿。作为生命的基础，细胞里包含众多种类的分子，如核酸、蛋白质、糖类和脂类等，这些生物分子相互作用，组装成结构与功能不一的复合体机器或亚细胞结构，并在细胞的基本生命活动过程中展现了纷繁复杂的各种生命现象。绘制细胞基本生命活动的分子调控图谱是生命科学研究的终极目标之一，而要真正实现这一目标，除了运用传统的分子生物学和细胞生物学研究方法，还需要进行多学科交叉，借助和发展不同于经典实验生命科学的新技术和新方法，系统地研究细胞的生命本质与活动规律。

分子细胞科学是以分子生物学和细胞生物学为基础，与数学、物理、化学、信息科学、医学等交叉融合，研究细胞生命活动规律及其分子机制的前沿性、交叉性新兴学科。利用学科交叉新技术、新手段，对生物分子、功能元件和细胞进行定性到定量、延滞到实时、静态到动态、离体到原位、单一到网络、局部到整体的研究，揭示其数量、形态、结构和功能的因果关系，为实现对细胞的操控与人工改造提供理论基础和技术手段。

发展分子细胞科学具有极大的科学意义与战略价值。在推动基础科学发展方面，分子细胞科学的发展将助力提高探索生命本质的深度，使对细胞结构和功能的探索实现从“对分子机制的单一表象分析”到“从多个维度深度阐释机理”的转变；同时为生命的改造提供更加优化的技术策略，进一步推动对细胞内分子表达和细胞功能的精准操控；此外，分子细胞科学的发展还为破解生命科学各学科中的重大科学问题提供更全面、更深入的支撑，将推进生命科学领域各学科的快速发展。在推动医药健康、农业和工业领域发展方面，分子细胞科学也具有重要的战略价值。第一，分子细胞科学将推动实现更精细、更真实地揭示人体生长发育和疾病发生发展机制和规律，可为疾病精准诊断和先进治疗的研发应用提供源头理论和创新策略。第二，分子细胞科学将进一步推动生物育种技术的优化和升级，推进高效、精准农业育种，同时也将为绿色农药创制和“细胞农业”产品开发等前沿方向的发展带来巨大的机遇。第三，分子细胞科学的发展还将为工业产品的绿色生物合成奠定理论和技术基础，助力设计构建酶分子、细胞和多细胞体系，优化调控代谢通路，最终实现对合成过程的人工控制。

分子细胞科学的发展主要经历了两个阶段。其一，分子生物学与细胞生物学交叉融合推动对细胞认识深度的不断跃迁：在传统细胞生物学的基础上，随着以基因组学为代表的组学研究技术促进细胞生命活动规律的全面解析、细胞工程技术的出现使细胞改造成为现实，遗传信息表达和调控、细胞物质运输、细胞信号转导、细胞命运调控等领域的研究逐渐勃兴，细胞生物学开始向分子细胞生物学的方向发展。其二，多学科交叉引领分子细胞科学前沿学科的出现及快速发展：单细胞、单分子技术，分子原位分析技术，成像技术，分子影像学，活细胞成像技术，多组学技术的出现和迭代进步使得这一时期的研究从平均化研究迈向单细胞、单分子差异化研究，从离体研究迈向携带空间信息的原位研究，从生物大分子的静态结构研究向动态分子电影制作发展，从断层式的分子细胞研究迈向无创、实时的观测，从单一分子的研究尺度向多层网络的整合分析发展，推动分子细胞生物学迈入分子细胞科学新阶段。

分子细胞科学作为生命科学的前沿领域，是全球科研的热点。在科研资助布局方面，美国国家科学基金会、英国医学研究理事会等国家科研资助机构均设置了专门的部门管理对分子细胞领域的资助，重点资助生物学与物理、化学、数学、计算科学等多学科交叉，增进对分子细胞基本功能、分子调控网络及机制的全面认识，并进一步将相关认识与生理学、病理学、医学相结合，以加深对疾病发生与治疗过程中应答机制的理解，进而推进药物靶标的发现和验证，以及新疗法的开发。美国国立卫生研究院也非常重视分子细胞科学相关基础研究，资助了4D细胞核组（4D Nucleome）计划和人类生物分子图谱计划（Human Biomolecular Atlas Program，HuBMAP）等专项计划，旨在推进活细胞追踪技术、单细胞测序技术、成像技术等研究单细胞活动的技术的发展，并在此基础上探究分子细胞特征及其相关变化对发育和各类疾病的影响。此外，美国、日本、欧盟等科技发达国家/组织相继出台了分子细胞科学领域的专项发展规划，包括全球大型国际合作项目人类细胞图谱（Human Cell Atlas，HCA）计划、日本活细胞图谱战略建议、日本4D细胞组战略建议、欧盟新“FET旗舰计划”的筹备行动——疾病发生发展机制研究（LifeTime）等。我国主要通过国家自然科学基金、国家重点研发计划、中国科学院战略性先导科技专项等对分子细胞科学领域进行资助。在科研论文产出方面，通过对作为分子细胞科学基础和重要组成部分的分子细胞生物学进行定量分析发现，1980 ～ 2020 年全球共发表研究论文 680 055 篇，始终维持较快的发展速度。2011 ～ 2020 年是我国分子细胞生物学研究领域快速发展阶段，论文数量从 2011 年的 2441 篇增长至 2020 年的 12 560 篇，年均增长率达到 19.96%。这 10 年期间，我国共发表分子细胞生物学论文 70 756 篇，占该时间段全球发表的分子细胞生物学论文总量的 24.93%，位居全球第 2 位；在全球分子细胞生物学论文数量排名前 10 位的国家中，我国论文的总被引频次居第 2 位，篇均被引频次居第 10 位。

本报告围绕分子细胞科学的 9 个方面（即：①基因组的结构与演化；②生物分子的结构、功能与设计；③生物分子模块的组装与功能性机器；④亚细胞结构的形成与相互作用；⑤细胞类型的区分与确定；⑥细胞命运的决定及其可塑性；⑦细胞间通信及细胞与微环境的互作与功能；⑧分子生物网络与“数字化”细胞；⑨细胞的人工改造），研究和阐述了各自的关键科学问题、发展现状与趋势分析，并提出我国该学科发展战略与重点方向。

一、基因组的结构与演化

基因组包含的遗传信息是生命活动的基础，而染色体是真核细胞遗传信息储存和传递的载体。在生物发育的过程中，遗传信息程序性地操控具有细胞特异性的基因表达，进而形成复杂的生命体。基因组线性编码如何决定多彩的细胞生命活动、生命体如何解决基因组序列的完整性维持与变异之间的矛盾以实现持续发展、能否通过构建人工合成生命探索生命奥秘并创造出自然界不存在的生命形式，这些都是基因组的结构与演化研究的重点和前沿。

在基因组结构与功能之间的关系方面，从化学结构、三维立体结构及空间分布（区室化）三个不同视角出发并全面揭示不同维度下的基因组分子行为是未来研究趋势。其中，对化学修饰的探索是揭示基因组表观动态调控机制的关键，染色质高级结构三维构象转化的生物学作用机制亟待明确，染色质元件区室化机制与功能的研究对于解开更多基因组结构谜题必不可少。在基因组序列的完整性维持与变异的分子机制方面，对DNA损伤及修复机理的理解仍不够深入，对RNA等生物大分子参与的DNA修复过程及其他新因子的新机制研究刚刚起步。探索长期“随机”突变及发育过程中程序性DNA损伤的运行机制及其生物学功能，有望揭示癌症深层次的致病机理并解答生命体发展进化的机制与走向。在构建人工合成生命方面，研究者已发现很大程度简化的基因组也可以维持生命运转，高度可控的人工生命系统的建立不仅可能极大地加速生命机制的揭示，也使创造新生命成为可能。目前，研究者已经成功扩充密码子，并将非天然碱基融入遗传信息的储存、解读、传代等功能中，未来有可能实现“镜像生命”等的人工合成。

建议我国通过整合多学科理论和技术，在“机制解析”与“人工合成”两个研究方向并进，为疾病诊疗提供新的思路，提升我国在该研究领域的国际影响力。

二、生物分子的结构、功能与设计

生物分子是自然存在于生物体内的分子总称，是生命的承载体和生命活动的执行者。生物分子包括核酸、蛋白质、糖类、脂类等大分子，也包括代谢物等小分子，不同类别的分子通过相互作用，协同发挥功能。生物分子的特异性识别是其发挥功能的基本前提，高精度解析生物分子的三维结构是本领域的重点之一。另外，研究生物分子代谢中的各个过程及其相互之间的耦联机制，从而解析生物分子的功能是本领域的又一核心。基于对生物分子的结构与功能的理解，根据需求设计和改造生物分子正在成为可能。

在核酸研究方面，大量新型RNA的发现及对其功能的深入理解正不断刷新研究者的认知。基因沉默、基因编辑和RNA测序等新技术，使核酸研究成为生命科学领域发展最快的方向之一。RNA操纵与编辑正在成为治疗遗传疾病的最有前景的手段之一。在蛋白质研究方面，近年来冷冻电镜技术的发展加速了对蛋白质活化过程中构象变化的认识，解析了其发挥生物学功能的分子机理。计算机模拟蛋白质结构与功能、人工设计蛋白质成为目前研究蛋白质的热门方向，并正将其应用于药物研发、疫苗研发、化学催化等多个领域。在糖类研究方面，复杂多变的糖基化修饰可通过与不同蛋白质、脂质甚至RNA结合，由体内的识别系统呈递信息。由糖链介导的生物分子相互作用在疾病诊治中有巨大的应用前景。在脂类研究方面，脂类代谢调控和代谢酶的新机制、新型及经典细胞器在代谢营养感知和物质运输中的作用、代谢器官的生理病理意义和物质交互方式等是重要的研究方向。

建议我国以解析各种生物分子结合靶标及其结构为基础，以生物分子加工、代谢及识别等为线索，以生物分子调控机制和细胞功能为核心，通过多学科交叉，整合多组学信息和关键技术，研发生物分子检测新技术、新方法，阐明生物分子代谢与调控的基本机制和功能。

三、生物分子模块的组装与功能性机器

细胞中纷繁芜杂的生物学功能并不是由单个生物大分子独立完成的，而是由成千上万种生物大分子通过相互作用、动态组装形成的生物大分子机器来执行的。生物大分子机器在能量代谢、物质运输、信号感知与应答、遗传信息的传递与维护、蛋白质的合成与降解，以及微生物-宿主互作与免疫应答等过程中发挥着重要而多样化的生物学作用。

揭示生物大分子机器的组装规律、修饰和动态调控机理，以及其在细胞内的定位和转运过程，可以阐明生物大分子机器在体内高度协同工作的规律，为实现生物大分子机器的设计、模拟及操控奠定基础。近年来迅速发展的“从头设计”生物大分子机器，不仅可以用于检验现有关于生物大分子机器功能和分子机制解释的正确性，还可以在很大程度上拓展大分子机器功能上的多样性。基于这些基础研究所获得的关于生物大分子机器的功能和机制的知识和研究技术，有助于揭示生物大分子机器的功能失调在疾病发生发展过程中的作用和机制，以及细菌和病毒与宿主细胞相互作用过程中生物大分子机器的功能和作用机制，为疫苗和药物的设计和研发，以及疾病的预防、诊断和治疗奠定基础。

鉴于目前国内外对生物大分子机器的研究进展和面临的挑战，建议我国积极鼓励开展生物大分子机器原位结构和功能的新技术和新方法研究，建立和发展可对细胞“天然分子景观”进行原位、高分辨率、实时动态结构描述的核心技术。这些新技术将能极大地促进人们对生理条件和病理条件下生物大分子机器发挥生物学功能的机制的理解。

四、亚细胞结构的形成与相互作用

亚细胞结构包括细胞膜、各种细胞器、细胞骨架、细胞核等。细胞内的细胞器除了经典的膜性细胞器（如内质网、高尔基体、溶酶体、内吞体、线粒体、自噬体等），还存在大量无膜细胞器——生物大分子凝聚体。细胞器通过精细分工、相互协作和密切接触，形成细胞器互作网络，实现快速的物质交换和信息交流，执行多种生物学过程。细胞器互作的紊乱与疾病发生发展密切相关。

在内膜系统方面，目前我国在膜转运的精细分选调控、内膜细胞器互作的生理功能和内膜完整性的维持等方向上已形成特色，未来将由单个细胞器研究转向多个细胞器的集成研究，聚焦内膜系统协同互作的精细调控机制和生理功能及意义。同时，内膜系统膜结构的完整性维持、修复与降解也将成为新的研究热点和前沿。在无膜细胞器方面，理解无膜细胞器的形成、超微结构、动态组装以及耦联的功能调控，对认识无膜细胞器发挥功能的分子机制，对多种疾病的病理机制研究及相关诊疗有着重要的意义。未来需要从生物大分子“相分离”和“相变”切入研究其形成和动态平衡调控，在更高的时空分辨率尺度下解析细胞核无膜细胞器的超微结构、分子组成，从分子层面理解细胞核无膜细胞器组装过程与功能调控的机制。

建议我国发展在体高时空分辨率解析技术以提升亚细胞结构的研究水平，包括发展基于光学工程、计算机、自动化、物理、生物等多学科交叉融合的生物显微成像技术。

五、细胞类型的区分与确定

细胞是生命的基本组成单元。多细胞生物在个体发育过程中，随着细胞数量的增加，细胞的类型通过细胞分化的方式不断增加。体细胞按照其形态和功能特点可以分为相应的细胞类型，如在发育成完整个体的人体中，估计有200多种不同的体细胞类型。如果想要了解多细胞生物不同的生命活动与功能，则需要认识其相应的细胞类型。

在细胞类型的遗传学基础方面，要区分和鉴定多细胞生物的细胞类型的一个重要任务就是要确定细胞谱系。如何构建复杂生物体的细胞谱系，尤其是人类的细胞谱系，是当前研究者面对的重大挑战。在细胞类型与环境之关系方面，基因型与环境之相互作用形成表型，细胞类型的形成也受到环境的影响，环境还能够导致体细胞基因组序列产生随机变异。如何在区分和确定细胞类型的研究中鉴定出环境引发的基因组异常影响是研究者面临的另一个重大挑战。

在细胞类型的分型技术与标准方面，过去研究者按照体细胞的形态和功能特点来进行细胞类型的划分，如今则按照分子特点来进行细胞分子分型。单细胞转录组测序是目前应用最广的分子分型技术，近年来国际上已经通过单细胞转录组测序技术获得了大量的人类细胞图谱信息。细胞谱系示踪可以追踪细胞在体的类型演化过程，弥补单细胞转录组测序技术丢失的谱系转化信息，这也是国际上细胞图谱研究的重要内容。在病理（如衰老、肿瘤）条件下的细胞图谱变化是近年来的研究热点。同时，本研究领域正从强调离体的、静态的单细胞分析发展到在体的、动态的细胞类型研究，可以使研究者以前所未有的时空分辨率观察组织的三维结构。

建议我国开展基于组学技术的细胞类型鉴定与分子分型研究、个体发育过程中细胞谱系的确定与追踪研究、复杂性疾病中的细胞类型的分子辨识及其演化研究，同时注重细胞的分子分型新技术和新方法的建立。

六、细胞命运的决定及其可塑性

细胞命运的决定及其可塑性是生命科学的重大问题。在机体的生命过程中，细胞每时每刻都在面临选择——增殖、分化、迁移或者死亡，这些选择使细胞维持现有状态或者改变为另一种状态，这个过程称为细胞命运的决定，也是细胞生命活动的核心事件。

在生殖细胞与成体干细胞研究方面，目前针对哺乳动物减数分裂的启动和调控机制、精原前体细胞向精原干细胞的动态变化机理、微环境对精原干细胞的调控等的研究均十分有限，对新型成体干细胞的发现、成体干细胞的自我更新以及微环境（干细胞巢）、成体干细胞在体内修复组织损伤的机理与应用等的研究亦需要深入开展。在细胞属性转换研究方面，终末分化的细胞可通过转分化或去分化转变为其他细胞类型，这可能是组织中广泛存在的非干细胞依赖的修复损伤与组织再生方式。DNA 甲基化、组蛋白修饰、非编码 RNA、染色体空间结构、微环境细胞或者因子对细胞属性转换的调控机制有待继续解析。在细胞信号网络方面，细胞通过感知细胞内外的信号变化，调控并整合关键信号通路调控特异性基因的表达，从而精准调控细胞的命运。Hippo、Wnt、mTOR、Notch、TGF-β 等信号通路参与调控细胞生长、增殖、分化与凋亡，在组织稳态和干细胞自我更新与分化中也发挥重要作用，并与体内免疫反应、代谢异常、肿瘤细胞增殖等密切联系。在研究技术手段方面，近年来建立的新型单倍体干细胞有单倍体性、自我更新、多能性和“受精”能力，可利用其开展细胞水平的正反向遗传筛选、构建异源二倍体、研究胚胎发育印记基因、高效制备复杂遗传修饰小鼠并在个体水平进行遗传筛选等。类器官是当下的研究热点，目前已经培养出肾、肝、肺、肠、脑、前列腺、胰腺、视网膜等组织的类器官。类器官能够在体外培养，能较好保持原位组织的异质性，在器官层面的生理病理研究等方面有巨大应用价值。

建议我国集中基于成体干细胞的基础应用、重大疾病与机体损伤修复中的细胞转分化、调控细胞命运决定的动态信号网络等方向的研究，并发展促进上述研究的新技术、新方法。

七、细胞间通信及细胞与微环境的互作与功能

细胞通信是细胞发出信息通过介质传递到另一个细胞产生相应反应的复杂过程。多细胞生物的细胞间形成复杂的细胞通信网络与反馈机制，精确、高效地发送与接收信息，协调细胞行为，完成复杂生命活动。细胞处于复杂、动态变化的微环境之中，细胞感知微环境并做出反馈，进而影响微环境，并与微环境进行交互调控。

细胞通信研究包括信号分子、分泌方式和通信模式的选择与调控等多个层面的内容。信号分子可以是水溶性或者脂溶性的小分子

化合物和多肽、蛋白质等大分子化合物；分泌方式可以是自分泌、旁分泌、内分泌；通信模式可以是信号分子与受体一对一的作用模式，也可以是细胞间孔道与细胞外囊泡的批量作用模式。细胞所处的生物、化学与物理的微环境存在高度的异质性和可塑性。细胞选择性接受并整合高度复杂、异质的微环境信号，并主动重塑其所处微环境。在发育以及成体组织器官稳态建立、维持过程中，细胞与微环境构成时空特异性的互作网络。细胞通信、细胞－微环境相互作用的异常改变导致发育异常、代谢性疾病、癌症、神经退行性疾病等疾病的发生。现代遗传学和合成生物学的研究方法和手段提升了细胞通信研究水平和深度，单细胞测序技术的发展使得在单细胞水平上开展细胞通信研究成为可能，在此基础上进一步发展立体、实时、动态、在体的研究体系和技术方法将有助于全面研究生理病理条件下细胞通信的关键节点与时空特异性调控网络。

建议我国以新型细胞通信信号分子与通信方式为抓手，以细胞－微环境互作与跨器官细胞通信为核心，阐明细胞通信时空特异性调控网络的调控机制及其功能，发展实时、动态研究新体系与新技术，进而开发基于细胞通信的复杂疾病诊治新策略、新技术、新药物。

八、分子生物网络与“数字化”细胞

细胞生命活动中多样化的功能是通过生物分子之间的相互作用即分子生物网络实现的，因而研究分子生物网络的结构和动力学是在系统水平上深入认识细胞功能的基础。网络重构、网络分析以及利用网络解决生物学问题是分子生物网络研究的三个基本问题，基于分子生物网络虚拟产生“数字化”细胞为分子细胞科学带来全新研究视角。

在网络重构方面，基于研究者开发的多样化的高通量测试方法既可以在基因组水平上对特定生物分子进行测量，产生有关基因组、转录组、蛋白质组、代谢组等组学数据，也可以表征蛋白质-DNA、蛋白质-蛋白质等大分子相互作用。利用这些信息可以建立描述基因调控、蛋白质相互作用、代谢、信号转导、表观调控及大小分子相互作用的分子生物网络，为从系统生物学视角理解生命过程奠定基础。在网络分析方面，通过整合巨量的生物学数据并进行高效的处理和分析，可能有助于研究者揭示复杂分子生物网络的共性运行规律，在静态和动态两方面揭示网络变化、反馈及调控机制。在利用网络解决生物学问题方面，网络标志物和动态网络标志物能够更准确地反映疾病相关基因间的关联，可稳定地诊断或表征疾病状态及生物系统状态，特别是生物系统的临界状态，从而实现疾病状态的预警。此外，基于分子生物网络设计的多靶点药物相较于传统的药物具有更好的疗效。在“数字化”细胞方面，DNA 折纸术、人工合成细胞及 DNA 信息存储技术已经充分展示了数字化生物技术的无限可能，利用虚拟的“数字化”细胞可以揭示细胞中复杂分子作用过程之间的深层次联系并最终预测细胞的行为。细胞的数字化有可能变革生命机理、病变细胞修复和生物存储等领域的研究范式。

建议我国优先建立单细胞层面基因调控网络图谱和蛋白质互作网络图谱及生物系统状态及临界变化的网络标志物理论，这对于揭示复杂生物现象的分子机制具有重要意义；同时积极发展将生物学信息、分子生物网络构建及计算机仿真技术整合的技术，以期实现全面真实地模拟细胞。

九、细胞的人工改造

细胞是生物体结构和功能的基本单位，细胞损伤导致的组织和器官的功能失调会导致重大疾病。通过细胞的人工改造，有望推进医疗技术从化学治疗向细胞治疗的革命性转变。例如，基于细胞改造的免疫疗法，如CAR-T和TCR-T细胞疗法，近年来在肿瘤临床治疗上取得了重要的进展。鉴定优化细胞功能的靶基因并深入阐明细胞疗法机制、拓展新型免疫细胞来源及改造方法、开发能感知微环境的“智能细胞”是未来发展细胞人工改造、实现疾病细胞治疗的重点研究方向。

在细胞疗法机制方面，大规模功能筛选为系统性解析免疫细胞应答中的关键基因及其功能提供了有效手段。找出可以帮助免疫细胞克服免疫抑制微环境的靶基因或通路并对细胞加以细致筛选及改造，将有可能发展出更具临床应用价值的人工免疫细胞疗法。在免疫细胞来源方面，针对自然杀伤细胞、树突状细胞、巨噬细胞等免疫细胞的功能靶向进行改造是开辟新型免疫细胞疗法的新方向。由人体外周血单核细胞经过重编程产生的诱导多能干细胞（iPSC）的可控分化是获取免疫细胞的重要途径，目前已经在免疫治疗方面得到越来越广泛的关注。此外，可以实现同种异体、可现货（off-the-shelf）、可规模化生产的通用型免疫细胞疗法，打破了传统自体免疫细胞在分离、改造和扩增等方面的限制，极大地拓展了免疫疗法的应用。在智能细胞构建方面，通过合成生物学改造，赋予某些功能性底盘细胞（尤其是免疫细胞）新的功能是目前研究的热点。利用高通量筛选技术和合成生物学技术，可以在功能细胞（尤其是治疗性免疫细胞）中导入基因线路来实现治疗功能的时空调控，使其智能地感知微环境中抗原表达、趋化因子浓度等各种信号，从而达到精准治疗的目的。此外，通过设计精确的筛选体系或靶向改造环路，建立高度还原人类疾病部位的疾病模型评价体系和筛选体系，将有可能解决目前细胞疗法在靶向性、持久性、精确打击性及通用性等方面存在的问题。

建议我国优先建立并发展寻找免疫细胞等功能性细胞中关键调节因子的新技术和新方法，积极探索新型的免疫细胞改造技术，重点关注和培育一批具有临床应用前景和自主知识产权的新型干细胞或免疫细胞改造技术和产品，并将其推向临床应用。

总而言之，分子细胞科学与技术前沿领域的突破不仅有助于阐释细胞的生命本质与活动规律，而且可为人工改造细胞提供理论基础、为生物医药和健康产业提供源头性应用基础成果，将引领生命科学进入新的时代，并对我国提升原始创新能力、掌握核心竞争力、实现科技自立自强、迈向世界科技强国发挥重要的战略作用。

Abstract

Cells are the basic unit of structure and function in organisms. All life activities of organisms, such as growth, development, reproduction and evolution, are based on cells. Multicellular organisms contain a variety of cells with different shapes and functions, all of which are derived from the division, proliferation, and differentiation of a fertilized egg. Why can a fertilized egg produce all types of cells? Why do cells change the shapes and functions during ontogeny? Why do the cells such as neurons and muscle cells in a tissue stably maintain their properties once they are formed? Why do normal cells in an organism turn into cancer cells or other diseased cells? These unsolved mysteries about cells are the hot frontiers of basic research in modern life sciences. As the basis of life, cells contain many kinds of molecules, such as nucleic acids, proteins, carbohydrates, and lipids. These biomolecules interact and assemble into complex biomolecular machines or cellular substructures with different structures and functions, which is the basis of the cell's behavior and further leads to various biological phenomena. Mapping the molecular regulation map of the basic cellular activities is one of the ultimate goals of life sciences research. To achieve this goal, in addition to using traditional molecular biology and cell biology research methods, it is necessary to carry out interdisciplinary research and develop innovative technologies and methods to systematically study the nature and the roles of behavior of cells.

Molecular cell science is a cutting-edge, interdisciplinary and emerging discipline, which is based on the molecular biology and cell biology, and cross-integration of mathematics, physics, chemistry, information science, as well as medicine to study cellular activities and their molecular mechanisms. With interdisciplinary new technologies and methods, molecular cell science makes the research on biomolecules, functional elements and cells from qualitative to quantitative, delayed to real-time, static to dynamic, *in vitro* to *in situ*, single to network, and local to global, reveals the causal relationship between the quantity, morphology, structure and function, as well as provides the theoretical basis and technical means for the manipulation and artificial transformation of cells.

The development of molecular cell science has great scientific significance and strategic value. In promoting the development of basic science, molecular cell science will increase the depth of exploration of the nature of cells, enabling the exploration of cell structure and function to transform from "surface analysis of molecular mechanisms in single perspective" to "deep interpretation of mechanisms from multiple dimensions". Meanwhile, it will provide more optimized technical strategies for the transformation of life, and further promote the precise manipulation of intracellular molecular expression and cell function. Furthermore, it will also provide more comprehensive and in-depth support for solving major scientific problems in various disciplines of life sciences, and promote the rapid development of life sciences. In promoting the development in the fields of human health, agriculture and industry, the development of molecular cell science also has important strategic value. First, it will promote the realization of the more precise and more realistic reveal of the mechanisms of human growth and development as well as disease occurrence and development, which can provide source theories and innovative strategies for the R&D and application of disease precise diagnosis and advanced treatment.

Second, it will further promote the optimization and upgrading of breeding technology, which promotes the realization of efficient and precise agricultural breeding. It will also provide huge opportunities for cutting-edge directions, such as the R&D of green pesticides and cellular agriculture products. Third, it will lay a theoretical and technical foundation for the green biosynthesis of industrial products, assist in the design and construction of enzyme molecules, cells and multicellular systems, as well as optimize and regulate metabolic pathways, thereby realizing manual control of the synthesis process.

The development of molecular cell science has mainly gone through two stages. First, the cross-integration of molecular biology and cell biology promoted the continuous transition of the depth of understanding of cells. On the basis of traditional cell biology, as the omics technology represented by genomics promoted a comprehensive analysis of the cell activities, and the emergence of cell engineering technology made cell transformation a reality, the research in the fields of gene expression regulation, cell material transport, cell signal transduction, and cell fate regulation has gradually flourished, and cell biology has begun to transform to molecular cell biology. Second, the multidisciplinary cross has led the emergence and rapid development of molecular cell science. The development and iterative refinement of single cell and single molecule technology, *in situ* analysis technology, imaging technology, molecular imaging, live cell imaging technology and multi-omics technology has made the research in this period transform from cell cluster average analysis to single cell or molecule differentiation analysis, from *in vitro* to *in situ* analysis with spatial information, from static structure analysis of biological macromolecules to the development of dynamic molecular movies, from tomographic molecular cell analysis to non-invasive and real-time observation, from single molecule scale analysis to the multi-layer networks integrated analysis, which lead molecular cell biology to enter into a new stage of molecular cell science.

As a frontier field of life sciences, molecular cell science is a hot spot of global scientific research. In terms of the layout of scientific research funding, national scientific research funding agencies such as the National Science Foundation of the United States and the Medical Research Council of the United Kingdom have set up special departments to manage the funding of molecular cells. The funding encourages the interdisciplinary research combining biology with physics, chemistry, mathematics as well as information science to improve the understanding of molecular cellular functions, molecular regulatory networks and mechanisms, and further integrate with physiology, pathology, and medicine to get a deeper understanding of disease occurrence and treatment response mechanisms, and then to improve the discovery and validation of drug targets and the development of new treatments. The National Institutes of Health of the United States also attaches great importance to basic research related to molecular cell science, and has funded special programs in this field, including the 4D Nucleome Program and the Human BioMolecular Atlas Program, which aim to promote the development of live cell tracking technology, single-cell sequencing technology, imaging technology and other technologies to study single cell activity, and further explore the impacts of molecular cell characteristics and related changes on human development and various diseases. In addition, the United States, Japan, European Union and other technologically advanced countries/regions have successively launched specific initiatives in the field of molecular cell science, including the Human Cell Atlas Project, the strategic proposal of Live Cell Atlas and 4D Cellome in Japan, and preparatory action for FET Flagships "LifeTime" in European Union. China has mainly funded the field of molecular cell science through research programs such as the National Natural Science Foundation of China, the National Key R&D Program, and the Strategic Priority Research Program of the Chinese Academy of Sciences. In terms of the output of scientific research papers, through the quantitative analysis of molecular cell biology, which is the basis and important part of molecular cell science, it was found that researchers published 680,055 papers between 1980 and 2020, maintaining rapid development speed. China has developed rapidly in the field of molecular cell biology in the past ten years. The number of papers in that field has increased from 2,441 in 2011 to 12,560 in 2020, with a compound annual growth rate of 19.96%. In these ten years, China has published 70,756 papers in molecular cell biology, ranking second in the world, accounting for 24.93% of the total number of papers worldwide. Among the ten countries with the most papers in molecular cell biology, the total cited frequency and average cited frequency per paper of China ranked second and tenth.

This report focused on nine aspects of molecular cell science, i.e., (1) genome structure, integrity and synthesis, (2) structure, function, and design of biomolecules, (3) assembly, functions and mechanisms of bio-macromolecular machines, (4) formation and interaction of subcellular structures, (5) identification and classification of cell types, (6) cell fate determination and plasticity, (7) cell communication and its interaction with the microenvironment, (8) biomolecular networks and "digital" cells, (9) artificial modification of cells. It not only studied and expounded the respective critical scientific issues, development status and trends, but also proposed our strategies and key directions in future development.

I. Genome structure, integrity and synthesis

The genetic information coded in the genome is the basis of life, while the eukaryotic chromosome is the carrier for information storage and transmission. The encoded genetic information is programmatically decoded into cell-specific gene expression to form complex organisms. How does the linear coding of the genome determine the diverse cell life activities? How does the cell solve the contradiction between integrity maintenance and sequence variation? Whether it is possible to explore the mysteries of life and create life that does not exist in nature through artificial synthetic life? These are all the frontiers that could be targeted in the research direction of genome structure and evolution.

In terms of genetic information en- and de-coding, the future research trend is to comprehensively reveal the folding of the genome in different dimensions, including chemical basis, hierarchical folding and compartmentalization. The basis of chromatin chemical modification is the key to revealing the dynamic regulations genome. The hierarchical folding mechanisms of chromatin in 3D remain to be clarified. The concept and molecular basis of chromatin compartment need further validation. In terms of genome integrity, the knowledge map of DNA damage repair still has many gaps. New factors and new mechanisms need to be revealed, such as the role of RNA. Rare low-frequent DNA alteration needs more mechanistic studies including the "random" mutation in aging cells. In the field of artificial synthetic life, researchers have found that a greatly simplified genome can also maintain the operation of eukaryotic life. This type of research not only greatly accelerates the revelation of life mechanisms, but also makes it possible to create new ones. Codon expansion, unnatural base/amino acid, and mirror molecules are all at the cutting-edge.

By integrating multi-disciplinary theories and technologies, we should move forward in two directions: "mechanism analysis study" and "genome artificial synthesis", which will provide new ideas for disease diagnosis and treatment, and enhance our international impacts in this field.

II. Structure, function, and design of biomolecules

Biomolecules are the general term for molecules that naturally exist in living organisms and are the carriers and performers of life activities. They include large molecules such as nucleic acids, proteins, carbohydrates, lipids, and small molecules such as various metabolites. The different classes of molecules exert their functions synergistically through interactions. The specific recognition of biomolecules is the basic premise for their functions, and the high-precision resolution of the three-dimensional structures of biomolecules is one of the focuses of this field. In addition, studying the various metabolic processes and coupling mechanisms and the functions of biomolecules is another core of this field. It becomes possible to design and modify biomolecules after understanding their structures and functions.

In nucleic acid research, the discovery of a large number of new types of RNAs and an in-depth understanding of their functions are constantly refreshing researchers' knowledge. New technologies such as gene silencing, gene editing, and RNA sequencing have made nucleic acid research one of the fastest emerging directions in life sciences. RNA manipulation and editing is becoming the most promising therapeutic tool for treating genetic diseases. In protein research, the recent development of cryo-electron microscopy technology has greatly accelerated the understanding of conformational changes in protein activation, and helped to resolve the molecular mechanisms of their biological functions. Computer simulation of the protein structures and functions, as well as the artificial design of proteins, has become popular in both protcin science and application in drug development, vaccine development, chemical catalysis, and many other fields. In carbohydrate research, glycosylation modifications including complex and variable glycan chain structures can deliver information *in vivo* by binding to different proteins or even ribonucleic acids. The biomolecular interactions mediated by glycan chains have great application potential in disease diagnosis and treatment. In lipid research, the regulation of lipid metabolism and new mechanisms of metabolic enzymes, the role of novel and classical organelles in metabolic nutrient perception and material transport, the physiopathological significance of metabolic organs, and the way of material interaction are important directions.

In the future, we could start with identifying the binding targets and solving the structures of various biomolecules, meanwhile following the lines of biomolecular processing/metabolism/recognition, and emphasizing biomolecular regulatory mechanisms and cellular functions. Moreover, we should develop new technologies for biomolecular detection through multidisciplinary integration. We anticipate illustrating the basic mechanisms and functions of biomolecular metabolism and regulation.

III. Assembly, functions and mechanisms of bio-macromolecular machines

The complicated biological functions in cells are usually executed by bio-macromolecular machines, which are formed by dynamic interactions and assembly of various bio-macromolecules, rather than simple individual biomolecule. Different bio-macromolecular machines play important and diverse biological functions in energy metabolism, material transport, signal sensing and response, transmission and maintenance of genetic information, protein synthesis and degradation, and microbial host interactions and immune responses.

Research on the molecular mechanisms of biological processes, including the assembly, modification and dynamic regulation, cellular localization, and transport of bio-macromolecular machines, is essential for understanding the synergistic functions and underlying mechanisms of bio-macromolecular machines *in vivo*. The achievements from the research could provide the molecular and theoretical basis for the design, simulation and manipulation of bio-macromolecular machines. The rapid development of the *de novo* design of bio-macromolecular machines in recent years can be used in the verification of existing interpretations on the functions of bio-macromolecular machines, as well as in the expansion of the functional diversity of bio-macromolecular machines to a great extent. From the aspect of applications, the gained knowledge and technology from the research would be able to help reveal the molecular mechanisms for the pathogenesis and development of various diseases, as well as the host-microbiome interactions, and lay the foundation for the design and development of vaccines, biomarkers and drugs for the prevention, diagnosis and treatment of diseases.

Based on the current progress and challenges in the research of bio-macromolecular machines, we suggest promoting and enhancing the research on establishing and developing new technologies and methodologies that can carry out *in situ* structural studies of bio-macromolecular machines, especially those that allow *in situ*, high-resolution, and real-time structural imaging of the "natural molecular landscape" of bio-macromolecular machines in cells. These technologies will greatly contribute to the understanding of the molecular mechanisms of bio-macromolecular machines and biological processes under normal physiological and abnormal pathological conditions.

IV. Formation and interaction of subcellular structures

Subcellular structures include cell membranes, various organelles, cytoskeleton, nucleus, etc. In addition to the classical membranous organelles, such as endoplasmic reticulum, Golgi apparatus, lysosomes, endosomes, mitochondria, autophagosomes, a large number of membraneless organelles, i.e., biomolecular condensates, are also found in cells. Through fine functional assignment, mutual cooperation and close contact, organelles network their interactions to achieve rapid material and information exchanges to execute a variety of biological processes. The disorder of organelle interactions is closely associated with disease development.

In terms of the membranous organelles, numerous advances have been made in each individual organelle, such as fine sorting regulation of membrane transport, the physiological functions of membranous endosomal organelle interactions, and maintenance of endosomal integrity. In the future, the research paradigm could be shifted from single organelle to integrated multiple organelles, focusing on the delicate regulatory mechanisms, the significance of synergistic interactions of membranous organelles and physiological functions. Additionally, the integrity, repair and degradation of the membrane structure of the membranous organelles will also be a new research hotspot and frontier. In terms of membraneless organelles, the components, ultrastructure, dynamic assembly and functional regulation of membraneless organelles are not well studied. Elucidating the molecular regulation and function of membraneless organelle in cells will be important for the diagnosis and treatment of associated diseases. In the future, we could start with "phase separation" and "phase transition" to investigate the formation and dynamic equilibrium regulation of bio-macromolecules, analyze the ultrastructure and molecular composition of membraneless organelles at a higher temporal-spatial resolution, and eventually understand the assembly process and functional regulation of membraneless organelles.

We suggest developing *in vivo*, high spatial and temporal resolution analytical techniques to improve and strengthen the research of subcellular structures, including the development of biomicroscopy imaging techniques based on cross-integration of multiple disciplines such as optical engineering, computer, automation, physics, and biology.

V. Identification and classification of cell types

Cells are the basic building blocks of life. Both cell numbers and types are constantly increased during the development of an individual. Somatic cells can be divided into corresponding cell types according to their morphological and functional characteristics. For example, there are estimated to be more than 200 different somatic cell types in a completely developed human body. It is a prerequisite to identify and understand each cell type before we reveal different cell activities and functions in multicellular organisms.

Regarding the genetic basis of cell types, an important task is to determine the cell lineage, which enables us to distinguish the history of various cell types in multicellular organisms. Currently, constructing cell lineages of complex organisms, especially human cell lineages, is a major challenge. In terms of the relationship between a certain type of cells and its environment, it is well known that phenotypes have resulted from the interactions of genotypes and environments. Accordingly, the formation of a cell type is influenced by the environment as well. Therefore, the identification of influences of environments on cell type formation is another major challenge.

In terms of the technology and standards for cell type classification, traditionally, cell types have been used to be classified according to morphological and functional characteristics. But they have now been categorized by molecular characteristics at different levels. First, single-cell transcriptome sequencing is the most widely used molecular typing technology at present. A large number of human cell chromatograms have been obtained in recent years. Second, cell lineage tracing, an important part of cell atlas research internationally, can track the development of cell types *in vivo*, compensating for the information loss that occurred in single-cell transcriptome sequencing. Third, the changes of the cell types under pathological (e.g., senescence and tumor) conditions have been a research focus in recent years. Finally, the field of cell typing is evolving from an emphasis on *in vitro*, static single-cell analysis to a higher level of *in vivo*, dynamic cell type studies, enabling researchers to observe the three-dimensional structure of tissues with unprecedented temporal-spatial resolution.

We should employ the combination of omics technologies to carry out studies on cell type identification and molecular typing, cell fate determination and cell lineage tracing during ontogeny, and cell atlas identification and evolution in complex diseases. Therefore, it is important to develop new techniques and methods for the molecular typing of cells.

VI. Cell fate determination and plasticity

Cell fate determination and plasticity are important issues in life sciences. During the life cycle of an organism, cells face choices at every moment, e.g., proliferation, differentiation, migration or death. These choices make cells maintain their existing state or change to another one. This process, namely cell fate determination, is a key event of the cellular life cycle.

In terms of research directions on germ cells and adult stem cells, there is limited progress on the initiation and regulation mechanisms of mammalian meiosis, the transformation mechanism of spermatogonial precursor cells to spermatogonial stem cells, and the regulatory impact of microenvironment on spermatogonial stem cells, etc. Several subjects in adult stem cells, such as novel adult stem cells, self-renewal and microenvironment (stem cell niche), underlying mechanism and application in repairing tissue damage *in vivo*, need to be further developed. In the research on cell identity change, terminally differentiated cells can change into other cell types by transdifferentiation or dedifferentiation, a non-stem cell-dependent process that repairs tissue damage. The epigenetic mechanisms (including DNA methylation, histone modification, non-coding RNA, chromosomal compartmentation and microenvironmental niches that regulate cell identity transition) need to be further elucidated. In terms of cell signaling networks, cells can regulate and integrate various signaling pathways to affect gene expressions by sensing intra- and extra-cellular signals, precisely regulating cell fate. The signaling pathways (such as Hippo, Wnt, mTOR, Notch, and TGF-β) regulate cell growth, proliferation, differentiation and death. They play an important role in tissue homeostasis and stem cell self-renewal and differentiation, and are also closely related to the immune response, metabolic abnormalities, and tumor proliferation. Regarding research technical development, researchers in China have established a new type of haploid stem cells with self-renewal, pluripotency, and "fertilization" ability. The cells can be used to carry out forward and reverse genetic screening at the cellular level, to construct heterologous diploids, to study imprinted genes in embryonic development, to efficiently prepare genetically modified mice and genetically screening *in vivo* at the organismal level. Organoid is a current research hotspot. Organoids have been derived from various organs, such as kidney, liver, lung, intestine, brain, prostate, pancreas, and retina. Organoids cultured *in vitro* can maintain the heterogeneity of the corresponding tissues/organs, and has great application potential for physiopathological study.

We should concentrate our effort on adult stem cells, particularly in the following application-oriented directions: cell transdifferentiation in major diseases and organ damage repair, and dynamic signaling networks for cell fate determination. The development of new technologies and methods will promote the investigations mentioned above.

VII. Cell communication and its interaction with the micro-environment

Cell communication is a complex process by which one cell transmits information to another to produce a corresponding response. The cells of multicellular organisms form complex cell communication networks and feedback mechanisms, which accurately and efficiently exchange information between cells and coordinate cell behavior to accomplish complex life activities. Cells reside in the complex and dynamic microenvironment. Cells interact with microenvironment, which perceive and respond to the microenvironment and influence the microenvironment in turn.

Cell communication research includes the selection and regulation of signaling molecules, secretion modes, and communication patterns. The signals, which can be water-soluble or lipid-soluble small molecules and large molecules such as peptides and proteins, are transmitted in autocrine, paracrine, or endocrine manner, depending on the nature of the receiver cells. Signals can be one-to-one transmitted by specific signal molecules and receptor recognition. Multiple signals can also be simultaneously transmitted in batch mode via cellular pores and extracellular vesicles. The physical, chemical and biological microenvironment of a cell is highly heterogeneous and dynamic. Cells not only selectively receive and integrate highly complex and heterogeneous microenvironmental signals, but also actively remodel their microenvironment. Cells and the microenvironment form a spatial and temporal-specific interactive cell-microenvironment network during the development and homeostasis of adult tissues and organs. Abnormal changes in cell communication and cell-microenvironment interactions lead to developmental abnormalities, metabolic diseases, cancer, neurodegenerative diseases, and other diseases. The methods and approaches of modern genetics and synthetic biology have improved the level and depth of cell communication research. The development of single-cell sequencing technology has made it possible to conduct cell communication research at the single-cell level. Further development of dynamic real-time and three-dimensional *in vivo* research systems and techniques will help fully discover the key nodes of cell communication regulatory networks under physiological and pathological conditions.

It is suggested that we take the discovery of new cell communication signal molecules and new communication modes as the breakthrough point in cell communication research to elucidate the regulatory mechanisms and functions of spatiotemporal-specific regulatory networks of cell-microenvironment interaction and inter-organ cell communication, which will be facilitated by the development of new real-time and dynamic technical platforms and methods. Furthermore, based on cell communication, we will invent cell communication-based new strategies, new technologies and new drugs for the diagnosis and treatment of complex diseases.

VIII. Biomolecular networks and "digital" cells

The diversified functions of cellular activitics are realized through the interactions between biomolecules, namely biomolecular networks. Therefore, the study of the structure and dynamics of biomolecular networks is the basis for an in-depth understanding of cellular functions at the systemic level. Network reconstruction, network analysis, and the use of networks to solve biological problems are the three basic issues in biomolecular network research, and the virtual "digital" cells generated from biomolecular networks will bring a new research perspective to molecular cell science.

In terms of network reconstruction, researchers have developed diverse high-throughput sequencing methods that can not only measure specific biomolecules at the omic level (e.g., genomic, transcriptomic, proteomic, and metabolomic), but also characterize molecular (such as protein-DNA and protein-protein) interactions. Using this information, researchers are able to establish biomolecular networks that describe gene regulation, protein interactions, metabolism, signaling transduction, epigenetic regulation, and large and small molecule interactions, laying the foundation for understanding various life processes from the perspective of systems biology. In terms of network analysis, integrating a huge amount of biological data, combined with efficient processing and analysis, may help researchers to reveal the common operating rules of complex biomolecular networks, as well as network changes, feedback, and regulatory mechanisms in both static and dynamic aspects. In terms of solving biological problems, the network markers and dynamic network markers can accurately reflect the association between disease-related genes. They can also stably diagnose or characterize disease states and homostatic states of biological systems, especially the critical states of biological systems, enabling early warning of disease states. In addition, multi-target drugs designed according to biomolecular-network information have better efficacy than traditional drugs. In terms of "digital" cells, the current DNA origami, synthetic cells, and DNA information storage technology have fully demonstrated the infinite possibilities of digital biotechnology. The use of virtual "digital" cells can reveal the deep connection between complicated molecular processes in cells, and eventually predict cell behavior. Cell digitization has the potential to change the paradigm of research in the fields of life mechanisms, repair of diseased cells and biostorage.

It is suggested that the research priorities in this field are the followings: mapping gene regulatory network at single-cell level, mapping protein interaction network, and developing the network biomarker theory of biological system state and its critical change. They are important for revealing the molecular mechanisms of complex biological phenomena. At the same time, China should actively pursue the integration of biological information, biomolecular networking, and computer simulation technology, achieving the goal of comprehensive and accurate simulation of cells.

IX. Artificial modification of cells

Cells are the basic structural and functional units of organisms. Dysfunction of tissues and organs caused by cell damage can lead to major diseases. Artificial modification of cells could be a revolution for medicine, i.e., from chemotherapy to cell therapy. For example, cell-modification-based immunotherapies, e.g., CAR-T and TCR-T cell therapies, have made significant progress in the clinical treatment of cancer over recent years. Under the theme of artificial cell modification and cell therapy, there are a few promising directions in the future: identify target genes for optimizing cellular functions, elucidate in-depth mechanisms of cell therapy, expand sources of immune cells, create new types of immune cells, and develop "intelligent cells" that can sense the microenvironment.

In terms of cell therapy mechanisms, large-scale functional screening appears to be an effective means for systematically identifying the key genes in immune cell response. Artificial immune cells with more therapeutic value will be a direction: discovering target genes or pathways that can help immune cells overcome the immunosuppressive microenvironment, and then genetically modifying the immune cells. In terms of immune cell sources, the functional modification of immune cells, such as natural killer cells, dendritic cells, and macrophages, is a new tool and direction to open up new models of immunotherapy. The controlled differentiation of induced pluripotent stem cells (iPSC), which is generated by reprogramming human peripheral blood mononuclear cells, is an important approach to obtaining immune cells, and has attracted more and more attention in immunotherapy. In addition, it is possible to achieve allogeneic, off-the-shelf, scalable production of universal immune cell therapies, which can solve the limitations of the isolation, modification, and amplification of traditional immune cells, and greatly expand the applications of immunotherapy. In terms of intelligent cell construction, it is currently a hot research topic to endow certain functional chassis cells, especially immune cells, with new functions through synthetic biology modification. By using high-throughput screening technology and synthetic biology technology, gene circuits can be introduced into functional cells, especially therapeutic immune cells, to achieve spatiotemporal regulation of therapeutic functions and intelligently sense various signals in the microenvironment (including antigen expression and chemokine concentration) for precision therapy. Moreover, by designing a precise screening system or targeted modification loop, researchers could establish a disease-model evaluation system and screening system that highly mimic the focal area of human disease, which will help to address current challenges (such as targeting, persistence, precise striking, and universality) in cell therapy.

It is suggested that the research priorities in this field are the followings: establish and develop new technologies and methods for finding key regulatory factors in functional cells (e.g., immune cells), actively explore new modification technologies for immune cells, focus on and cultivate a number of new "stem cell or immune cell modification technologies and products" with clinical application prospects and independent intellectual property rights for clinical application.

All in all, breakthroughs in the frontier of "molecular cell science and technology" not only help to explain the nature and the law of activity of cells, but also provide theoretical basis for artificial modification of cells, source application results for the bio-pharmaceutical and health industry, and lead to a new era of life science. They will contribute to improve nation's original innovation ability, grasp the core competitiveness, achieve self-reliance in science and technology, and process China toward a world technology powers.

目　录

第一篇

分子细胞科学领域总论

第一章

分子细胞科学领域的科学意义与战略价值

第一节　分子细胞科学的概念内涵

分子细胞科学是以分子细胞生物学为基础，与医学、数学、物理、信息科学、化学等交叉融合，研究细胞生命活动规律及其分子机制的前沿性、交叉性新兴学科。利用学科交叉新技术、新手段，对生物分子、功能元件和细胞进行定性到定量、延滞到实时、静态到动态、离体到原位、单一到网络、局部到整体的研究，揭示其数量、形态、结构和功能的因果关系，为实现对细胞的操控与人工改造提供理论基础和技术手段（图 1-1）。

从这个概念出发，分子细胞科学体系中，分子细胞生物学是最核心的基础，也是最重要的组成部分，细胞仍然是分子细胞科学的研究对象；与分子细胞生物学不同的是，分子细胞科学更强调生物学与其他学科的交叉融合，通过融合，使原有的分子细胞生物学的体系得以拓宽，研究深度得以加深，进而对细胞内分子的作用机制获得更透彻的认识，也为相关理论知识的应用带来了全新的契机。

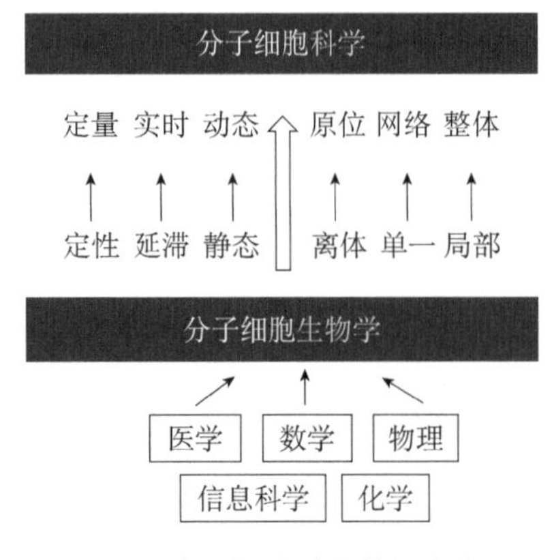

图 1-1 分子细胞科学的概念内涵

第二节 发展分子细胞科学领域的科学意义

一、分子细胞科学为认识生命和改造生命带来理论的深化和技术的升级

一方面，分子细胞科学的发展进一步加深了人类探索生命本质的深度，使对细胞结构和功能的探索实现了从“对分子机制的表象分析”到“从多个维度深度阐释机理”的转变。随着生命科学各类高通量分析平台的逐步建立，单细胞、单分子水平高灵敏度分析技术的飞速发展，以及成像技术空间分辨率的不断提升，分子细胞科学对生物分子、功能元件和细胞的数量、形态、结构和功能的研究发展到一个全新的层面，实现了从实时、动态、原位等更多的维度揭示细胞内基因表达、蛋白质修饰、物质运输、细胞运动、信号转导等分子调控过程，以及细胞增殖、分化、死亡等细胞功能的调控机制，进一步加深了对生命本质的认识。

另一方面，分子细胞科学的发展为生命的改造提供了更加优化的技术策略，进一步推动了对细胞内分子表达和细胞功能的精准操控。分子细胞科学研究使得对细胞内的基因表达及其调控机制、细胞命运可塑性等关键科学问题的认识更加全面、系统、深入，在此基础上进一步通过技术优化提高了分子和细胞改造工具的稳定性和靶向性，为人工改造细胞以及进一步设计合成细胞，实现细胞在医学、农业、工业等各个领域的应用奠定了坚实的基础。

二、分子细胞科学将进一步推动生命科学各学科的发展

生命科学是研究生命现象，生命活动的本质、特征和发生发展规律，以及各种生物之间和生物与环境之间相互关系的科学，其分支学科包括生理学、遗传学、免疫学、发育生物学、进化生物学等，而所有这些学科的发展都离不开在分子细胞水平上对相关生命现象发生机制的深度阐释。

分子细胞科学的发展将为破解生命科学各学科中的重大科学问题提供更全面、深入的理论支撑，进一步推进生命科学领域各学科的快速发展。例如，理解中枢神经系统神经递质的释放和传递过程是生理学研究的重要问题，分子细胞科学的发展可助力获得各种神经递质更高的时间和空间分辨率的动态信息，为深度解释神经递质生理学功能以及相关调控机制提供重要支撑；细胞免疫学是免疫学研究的重要组成部分，其核心研究内容是免疫细胞的发生、分化和介质功能以及免疫细胞间的相互作用，分子细胞科学的发展使得细胞免疫学研究可实现动态、定量、在体分析，为细胞免疫学理论发展提供了更多证据；发育生物学研究是对生物体从生命的产生、发育、生长到衰老、死亡全过程开展研究的学科，分子细胞科学的发展推动了在单细胞水平对细胞谱系进行追踪研究，重建了单细胞胚胎发育成拥有多种器官和亿万细胞的成体过程中细胞和基因表达的时间与空间信息。

第三节　发展分子细胞科学领域的战略价值

一、医药健康领域

人体的正常生理功能和体内的稳态是依靠细胞内各类分子之间的有机调控来实现的，而人体各类疾病的发生在很大程度上是分子层面的调控紊乱导致的。因此，开展分子细胞科学研究是了解人体生长发育和疾病发生发展机制，以及开发疾病预防和治疗方法的重要基础。

首先，分子细胞科学研究使人们对细胞结构和功能的认识更加深刻，可在分子细胞水平上更全面、更精细、更真实地揭示人体生长发育和疾病发生发展的机制与规律。深层次地理解与出生缺陷、人体生长发育异常、衰老和疾病相关的基因表达、信号转导、分子互作等的调控机制，可为疾病的早期发现、预防和治疗，以及预防出生缺陷和实现健康衰老等提供路径。

其次，分子细胞科学研究是发现新型生物标志物的主要途径，可助力实现疾病的精准诊断和治疗。一方面，通过分子细胞科学研究能够发现大量疾病相关的生物标志物，为基因水平的单核苷酸多态性（single nucleotide polymorphism，SNP）筛查、点突变基因诊断、蛋白质和代谢物水平的各类生物标志物检测，以及细胞水平的循环肿瘤细胞的检测等多种分子诊断新技术的发展和应用提供了基础支撑，可助力实现疾病的早期诊断；另一方面，通过分子细胞科学研究探索疾病特异性生物标志物，有助于解析疾病发生的分子机制，促进疾病的精准分型和患者分层，进而推动精准医学的实现。

最后，分子细胞科学是先进疗法研发和应用的重要理论基础。第一，通过分子细胞科学研究，可解析疾病发生、发展的分子机制及调控原理，指导研发针对特定分子进行靶向攻击的方法，促进分子靶向治疗技术的发展。第二，分子细胞科学研究促进了人们对分子和细胞认识的加深，基因编辑技术、基因重组技术、基因转移载体等基因操作技术的精准性和安全性得到进一步提高，推动基因治疗逐渐进入临床试验阶段。第三，通过分子细胞科学研究，可明确干细胞多能性和可塑性机制，发现免疫治疗靶点，进而推动干细胞治疗、免疫细胞治疗等一系列革命性细胞治疗技术从理念成为现实，为重大慢性疾病、恶性肿瘤的治疗带来了新希望。

二、农业领域

在农业领域，作物和畜禽的培育和繁殖都已经全面进入“分子时代”，分子细胞科学的发展将进一步推动生物育种技术的优化和升级，推进高效、精准农业育种，实现产量、品质和抗性等性状指标的优化，同时也将为绿色农药创制和“细胞农业”产品开发等前沿领域的发展带来巨大的机遇。

首先，分子细胞科学研究为农业育种提供了重要理论。研究人员能够利用全基因组筛选、基因工程、人工智能等技术，挖掘协同调控重要农艺性状的基因，突破常规育种局限，实现从经验育种到定向高效设计育种阶段迈进，大幅提高了育种效率。

其次，分子细胞科学研究为农药分子设计和施用提供了新的思路。通过分子细胞科学研究，厘清农药分子与受体相互作用机制，发现关键靶标，进一步利用计算机分子设计、化学合成等方法实现绿色农药创制，可实现农业病虫害精准防控关键目标。

最后，分子细胞科学研究也为“细胞农业”产品（即从细胞培养物中生产农产品）创制铺平了道路，指导细胞系、培养基、组织支架的选择和设计。与传统农业相比，“细胞农业”对环境的影响较小，产品更加安全纯正，并且供给更加稳定。

三、工业领域

工业产品生物制造是以工业生物技术为核心技术手段，利用生物质、CO_2等可再生原料生产能源、材料与化学品等，实现原料、过程及产品绿色化的新模式，相关产品产量、质量、经济性依赖于分子细胞水平的精细调控。分子细胞科学的发展为工业产品的绿色生物合成奠定了理论和技术基础，助力设计构建酶分子、细胞和多细胞体系，优化调控代谢通路，将推动酶促反应速率、底物转化率、产物产量、菌株生长速度和稳定性的提升，产物提取难度的降低，进而实现对合成过程的人工控制。

第二章

分子细胞科学领域的发展现状

第一节　分子细胞科学领域的发展历程

20 世纪 80 年代以来，从分子水平揭示细胞生命活动现象的本质逐渐成为细胞研究的核心内容，在细胞生物学与分子生物学不断实现深度融合的同时，物理、化学、信息等其他学科不断渗透，推动分子细胞科学理论体系的形成和不断完善。在这个新阶段中，对生物分子、功能元件和细胞结构与功能研究的能力逐渐在时间维度和空间维度实现了升级，实现了实时、动态、原位等层面的研究，更能够从全局的角度揭示分子细胞互作和调控网络，进而更好地揭示细胞生命活动的机制和规律。

一、分子生物学与细胞生物学交叉融合推动对细胞认识深度的不断跃迁

分子细胞科学发展的第一个阶段是分子逐渐成为认识细胞的核心，分子生物学的不断发展，以及其与细胞生物学的不断融合，推动了对细胞认识深

度的不断跃迁。

（一）细胞学向细胞生物学的发展

分子生物学研究方法的应用打破了通过显微观察，研究细胞形态、结构的局限，实现了在显微、亚显微，甚至分子层面观察和研究细胞，推动了细胞学逐渐发展成为细胞生物学。在这一阶段，科学家相继发现了 DNA 和蛋白质的结构和功能，为认识细胞提供了新的研究方向。1944 年，美国生物学家奥斯瓦德·艾弗里等在肺炎双球菌的体外转化实验中证明了 DNA 是遗传物质；1950 年，美国科学家莱纳斯·鲍林等首次发现了蛋白质的二级结构 α- 螺旋；1953 年，美国科学家詹姆斯·沃森和英国科学家弗朗西斯·克里克用 X 射线衍射法发现了 DNA 的双螺旋结构；同年，英国科学家弗雷德里克·桑格测定了胰岛素中氨基酸的排列顺序，开启了蛋白质一级结构的研究。

（二）细胞生物学向分子细胞生物学的发展

1. 技术发展历程

在上述对细胞获得更深层次认识的基础上，科学家进一步阐明了 DNA 复制、RNA 转录合成的机理，破译了 RNA 编码合成蛋白质的遗传密码，认识了蛋白质合成过程，并发现了以 RNA 为模板合成 DNA 的反转录酶，建立并补充完善了遗传信息传递的中心法则。之后，随着分子生物学和细胞生物学进一步交叉融合，分子细胞生物学的概念和体系于 1980 年进入人们的视线，自此，对细胞生命活动的研究真正从显微结构、亚显微结构观察阶段迈向了分子水平研究阶段。在这一阶段，科学家开发了一系列新技术和新方法，在分子水平研究细胞结构和功能，解析细胞生命活动的调控机制，并实现了分子 / 细胞的人工改造和合成。

1）组学研究技术促进细胞生命活动规律的全面解析

基因组、蛋白质组、代谢组等组学理论和技术的提出，推进了对细胞生命活动分子调控规律的全面认识。

在基因组方面，1977 年以双脱氧测序和化学降解法为代表的第一代测序技术的提出，标志着基因组学研究的开端。1977 年噬菌体单链完全测序完成，这是人类历史上首个完成的基因组测定。1985 年，聚合酶链反应（polymerase chain reaction，PCR）的成功发明，更是为基因组研究提供了强有力的工具。PCR 技术的发明者美国科学家凯利 • 穆利斯因此于 1993 年获得了诺贝尔化学奖，直到如今，这一技术仍然在基因组研究中广泛应用。进入 21 世纪，基因组测序技术开始步入快速革新阶段，2005 年之后出现了第二代测序技术，以罗氏（Roche）公司的 454 技术、Illumina 公司的 Solexa 技术和美国应用生物系统公司（ABI）的 SOLID 技术为代表。这类技术基于边合成 / 连接边测序的核心思想，显著增加了测序通量，引领基因组研究进入高通量时代。2008 年，基因组测序技术进一步"进化"到第三代，以美国螺旋生物科学公司（Helicos BioSciences）的真正单分子测序技术（true single molecule sequencing，tSMS）、美国太平洋生物科学公司（Pacific Biosciences）的单分子实时测序技术（single molecule real-time，SMRT）和英国牛津纳米孔技术公司（Oxford Nanopore Technologies）的纳米孔单分子测序技术为代表，有效解决了第二代测序技术读长短和系统偏向性的问题。

在基因组研究发展的基础上，蛋白质组和代谢组的概念于 1994 年和 1998 年相继提出，人们开始进一步在蛋白质和代谢物水平对基因的功能进行大规模研究。在技术层面，由 2002 年获得诺贝尔化学奖的美国科学家约翰 • 芬恩和日本科学家田中耕一发明的生物大分子质谱分析技术为蛋白质组分析铺平了道路。

2）细胞工程技术的出现使细胞改造成为现实

细胞工程是指应用细胞生物学、遗传学和分子生物学技术，按照预定的设计来改变或创造细胞的某些生物学特性，包括分子水平（如基因工程、蛋白质工程）、细胞器水平（如核移植）、细胞整体水平（如细胞融合）等不同层次的改造。

在分子水平的细胞工程技术中，基因工程技术是发展最快、应用最广泛的一项技术。DNA 连接酶（1967 年）和限制性内切核酸酶（1970 年）的发现，为 DNA 的人工改造提供了重要工具。利用这一工具，1972 年，美国科学家保罗 • 伯格构建了第一个体外重组 DNA 分子，该项研究成果成为基因工程领域发展的里程碑，保罗 • 伯格也因此获得了 1980 年诺贝尔化学奖。1974 年，研究人员将金黄色葡萄球菌质粒上的抗青霉素基因转到大肠杆菌体内，拉开了转基因技术应用的序幕。之后，基因工程技术日趋成熟，在多个领域实现了应用。首个转基因动物（1980年）、基因工程药物（1982年）、首个转基因植物（1983年）相继问世，首个人体细胞基因治疗临床试验也于1990年获得美国食品药品监督管理局（FDA）的批准。此外，早期的基因工程技术只能将外源或内源遗传物质随机插入宿主基因组，基因编辑技术的出现使基因的定点编辑成为可能。最早的基因编辑是利用机体自身的同源重组，1987年多家实验室在小鼠胚胎干细胞上利用同源重组实现基因替换和引入突变，并称之为基因打靶技术。之后，锌指核酸酶技术（1996年）、转录激活因子样效应物（transcription activator-like effector，TALE）核酸酶技术（2010年）、CRISPR-Cas9技术（2012年）等基因编辑技术相继出现，进一步提高了基因编辑的效率和安全性。其中CRISPR-Cas9技术的发明者——法国科学家埃玛纽埃尔·沙尔庞捷和美国科学家珍妮弗·杜德纳获得了2020年诺贝尔化学奖。目前，基因工程技术已广泛应用于生物医药、生物能源、生物基化学品、动植物育种、基因治疗等领域。在基因工程技术快速发展的同时，随着对蛋白质结构解析的不断深入，蛋白质工程的概念于1983年提出——通过对蛋白质化学、蛋白质晶体学和蛋白质动力学的研究，获得有关蛋白质理化特性和分子特性的信息，在此基础上对编码蛋白质的基因进行有目的的设计和改造。

在细胞器水平的细胞工程技术中，核移植技术最具代表性。该技术最早由德国胚胎学家汉斯·施佩曼于1938年提出，旨在研究细胞核全能性机制。1952年，美国科学家罗伯特·布里格斯和托马斯·金利用该技术首次在美洲豹蛙（*Rana pipiens*）核移植实验中成功获得正常克隆后代。之后随着去核技术、显微操作技术等各项技术的发展和完善，核移植实验由在两栖类和鱼类中开展研究逐步发展为在哺乳动物中开展研究。科学家先后以不同哺乳动物细胞作为供体，培育出了克隆羊、克隆牛、克隆小鼠、克隆家兔等。因核移植技术可短期内批量生产遗传背景一致且无嵌合现象的动物模型，其逐步发展为构建基因修饰动物模型的最佳方法。2018年，中国科学家首次基于体细胞核移植技术成功克隆出猕猴，进一步开创了使用非人灵长类动物作为实验模型的时代（Liu et al.，2018）。

在细胞整体水平的细胞工程技术中，细胞融合技术的发明和发展使不同种类的两种生物细胞能够实现直接融合，产生能够同时表达二者有益性状的杂合细胞。1838年，科学家首次观察到了细胞自发融合现象。之后，基于对细胞融合现象的观察和机制研究，1958 年科学家首次发现了仙台病毒能够触发细胞融合，1974 年科学家开发了聚乙二醇（PEG）化学融合技术，这两项技术的成功开发成为细胞融合技术发展的里程碑，标志着人工触发细胞融合的全面实现。此后细胞融合技术不断得到优化和创新，例如在 20 世纪 80 年代电融合技术、激光诱导技术、离子束细胞融合技术等技术相继创立。

3）DNA 从头合成技术的发展推进基因组与细胞的人工设计与合成

DNA 从头合成技术主要包括寡核苷酸合成、基因合成、基因组装等几大关键步骤。在寡核苷酸合成方面，1955 年英国科学家亚历山大・托德（Alexander Todd）等第一次基于固相化学合成法成功合成寡核苷酸；1987 年美国科学家马文・卡拉瑟斯（Marvin Caruthers）发明了目前广泛应用的固相亚磷酸三酯法；进入 20 世纪 90 年代之后，高通量寡核苷酸合成的方法取得了重要进展，1991 年美国昂飞公司（Affymetrix）开发了基于微阵列合成的方法，运用有掩模光刻（mask-based photolithographic）技术选择性保护光不稳定核苷亚磷酰胺。在基因合成技术方面，从 20 世纪 80 年代起，连接酶介导技术和聚合酶循环组装技术被应用于基因合成。在基因组装技术方面，重叠序列拼接技术和酵母或枯草芽孢杆菌体内 DNA 拼接技术等一系列大片段 DNA 组装技术的发现使得基因组人工合成成为可能。伴随着相关技术的进步，2010 年第一个由化学合成的基因组控制的合成支原体细胞 JCV1-SYN1.0 诞生（Gibson et al.，2010）；2014 年，科学家成功合成第一个功能性酵母染色体（Annaluru et al.，2014）。

2. 领域发展历程

在上述技术的飞速发展下，科学家在遗传信息表达和调控、细胞物质运输、细胞信号转导、细胞命运调控等分子细胞生物学研究重要子领域不断取得突破，推动对细胞认识的不断深入。

1）遗传信息表达和调控

基因的功能并不是固定不变的，而是可以根据环境的变化进行调节。原核生物遗传信息表达和调控相对简单，1961 年，法国科学家雅克・莫诺（Jacques Monod）与弗朗索瓦・雅可布（Francois Jacob）在《蛋白质合成中的遗传调节机制》（Genetic Regulatory Mechanism in the Synthesis of Protein）

一文中提出了操纵子学说，开启了原核生物的基因表达和调控的研究序幕。在真核生物遗传信息表达和调控方面，1977年，英国科学家理察德·罗伯茨（Richard Roberts）和美国科学家菲利普·夏普（Philip Sharp）以DNA排列序列与高等动物很接近的腺病毒作为研究对象，发现它们的基因在DNA上的排列由一些不相关的片段隔开，是不连续的，进而提出了真核生物中存在断裂基因的概念，他们因此获得了1993年诺贝尔生理学或医学奖。1998年，美国科学家安德鲁·法厄（Andrew Fire）和克雷格·梅洛（Craig Mello）发现了RNA干扰（RNA interference, RNAi）现象，即双链RNA可以引起体内同源mRNA降解，从而导致基因沉默，他们因此获得了2006年诺贝尔生理学或医学奖。而2006年的诺贝尔化学奖授予了美国科学家罗杰·科恩伯格（Roger Kornberg），他在分子水平上首次阐明了真核生物转录的基本机制。

基因自身的稳定性也将影响遗传信息的表达。美国科学家巴巴拉·麦克林托克（Barbara McClintock）因发现可移动的遗传元件“转座子”获得1983年的诺贝尔生理学或医学奖。转座子的发现改变了人们对基因组序列稳定性的认识，打破了遗传物质在染色体上呈线性固定排列的传统理论。之后，日本科学家利根川进证实抗体基因可以发生突变和重组衍生出新的抗体，揭示了抗体多样性的遗传学原理，于1987年获得了诺贝尔生理学或医学奖。2015年，诺贝尔化学奖授予了在细胞基因组“维稳”的DNA修复机制研究方面做出贡献的三位科学家——托马斯·林达尔（Tomas Lindahl）、保罗·莫德里克（Paul Modrich）、阿齐兹·桑贾尔（Aziz Sancar），他们在分子水平描述了细胞如何完成DNA修复及保护遗传信息。

2）细胞物质运输

物质跨膜运输是细胞维持正常生命活动的基础之一，也是细胞膜的重要功能之一。

离子通道是一类跨膜糖蛋白，其在细胞膜上形成的亲水性孔道使带电荷的离子得以进行跨膜转运。20世纪七八十年代，德国科学家厄温·内尔（Erwin Neher）和伯特·萨克曼（Bert Sakmann）发明了膜片钳技术，并利用该技术阐述了单离子通道的功能，开启了该领域的研究。此后，随着膜片钳技术的广泛应用和生化技术的进步，离子通道研究逐渐进入飞跃发展时期。1988年，美国科学家彼得·阿格雷（Peter Agre）等从红细胞和肾小管中分离出一种功能未知的新的膜蛋白 CHIP28，并进一步证实其为水通道蛋白。1998 年，美国科学家罗德里克·麦金农（Roderick Mackinnon）获得了青链霉菌钾离子通道的高分辨率三维结构，首次在原子水平上揭示了离子通道的结构和机理。彼得·阿格雷和罗德里克·麦金农因此获得了 2003 年诺贝尔化学奖。

蛋白质等生物大分子无法直接穿过细胞中的膜结构以行使功能，需要依赖围绕在膜周围的囊泡进行传递运输。20 世纪 60 年代，乔治·帕拉德（George Palade）利用豚鼠胰腺腺泡细胞进行实验研究，发现细胞分泌的蛋白质需要先进入内质网，再到高尔基体，然后分泌到细胞外，证实了分泌蛋白的合成、运输及分泌途径，由此获得了 1974 年诺贝尔生理学或医学奖。20 世纪 70 年代，科学家通过显微镜观察发现蛋白质的运输需要一种囊泡结构来介导完成。1975 年，甘特尔·布洛贝尔（Günter Blobel）发现了蛋白质具有内在信号以控制其在细胞内的传递和定位，进一步提出了分泌蛋白进入内质网的信号肽学说，并因此获得了 1999 年诺贝尔生理学或医学奖。20 世纪 80 年代，迈克尔·布朗（Michael Brown）和约瑟夫·戈尔斯坦（Joseph Goldstein）因发现胆固醇进入细胞的机制，证明低胆固醇的饮食可以增加低密度脂蛋白受体的数量，降低血液里胆固醇含量，于1985年获得诺贝尔生理学或医学奖。美国科学家詹姆斯·罗斯曼、兰迪·谢克曼和德国科学家托马斯·苏德霍夫发现了编码调节囊泡运输关键蛋白的基因，解释了蛋白复合物在囊泡锚定和融合中的作用机制，还在神经细胞突触传导中证实了囊泡运输系统的时空调控性，这三位科学家因此获得了 2013 年诺贝尔生理学或医学奖。

3）细胞信号转导

20 世纪下半叶开始，随着对激素、神经递质等生物体内细胞外信号分子以及环磷酸腺苷（cAMP）等细胞内信号转导分子作用机理研究的深入，对细胞信号转导的认识有了很大的进步。

在激素、生长因子、神经递质等第一信使（细胞分泌的调节靶细胞生命活动的化学物质）的研究方面，1986 年诺贝尔生理学或医学奖授予了意大利生物学家丽塔·莱维·蒙塔尔奇尼和美国生物化学家斯坦利·科恩，以表彰他们在外周神经组织和脑内发现并证实了对神经细胞生长、发育和维持具有重要意义的物质——神经生长因子（nerve growth factor，NGF），以及另一种生长调节蛋白——表皮生长因子（epidermal growth factor，EGF）。美国科学家罗伯特·弗奇戈特（Robert Furchgott）、路易斯·伊格纳罗（Louis Ignarro）和费里德·穆拉德（Ferid Murad）因发现一氧化氮（NO）是心血管系统中的信号分子，证实NO在神经信号传递、血压控制、血液流量控制等方面起到了重要作用，而获得1998年诺贝尔生理学或医学奖。

在细胞内信号转导方面，1965年厄尔·萨瑟兰（Earl Sutherland）首次提出了第二信使学说，认为人体内各种含氮激素（蛋白质、多肽和氨基酸衍生物）都是通过细胞内的环磷酸腺苷（cAMP）来发挥作用的，同时首次将cAMP称作第二信使，将激素等称作第一信使。此后，科学家又相继发现了多个第二信使，并陆续阐明了其行使功能的机制。1978年，基于钙离子受体蛋白——钙调素的发现及其功能研究，霍华德·拉斯马森（Howard Rasmussen）提出Ca^{2+}第二信使学说。到了20世纪80年代，科学家进一步提出了肌醇磷脂代谢途径产生的另外两个第二信使肌醇三磷酸（inositol triphosphate，IP3）和二酰甘油（diacylglycerol，DG）。之后，美国科学家埃德温·克雷布斯（Edwin Krebs）和埃德蒙·费希尔（Edmond Fisher）发现了蛋白质可逆磷酸化过程关键酶——蛋白质激酶和磷酸酯酶，相关研究为揭示cAMP如何行使功能奠定了基础。这两位科学家因此获得了1992年诺贝尔生理学或医学奖。

在细胞信号跨膜转导方面，科学家发现，细胞质膜上最多也是最重要的信号转导系统由鸟苷三磷酸（guanosine triphosphate，GTP）结合蛋白（G蛋白）介导，依靠GTP的结合或水解产生的变构作用完成，相关研究获得重大进展。马丁·罗德贝尔（Martin Rodbell）等在20世纪70年代发现跨膜信号转导需要GTP存在，之后阿尔弗雷德·吉尔曼（Alfred Gilman）分别于1977年和1981年发现和纯化了G蛋白。吉尔曼和罗德贝尔因在G蛋白及其在细胞中的信号转导作用方面的贡献获得了1994年诺贝尔生理学或医学奖，这两位科学家揭示了细胞外信号如何转换为细胞内信号的真正机制，开辟了细胞信号跨膜转导研究的新时代。美国科学家罗伯特·莱夫科维茨（Robert Lefkowitz）和布莱恩·科比尔卡（Brian Kobilka）因进一步发现了G蛋白藕联受体（G-protein coupled receptor，GPCR）家族，并阐明了其结构和作用机理获得了2012年诺贝尔化学奖。

4）细胞命运调控

细胞是生命的基本结构和功能单元，细胞一旦产生就面临着分裂、增殖、运动、分化和死亡等各种不同的细胞命运，细胞命运调控机制研究是分子细胞科学研究的关键环节。

在细胞分裂、增殖研究方面，在发现细胞分裂和细胞周期的基础上，找到细胞周期的关键调控因子并确认相关分子机制，可为解开生命奥秘奠定重要基础。20 世纪 70 年代，科学家发现了有丝分裂促进因子（maturation-promoting factor，MPF），开启了细胞周期调控的研究之路。之后，周期蛋白（cyclin）和周期蛋白依赖性激酶（cyclin-dependent kinase，CDK）的发现成为该研究领域的里程碑。作为细胞周期的关键调控因子，二者共同驱动细胞从一个时相向另一个时相转换。相关研究者获得了 2001 年诺贝尔生理学或医学奖。

体细胞重编程是细胞命运调控领域的研究热点。19 世纪末期到 20 世纪初，科学家普遍认为未成熟细胞发展成特定成熟细胞是单向性的，不可能再恢复到多能干细胞的阶段。直到 20 世纪中期，科学家相继在美洲豹蛙和非洲爪蟾中证实，两栖类等低等动物中部分已分化的体细胞在卵母细胞质的作用下，可以恢复其全能性并重新发育到成体，即细胞发生了去分化的过程，打破了细胞分化不可逆理论思维。1996 年，科学家通过细胞核移植技术生成了克隆羊多莉（Dolly），进一步证实了成熟的、已分化的细胞可以通过去分化返回到未成熟、多能分化阶段。之后，随着干细胞调控机制研究的逐渐深入，诱导多能干细胞（induced pluripotent stem cell，iPSC）技术应运而生。该技术从成体细胞出发，通过体外人工诱导，实现“生命时钟”的逆转，使其重新具备多能性。2006 年，日本科学家山中伸弥等利用病毒载体将 4 个转录因子转入已分化的体细胞中，使其重编程而得到类似于胚胎干细胞的一种细胞类型，即诱导多能干细胞（Takahashi and Yamanaka，2006）。该项研究为寻找无免疫排斥反应、避开胚胎干细胞伦理限制的新型干细胞指明了方向，山中伸弥因此获得了 2012 年诺贝尔生理学或医学奖。随后，科学家陆续发现其他方法（如小分子化合物）同样也可以制造出诱导多能干细胞。在此基础上，科学家进一步提升了诱导多能干细胞重编程效率和安全性。

在细胞程序性死亡研究方面，常见的细胞程序性死亡形式包括细胞凋亡（apoptosis）、自噬（autophagy）、细胞坏死性凋亡（necroptosis）、细胞焦亡（pyroptosis）等。其中，细胞凋亡是为了维持内环境稳定由基因控制的细胞自主的有序的死亡，这个概念由美国病理学家约翰·克尔（John Kerr）等于1972年首次提出。之后，随着分子生物学技术的发展，研究人员进一步揭示了细胞凋亡的诱导和执行需要一系列分子的配合，包括信号分子、受体、酶和基因调节蛋白。2002年诺贝尔生理学或医学奖授予了美国科学家罗伯特·霍维茨（Robert Horvitz）、英国科学家悉尼·布伦纳（Sydney Brenner）和约翰·萨尔斯顿（John Sulston），以表彰他们在细胞凋亡研究中的突出贡献。这三位科学家以线虫作为研究对象，发现细胞凋亡由基因控制，并证实人体内也存在相应的基因。细胞自噬的概念于20世纪60年代提出，科学家观察发现细胞能破坏自身成分，用膜将这些成分包裹，形成袋状囊泡并转移给溶酶体进行降解回收。20世纪90年代，日本科学家大隅良典在利用酵母菌鉴别出细胞自噬发生的关键基因的基础上，进一步阐明了酵母菌自噬的基本原理，并证明类似的复杂机制也存在于人体细胞内，因此获得了2016年诺贝尔生理学或医学奖。细胞焦亡是一种新发现的细胞程序性死亡方式，表现为细胞不断胀大直至细胞膜破裂，导致细胞内容物的释放进而激活强烈的炎症反应。2001年，美国华盛顿大学布拉德·库克森（Brad Cookson）等首次使用“pyroptosis”（细胞焦亡）来描述在巨噬细胞中发现的胱天蛋白酶（caspase）依赖的细胞死亡方式（Cookson and Brennan，2001）；2015年，北京生命科学研究所邵峰课题组首次鉴定发现GSDMD蛋白是胱天蛋白酶的特异底物，是细胞焦亡发生的最终效应蛋白（Shi et al.，2015）。

二、多学科交叉引领分子细胞科学前沿学科的出现和快速发展

随着多学科交叉发展，一系列新兴前沿技术和研究理念不断涌现，推动分子细胞生物学迈入分子细胞科学新阶段。在这一阶段，分子细胞研究在时间、空间、研究尺度等多个维度实现了深度和广度的拓展，极大地推进了人类对细胞的认识。

1. 单细胞、单分子技术引领从平均化研究到差异化研究的发展

在单细胞和单分子技术出现之前，对细胞和分子的研究结果仅是多细胞或多分子水平上的“平均值”，无法反映单个细胞或分子的真实情况。因此，在微米－纳米水平上对单细胞、单分子开展研究对于解析生命本质具有重大意义。

细胞是生物学研究的基本单位，将细胞进行单个分离、研究和比较，揭示其异质性已成为生物学领域研究的热点和重点。随着单细胞分离技术、基因扩增技术以及高通量测序技术的飞速进步，单细胞分析的通量不断获得质的飞跃，逐渐进入高通量分析阶段。其中，基于生物、化学、医学、流体、电子、材料、机械等学科的交叉融合，单细胞分离技术从最初的显微吸取法、激光显微切割法等人工手动分离法，发展到整合集成化芯片的高通量集成流体电路技术、液滴微流控技术等，细胞通量从个位数跳跃到数万的数量级。与此同时，DNA 和 RNA 扩增技术也不断改进。结合高通量测序技术，2009 年，汤富酬等开发了首个单细胞 RNA 高通量测序技术（Tang et al., 2009）；2011 年，尼古拉斯·纳文（Nicholas Navin）等实现了单细胞 DNA 的高通量测序（Navin et al., 2011）；2013 年，汤富酬等进一步开发出了单细胞全基因组 DNA 甲基化检测技术——单细胞简化代表性重亚硫酸盐测序技术（single-cell reduced representation bisulfite sequencing, scRRBS）(Guo et al., 2013)。这一系列技术的发展标志着单细胞测序分析迈入了高通量时代，为从单个细胞的水平上更为精确地解析细胞的分化、再生、衰老以及病变机制奠定了重要基础，也将为肿瘤学、发育生物学、神经科学等领域的发展奠定重要基础。

研究单个分子（如蛋白质或 DNA）的行为能够揭示重要的生物学机制，这对于分子细胞研究来说至关重要。近年来基于多学科交叉融合，一些单分子技术逐步成熟，例如可以检测分子的结合、折叠或机械行为的单分子力谱术（single molecular force spectroscopy），以及能够在体外和体内对单分子进行追踪的荧光显微镜等。与此同时，一系列新兴的单分子技术和设备也不断出现，例如单分子纳米孔测序技术，无需标记就能检测单分子的光学设备和等离子设备等。这些技术和设备的出现，为科学家能够以空前的深度探索单分子的功能提供了可能。

2. 分子原位分析技术推动从离体研究到携带空间信息的原位研究的转变

分子、细胞在体内的空间位置信息对于理解分子与分子之间、细胞与细胞之间的相互作用，以及发育进化过程至关重要。近年来，研究人员综合生物学、物理、化学等多学科原理开发出了多种原位分析技术，引领分子细胞研究从离体分析发展到能够保留空间特征的原位分析。

在基因测序方面，一般的测序技术分析的样品是纯化后的核酸，丢失了核酸序列的空间背景；而传统荧光原位杂交能够原位检测单个 mRNA 分子，但却无法区分结构非常相似的序列，因此无法用于等位基因失活和剪接变体等研究。为了解决相关问题，研究人员提出了 RNA 原位测序的概念，对固定组织或细胞样本中的 mRNA 进行直接测序。2010 年，瑞典乌普萨拉大学的马茨·尼尔森（Mats Nilsson）团队首次提出基于锁式探针的原位测序方法，实现了转录物的原位检测，能够鉴定出多个基因中的单个突变，并极大地保留了 RNA 分子的位置信息（Larsson et al.，2010）。2013 年末，《自然 - 方法》（*Nature Methods*）期刊评选出 2014 年值得关注的技术，原位测序位列其中，这反映了细胞基因空间表达研究已经成为国际前沿。2014 年，美国哈佛大学医学院乔治·丘奇（George Church）团队进一步研发出非靶向原位测序策略——荧光原位 RNA 测序（fluorescent *in situ* sequencing，FISSEQ）技术，实现了在固定的细胞和组织中获得全基因组范围的基因表达图谱（Lee et al.，2014），解决了之前 RNA 原位分析方法限于分析单个样本的少数基因的问题。

在蛋白质、代谢物分析方面，质谱等技术的革新逐渐实现了保留空间原始信息的原位化分析，质谱离子源的升级是其中的关键环节。最早的质谱成像技术是基质辅助激光解吸电离（matrix-assisted laser desorption ionization，MALDI）质谱分子成像技术于 1997 年发明，该技术可提供分子在组织中的精确空间分布信息。但由于其设计仍为封闭的离子源结构，需要在真空环境中进行，无法实现即时原位分析。2004 年和 2005 年，在电喷雾电离（electrospray ionization，ESI）基础上研发出的解吸电喷雾电离（desorption electrospray ionization，DESI）技术，以及在大气压化学电离（atmospheric pressure chemical ionization，APCI）基础上研制出的实时直接分析技术（direct analysis in real time，DART）相继问世，正式开启了常压敞开式离子化技术时代，使得质谱分析发展到实时、表面、原位分析的新阶段（贺玖明等，2012），为蛋白质、代谢物的原位分析带来了全新的发展机遇。

3. 成像技术迭代促进生物大分子结构研究从静态结构到动态分子电影的发展

生物大分子在细胞功能实现中的作用在很大程度上是由它们的三维结构所决定的，因此生物大分子的动态结构变化以及结构的不均一性一直是结构生物学的研究重点之一。通过传统的 X 射线晶体学能够获得大量生物大分子的静态结构信息，但需要事先获取这种大分子的晶体，因而不适用于膜蛋白或动态的复合体等大分子的分析。直到 20 世纪 80 年代，得益于快速冷冻技术、三维重构技术的快速发展，不依赖于结晶的冷冻电镜技术出现，这使得在不同活性状态下对生物大分子瞬时构象的捕捉成为可能。冷冻电镜技术是先将生物大分子在毫秒时间尺度内快速冷冻在玻璃态的冰中，再应用低温透射电子显微镜收集生物大分子的二维投影，并利用三维重构技术得到生物大分子三维精细结构。2013 年，研究人员首次将直接电子探测器应用于记录电镜图像，突破了冷冻电镜技术最后一个障碍，迎来了“分辨率革命”，冷冻电镜解析精度达到近原子分辨率水平（Liao et al.，2013）。由此，冷冻电镜技术逐渐成为生物大分子动态结构解析最重要的手段。

除了冷冻电镜技术，X 射线自由电子激光（X-ray free electron laser，XFEL）技术的发明及其在结构生物学中的应用同样给生物大分子结构研究带来了革命。X 射线自由电子激光可以产生能够与单个生物颗粒作用的高能窄脉冲，以在不结晶的基础上对单个样品颗粒进行散射成像，进而利用全同的样品颗粒，在不同的空间取向探测样品结构信息，抓取生理环境下的结构特征（时盈晨和刘海广，2018）。X 射线自由电子激光技术与冷冻电镜技术相互补充，使得生物大分子结构解析实现了从静态结构到动态分子电影的飞跃。

4. 分子影像学和活细胞成像技术的发展推动分子细胞研究迈向无创、实时观测阶段

20 世纪，大部分分子细胞检测采用的是传统的终点法，这种方法无法得到分子和细胞的真正状态，也无法对细胞的生长过程以及其中分子表达水平进行动态监测和分析。为了解决上述问题，基于分子生物学、纳米材料学、

医学影像学、核医学、计算机学等学科的高度交叉融合，高特异性分子探针技术及成像技术快速发展。1999 年，哈佛大学的拉尔夫・维斯莱特（Ralph Weissleder）等首次提出分子影像学（molecular imaging，MI）的概念，即采用结合了检测基因或纳米材料的分子探针，通过多模态成像方法，最终对体内特定的靶点（活体组织、细胞或亚细胞结构）进行分子水平的成像。分子影像学可实现无创、实时地观察分子表达情况，起到连接分子生物学与临床医学的桥梁作用，能够更好地用于探索疾病的发生、发展、转归以及评价药物疗效（郭燕丽和范校周，2014）。由此，医学影像学进入了新的发展阶段。活细胞成像技术的发展则为即时、无创监测活细胞增殖、激活、死亡过程提供了支撑。其中，多光子显微技术、激光片层扫描显微技术等对活标本杀伤性更小的新技术，以及高灵敏度和高速图像采集技术等一系列技术的发展，使得对活细胞动态生长过程进行长时间、高时间分辨率和空间分辨率的动态观察成为可能，引领活细胞成像向更深、更快和更宽方向发展（Dance 和李楠，2019）。

5. 多组学技术的不断成熟促进分子细胞的研究尺度从单一分子向多层网络发展

作为生命体的基本单元，细胞是由复杂的生物大分子（复合体）和亚细胞结构（细胞器）组成的，因此如何整合多源数据并构建一个更能反映细胞内分子机制的多层生物分子网络模型是分子细胞科学一个亟待解决的关键问题。随着大规模组学数据的积累、信息理论的应用，以及化学和工程科学等多学科交叉与融合，基于高效的大数据信息分析技术实现了高水平的系统整合分析，系统、整合、跨尺度地研究细胞内不同组分和结构的功能与互作机制成为可能。多组学联合分析的提出和应用使得分子细胞科学的研究模式从以往的只关注某个分子扩展到对分子之间相互形成的多层生物分子网络的系统分析。传统多组学（即基因组学、转录组学、蛋白质组学、代谢组学等）开始向包括表观组学、免疫组学、微生物组学、生物影像组学等多个方向的全组学联合分析发展。

第二节　分子细胞科学领域发展现状

一、国际分子细胞科学领域规划与布局

（一）科研资助机构对分子细胞科学领域的资助方向

分子细胞科学作为生命科学的前沿领域，是全球科研的热点。美国国家科学基金会（National Science Foundation，NSF）、英国医学研究理事会（Medical Research Council，MRC）等国家层面的科研资助机构均设置了专门的部门管理对分子细胞领域的资助，重点资助方向是在生物学、物理、化学、数学、计算科学等多学科交叉的基础上，开发新技术，增进对分子细胞基本功能、分子调控网络及机制的全面认识，并进一步将相关认识与生理学、病理学、医学相结合，以加深对疾病发生与治疗应答机制的理解，进而推进药物靶标的发现和验证以及新疗法的开发。美国国立卫生研究院（National Institutes of Health，NIH）也非常重视分子细胞科学相关基础研究，资助了该领域的多个专项计划，旨在推进活细胞追踪技术、单细胞测序技术、成像技术等研究单细胞活动的技术的发展，并在此基础上探究分子细胞特征，以及相关变化对发育和各类疾病发生发展的影响。

1. 美国国家科学基金会对分子细胞科学领域的资助

美国国家科学基金会生物科学部设立了分子与细胞生物科学处（Division of Molecular and Cellular Biosciences，MCB），专门负责资助分子细胞科学相关的研究项目。MCB 的资助重点包括以下 4 个方向。

1）细胞动力学和功能

该方向重点支持针对各种细胞和亚细胞系统的理论研究。鼓励采用物理、化学、数学和计算科学等多学科的方法来开发新技术，以增进对细胞基本功能的全面认识。鼓励使用植物、微生物和非传统模式生物开展研究。

该方向优先资助以下3个领域。

（1）通过建模和实验相结合的方式研究活细胞行为。

（2）利用进化研究方法理解细胞功能调控规律。

（3）鼓励从时空角度将细胞功能与细胞特征进行关联。

2）遗传机制

该方向旨在了解细胞系统内遗传信息的进化、组织、动力学和功能的调控机制。鼓励综合使用理论方法、计算技术和实验方法来整合结构、生化、遗传和组学数据，以发现基因组稳定性维持和基因表达的机制。

该方向优先资助以下3个领域。

（1）染色质和RNA介导的调节机制。

（2）基因组复制、DNA修复、染色质修饰、转录和翻译的动态和时空协调机制。

（3）遗传聚合物（genetic polymers）的起源和进化，包括DNA、RNA和蛋白质。

3）分子生物物理学

该方向支持生物分子的结构、动力学和功能之间相互作用的基础研究。支持生物分子相互作用的基本原理研究、在分子水平上对生物功能的调节或基于这些原理设计分子新功能的研究。

该方向优先资助以下3个领域。

（1）开发大规模计算方法，解决不同实验技术带来的限制。

（2）基于方法学创新，在多个时间尺度和空间尺度上提供生物分子动力学信息，以破译它们在分子识别和功能中的作用。

（3）在原子或分子水平上确定大型生物相关组件装配体（assembly）的结构和相互作用。

4）系统与合成生物学

该方向支持利用系统和合成生物学工具来全面了解生物系统内的复杂相互作用。聚焦使用新兴的模型系统分析调控网络和代谢网络动态、控制复杂行为的基本机制、微生物群落及其相互作用。

该方向优先资助以下4个领域。

（1）建立调控、信号转导和代谢网络，以及网络之间相互作用的机理模型。

（2）生命的起源。

（3）开发新的实验技术和计算工具，包括开发有助于发现分子尺度机制的工具。

（4）微生物群落从分子到系统尺度的组装规律及功能。

2. 英国医学研究理事会对分子细胞科学领域的资助

英国医学研究理事会下设分子与细胞医学委员会（Molecular and Cellular Medicine Board，MCMB），该委员会提出了三个方面的优先发展方向。

1）了解动态生物系统

MCMB 致力于支持应用最新技术、定量分析工具和系统方法开展研究，从不同的层面（包括基因型与表型、单细胞、组织、生物体），研究健康和疾病的动态生物学机制。优先发展以下 4 个方向。

（1）开发新型跨尺度成像方法。

（2）开发新型实验系统（细胞、组织、动物模型），包括开发基质、结缔组织、脉管系统等组织模型。

（3）使用生物信息学、数学，以及机器学习和计算机建模技术来分析生物系统，并了解生物的复杂性和病理学信息。

（4）应用上述方法和知识，开发基因 / 细胞疗法，结合细胞 / 组织工程系统生物学方法和知识，进行药物靶标的发现、验证和开发，以及了解药物的作用、毒理和耐药性机制。

2）环境毒性因素暴露与疾病发生的关系及生物学机制

了解当前和新出现的环境暴露威胁（如空气污染、化学物质、纳米颗粒、混合物等）与人类疾病之间的因果关系是 MRC 的优先资助方向。该主题强调基于从流行病学研究中识别出的风险证据，加深对特定毒性因素暴露（包括所有类型的环境暴露因素和药物暴露）与疾病发生相关性机制（如分子引发事件、细胞应激反应、不良结局的机制等）的理解。优先发展以下 3 个方向。

（1）探索将毒性因素暴露与不良后果联系起来的因果关系和机制。

（2）分析多维数据以研究疾病的分子、细胞和生理信息，以及基因型对这些疾病的影响。

（3）开发新型计算模型和实验系统以研究暴露因素、暴露因素引发的生物学效应和暴露因素与疾病发生的因果关系。

3）放射生物学

当前，放疗仍然是癌症治疗中必不可少的方法。对放射生物学理解的不断深入将为多种肿瘤疗法的开发和优化提供支撑。未来还应进一步强化放射生物学和肿瘤研究的基础。优先发展以下 4 个方向。

（1）应用先进的成像技术、分子技术和细胞技术，在分子、细胞、组织和器官水平上，研究恶性肿瘤和健康组织对辐射的生物学应答及其潜在机制。

（2）将 DNA 损伤和修复研究与放射生物学方法相结合，在临床中研究机体发生损伤后分子层面的应答机制。

（3）研究基质、血管系统和宿主对放射线所产生的免疫反应的作用。

（4）改善放疗治疗癌症的效果，包括通过精准靶向癌变部位、设置合理的放疗流程、准确地给予辐射剂量；将放疗与药物和生物制剂结合治疗，以及开发应答生物标志物。

3. 美国国立卫生研究院对分子细胞科学领域的资助

美国国立卫生研究院（NIH）重视分子细胞科学相关基础研究，通过共同基金（Common Fund）资助了 4D 细胞核组计划（4D Nucleome Program）、人类生物分子图谱计划（Human BioMolecular Atlas Program，HuBMAP）等该领域的多个专项计划。同时，在脑科学计划和癌症登月计划（Cancer Moonshot）中设置了细胞图谱绘制任务。2017 年 10 月，NIH 宣布未来 5 年将提供 2.5 亿美元用于开展脑科学计划中的细胞普查网络（BRAIN Initiative Cell Census Network，BICCN）研究，该研究将整合不同类型细胞的分子、解剖学和生理学特性，创建小鼠脑细胞 3D 参考图谱，同时也将基于人类和非人灵长类脑组织样本，绘制脑细胞参考图谱。2018 年 9 月，作为 NIH 下属美国国家癌症研究所（National Cancer Institute，NCI）资助的癌症登月计划的一部分，人类肿瘤图谱网络（Human Tumor Atlas Network，HTAN）计划启动，该计划旨在构建反映从癌前病变发展为晚期癌症过程中肿瘤的细胞、形态学和分子动态特征的 3D 图谱。

1）美国 4D 细胞核组计划

2015 年 10 月 5 日，NIH 共同基金宣布资助 4D 细胞核组计划，旨在了解细胞核组的时空特征，研究细胞核组在基因表达和细胞功能中所发挥的作用，以及探究细胞核组的变化对发育和各类疾病的影响。该计划包括以下 6 个方面。

（1）细胞核组和功能跨学科联盟（Nuclear Organization and Function Interdisciplinary Consortium，NOFIC）：NOFIC 由多学科团队组成，这些团队将通过开发全基因组图谱绘制技术等新技术，研究细胞核组织结构的时空特征及其在基因表达调控中的作用。

（2）核酸组学工具。4D 细胞核组计划旨在通过开发化学和生化技术，分析哺乳动物细胞中基因组特定位置（基因座）之间或基因座与基因组组织和功能调控元件之间的三维相互作用。

（3）核质体和核区室的研究。核质体（细胞核内的结构和功能性亚单位）和核区室（细胞核内的特定子区域）共同形成了促进许多重要细胞核过程（nuclear process）发生的独特局部环境。4D 细胞核组计划支持工具和策略的开发，以研究与核质体的空间排列和基因表达调控分子机制有关的细胞核三维结构，特征不明显的细胞核结构、核区室的结构和功能，以及特定蛋白质和 RNA 在细胞核组织结构和功能中发挥的作用。

（4）成像工具。4D 细胞核组计划促进更高通量、更高分辨率和更高内涵成像技术的发展，以在活细胞中测量单细胞核组的变化。

（5）组织中心（4D Nucleome Organizational Hub，4DN-OH）。4DN-OH 的建立旨在促进 4D 细胞核组研究人员之间以及他们与其他科研人员之间就数据、试剂、标准和方案的共享与合作。同时，4DN-OH 还负责管理机会基金（Opportunity Pool），用于支持新项目和计划，以满足该计划实施全过程中产生的新需求。

（6）数据协调和集成中心（4D Nucleome Data Coordination and Integration Center，4DN-DCIC）。4DN-DCIC 的建立旨在跟踪、存储和展示所有 4D 细胞核组的数据。同时，4DN-DCIC 还计划建立一个数据分析中心，以协助研究人员进行数据综合分析，制定统一的衡量指标和标准，并提供可视化工具，以促进对复杂数据集的访问和理解。

2）美国人类生物分子图谱计划

2012年，NIH共同基金项目单细胞分析计划（Single Cell Analysis Program，SCAP）启动，旨在发展用于研究单细胞活动的技术。最终，该计划开发了可捕获单个细胞RNA的体内转录组分析标签技术、高通量单细胞转录组分析技术——液滴测序技术（indexing droplet，inDrop）、荧光原位测序技术、全基因组核测序技术等新技术，以及双视图倒置选择性平面照明显微术（dual-view inverted selective plane illumination microscopy，diSPIM）成像系统等新工具，并推动了用于RNA和蛋白质检测的单分子荧光原位杂交技术的发展，使得研究人员能够更好地研究单细胞。

基于相关研究基础，2018年，NIH启动了人类生物分子图谱计划（HuBMAP）。该计划旨在建立用于人体细胞图谱绘制的开放框架和技术，同时构建跨不同年龄段个体组织的细胞分子图谱。NIH此前曾资助过类似的项目，对1000人的组织样本中的DNA变异和表达模式进行了研究。此次HuBMAP希望能够利用更少数量的个体样本，获得单细胞水平、空间尺度的分子图谱。

HuBMAP计划通过两步路径获得空间上解析的单细胞图谱。

第一步，利用生命组学技术，获得单细胞的全基因组序列和基因表达图谱；同时，利用单细胞转录组和染色质可及性（chromatin accessibility）分析技术解析每个细胞的分子状态；将已经公开的转录因子结合域数据与基因表达数据相结合，解释不同细胞类型独特的基因表达调控模式。

第二步，利用荧光显微术、序贯荧光原位杂交（sequential fluorescence *in situ* hybridization，seqFISH）、质谱成像、成像质谱流式（imaging mass cytometry，IMC）等技术，获取各类分子的空间信息（冗余、一致性和定位）。

这种方法获得的大范围的单细胞图谱将助力解析细胞内分子的原位模式信息，进而为数百种分子靶标的研究提供空间信息。对细胞特异性的表观信息或转录组信息进行电子化管理，可解释细胞所处的不同的微环境状态，这些信息包括蛋白质在细胞质、细胞核和细胞表面的定位信息，磷酸化、复杂包装、细胞外环境和细胞表型信息等。

（二）分子细胞科学领域专项规划

近年来，美国、日本、欧盟等科技发达国家/组织相继出台了分子细胞科学领域的专项发展规划，重点资助方向涵盖了利用单细胞多组学技术、高分辨率成像技术、生物大数据技术等多学科交叉技术，旨在实现实时追踪分子 / 细胞空间分布特征，解析分子表达和细胞功能调控机制以及分子 / 细胞互作和通信网络等目标。

1. 人类细胞图谱计划

人类细胞图谱计划（Human Cell Atlas，HCA）于 2016 年在陈 – 扎克伯格计划（Chan Zuckberg Initiative，CZI）的资助下启动，目标是从细胞的基因表达、生理状态、发育轨迹和空间位置等维度来识别人类细胞的所有类型。HCA 计划最终将提供一个全面的涉及人类所有细胞类型和特征的参考图谱，助力解答人类生物学所有领域的问题，包括细胞结构、细胞发育、细胞命运和谱系、细胞生理、细胞内稳态等，同时也将使人们增进对健康的理解，加强对健康状况的监测，并提升疾病的诊断和治疗能力。

HCA 计划的主要研究方向包含以下 5 个方面。

（1）将人体的所有细胞类型（例如，免疫细胞、脑细胞）和子类型进行分类。

（2）将细胞类型映射到它们在组织和体内的空间位置。

（3）区分细胞状态（例如，比较一个尚未遇到病原体的原始免疫细胞与被细菌激活后的同一类型免疫细胞）。

（4）捕获细胞（如干细胞）在激活和分化等过程中的关键特征。

（5）通过谱系来追溯细胞的发育，如从骨髓中的干细胞到功能性红细胞。

HCA 计划的研究目标包含以下 9 个方面。

（1）为比较细胞和识别新的细胞类型提供参考图。

（2）识别基因在哪些细胞中发挥作用，助力解释遗传变异机制。

（3）辨识疾病和健康状态。

（4）确定可用于病理学研究、细胞分选和其他测量或测试的标签。

（5）增进对于细胞类型、细胞状态和细胞转型的理解。

（6）直观地展示人体内的生物学组织模式，消除细胞培养中存在的问题，促进遗留数据的分析。

（7）识别细胞分化、细胞之间相互作用、细胞状态维持的调节机制。

（8）发现疾病治疗干预的靶标。

（9）推动新技术和先进分析技术的发展。

2. 日本《活细胞图谱——通过多维分析揭示生命系统的动力学战略建议》

2019年3月，日本科学技术振兴机构（Japan Science and Technology Agency，JST）研发战略中心（Center for Research and Development Strategy，CRDS）发布了《活细胞图谱——通过多维分析揭示生命系统的动力学战略建议》。其中，活细胞图谱（Live Cell Atlas，LCA）将由全面的4D细胞参考图组成。该战略建议重点关注生物分子和细胞之间的互作网络的动力学，包括细胞间的通信。该战略建议将为广泛的生命科学研究奠定基础。

该战略建议提出了如下研究目标。

（1）开发新技术，在单细胞水平对多细胞系统中的生物分子进行全面定量分析：①开发细胞分离以及分类技术；②开发在完整的活细胞中分析生物分子（如RNA）的位置和表达状态的技术；③开发单细胞水平的蛋白质组和代谢组分析技术。

（2）开发能够在较高的分辨率下捕获生物分子/细胞时空分布特征的技术：①开发高通量原位测序技术；②开发多通道探针技术，同时对多种蛋白质代谢物进行时空分析；③开发在单细胞水平观察整个多细胞系统（宽视野和高分辨率）的成像技术；④开发和改进长时间活细胞成像技术，如观察发育过程所需的自动聚焦和细胞追踪技术。

（3）建立有助于以定量的方式理解生物分子和细胞动力学的数学模型：①建立综合数据库，以存储和利用各种组学与成像大数据；②开发多组学和成像数据的综合分析技术；③建立数学模型实现生物分子和细胞互动网络的定量可视化分析。

3. 日本4D细胞组战略建议

2020年，日本科学技术振兴机构（JST）研发战略中心（CRDS）发布《4D细胞组——面向细胞结构、动力学和功能研究的新阶段战略建议》（简称为《4D细胞组战略建议》）。

为了准确地构建4D细胞组（4D cellome），该战略建议除了着眼于特定的基因组和蛋白质组研究以外，还将促进创新技术的开发，以阐明细胞内功能元件（超分子复合体、细胞器等）的动态结构、空间定位、数量和功能的定量因果关系（图 2-1）。这些技术包括：细胞内功能元件的动态结构、空间定位、数量等时空信息的测量技术，用于数据解释和分析的数学、物理、计算化学、信息科学技术，基于化学生物学、生物技术等对细胞内功能元件进行操作和控制的技术等。

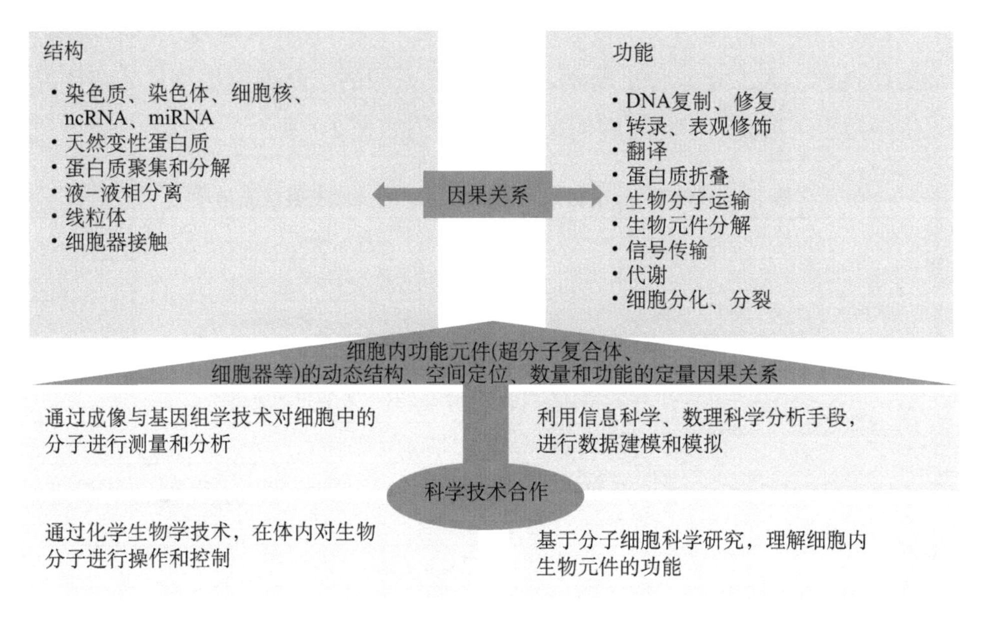

图 2-1 《4D 细胞组战略建议》（通过时空信息了解整个细胞的分子动态）的研究对象

4. 欧盟委员会新“FET 旗舰计划”的筹备行动——疾病发生发展机制研究

2019 年 3 月 1 日，欧盟委员会启动了 6 个“未来和新兴技术”（Future and Emerging Technologies，FET）旗舰计划的筹备行动项目，旨在解决未来数十年欧洲面临的主要挑战。每个项目资助 100 万欧元，在 1 年内完成具体科学和技术议程。这些项目将有助于在“地平线欧洲”（Horizon Europe）计划框架下发起新的研究活动，并有望成为新 FET 旗舰计划。其中之一的“疾病发生发展机制研究”（LifeTime）项目旨在更好地理解人体内疾病的发生与发展机制。LifeTime 项目将开发和整合多项突破性技术，包括单细胞多组学技术、成像技术、机器学习和人工智能（artificial intelligence，AI）技术，以及类器官等个性化疾病模型构建技术，以帮助人们了解基因组在细胞内如何发挥作用、细胞如何形成组织，以及在组织病变时细胞如何动态重塑组织活动。LifeTime 项目将对慢性和进行性疾病的早期发现、预防和创新治疗产生巨大影响。

2020 年 6 月，在综合了 100 多家机构和医疗中心、80 家公司的专家观点的基础上，LifeTime 项目发布战略研究议程，绘制了科技发展路线图。

LifeTime 项目包括 3 个核心的技术支柱，即：①单细胞多组学和成像；②数据科学、人工智能和机器学习；③患者来源的实验疾病模型。战略研究议程中指出了相关技术的发展优先事项，具体如表 2-1 所示。

表 2-1　LifeTime 项目战略研究议程中的技术发展优先事项

核心技术	发展优先事项
单细胞多组学和成像	1. 为不同大分子开发强大的高分辨率单细胞分析技术
	2. 基于新型单细胞多组学方法以探索疾病机制
	3. 开发基于高灵敏度、高分辨率成像和测序的空间组学分析技术
	4. 新型空间单细胞多组学分析和先进的成像技术
	5. 用于数据集成和分析的计算方法
	6. 将基因干扰筛选和谱系示踪分析与单细胞多组学分析相整合
	7. 建立技术标准和基准
	8. 提升临床样品采集和处理能力
数据科学、人工智能和机器学习	1. 大规模数据的存储和协调
	2. 开放和受控数据的访问
	3. 联合查询和分析
	4. 用于整合来自不同实验模式的数据的计算方法
	5. 将计算分析扩展到更高维度的数据
	6. 将多组学分子数据整合到电子健康记录中
	7. 根据患者细胞轨迹生成疾病预测模型
	8. 确立影响疾病进展的因果分子机制
	9. 验证和基准化分析计算工具
患者来源的实验疾病模型	1. 增加类器官的细胞、结构和功能复杂性
	2. 通过整合不同的类器官研究不同组织和器官间的相互作用
	3. 类器官培养和存储的标准化、自动化、规模化和基准化
	4. 用于研究人类病理生理反应和治疗应答的先进动物模型
	5. 基于大规模扰动工具在模式生物中确立疾病因果关系

LifeTime 项目还确定了 5 个关键的疾病领域，即：①癌症；②神经和精神疾病；③感染性疾病；④慢性炎症；⑤心血管和代谢性疾病。相关技术的整合和应用将对这些疾病患者的诊治和结局产生直接影响。战略研究议程中为如何在深入认识疾病发生发展的机制基础上，实现疾病的早期发现、预防和创新治疗提出了具体的资助方向，包括：利用单细胞多组学方法绘制患者细胞类型和状态纵向图谱，开发先进的患者来源的疾病模型；基于患者队列和先进的疾病模型进行纵向研究，以理解疾病发生发展机制并确定新的药物靶标；在新设计的临床试验中验证用于患者分层、疾病拦截和预防的生物标志物；等等。

二、我国分子细胞科学领域规划与布局

我国已通过国家自然科学基金、国家重点研发计划等对分子细胞科学领域进行了资助，旨在通过多学科交叉研究技术深度解析分子 / 细胞结构和功能，并进一步实现对分子 / 细胞的精准操控与人工设计合成。

（一）国家自然科学基金

以 2020 年国家自然科学基金的规划情况为例进行说明。

1. 2020 年国家自然科学基金面上项目

分子细胞科学作为由分子生物学和细胞生物学高度融合，并与物理、化学、医学、信息科学多学科交叉而衍生的前沿学科，在 2020 年国家自然科学基金面上项目中多个学科项目涉及该主题的资助布局（表 2-2）。

表 2-2　2020 年国家自然科学基金面上项目指南中涉及分子细胞科学领域的资助方向

学科	资助方向
遗传学与生物信息学	多组学数据分析与整合；分子模块和网络的建模、分析、重构与设计研究；等等
细胞生物学	细胞微环境与细胞命运决定生物膜及膜性细胞器的发生、重构、运输、清除机制与生物学意义；代谢物感应与细胞稳态维持；非膜性细胞器的相变；功能分区化的结构和调控；核质互作；细胞信号网络的时空调控与定量；细胞示踪与谱系；细胞衰老机制及干预；细胞间识别、互作与功能调控；等等

续表

学科	资助方向
生物物理与生物化学	生物活性分子的实时、动态和微量检测；超高分辨率显微镜；单分子成像；非荧光成像；亚细胞器在细胞内的时空动态变化监测；生物分子体内外定量体系；等等
分子生物学与生物技术	合成生物学；基因编辑；生物分子的原位与活体分析；复杂系统的单分子与单细胞分析；多尺度多模态成像；人工智能与计算生物学；等等
化学生物学	分子探针的发现、构建及其在生物重大事件和重大疾病中的分子机能与功能调控；等等

2. 2020 年国家自然科学基金重大研究计划

2020 年，国家自然科学基金设置了“细胞器互作网络及其功能研究”“生物大分子动态修饰与化学干预”等分子细胞科学领域相关的重大研究计划。其中，“细胞器互作网络及其功能研究”计划旨在以内质网、线粒体、溶酶体、高尔基体、叶绿体等细胞器（包括细胞质膜、核膜等膜性结构）之间的互作为研究主线，发现新的细胞器互作方式，阐明细胞器互作的分子机制和生物学效应，绘制细胞器互作网络图谱，阐明细胞器互作网络的建立、维持、动态变化及其调控机制，揭示细胞器互作网络的生理、病理功能。“生物大分子动态修饰与化学干预”计划旨在充分发挥化学、生命科学和医学的特点以及学科交叉的优势，引领生物大分子动态修饰与化学干预研究，为生物大分子动态修饰的机制研究提供具有化学特征的新工具和新模式，获得针对动态修饰的新药物靶标和相应的干预小分子；加速从基础研究到药物开发的转化，为认识生命体系调控的内在规律、为重大疾病的诊断与防治提供基础性和前瞻性的科学技术储备；促进化学与生命科学和医学研究的衔接和交叉集成，形成新的学科生长点，提升我国生物大分子动态修饰的基础研究和应用性研究的综合实力，以及其在国际化学生物学领域和生物医学前沿研究中的地位；同时，打造一支学科深度交叉、具有国际影响力的化学生物学科研队伍。

（二）国家重点研发计划

国家重点研发计划中的一些重点专项提出了与分子细胞科学研究相关的资助方向，旨在通过项目研究揭示生命发育过程中关键蛋白质、代谢调控机制，以及细胞命运调控机制，并在分子细胞人工设计合成中取得突破（表 2-3）。

表 2-3　2020 年国家重点研发计划在分子细胞科学领域的支持方向

重点专项	支持方向
蛋白质机器与生命过程调控	1. 细胞增殖与分化过程中关键蛋白质机器的功能机制研究；
	2. 黏膜免疫相关蛋白质机器的功能机制研究；
	3. 重要病原体感染和致病过程中的蛋白质机器研究；
	4. 调控重要植物、作物关键生命过程的蛋白质机器研究；
	5. 神经系统疾病发生发展或发育相关的蛋白质机器研究；
	6. 极端条件下外泌体的功能与调控研究；
	7. 畜牧或水产产品生殖发育、抗逆、抗病的关键蛋白质机器研究
发育编程及其代谢调节	1. 器官发育与稳态编程及其代谢调节；
	2. 营养与环境对器官发育和稳态的调节机制；
	3. 代谢和发育紊乱相关疾病的发生发展机制
变革性技术关键科学问题	1. 生物界面蛋白质冠主动精准调控与高效递送载体构建；
	2. 工程化细胞逆转重要器官纤维化的分子机制和临床转化研究
干细胞及转化研究	1. 多能干细胞的谱系分化过程与细胞命运决定；
	2. 细胞命运调控中的间质状态和上皮状态之间的转变；
	3. 胚层特异干细胞的建立与调控
合成生物学	1. 人工基因组合成与高版本底盘细胞构建；
	2. 人工元器件与基因回路；
	3. 特定功能的合成生物系统；
	4. 数字细胞建模与人工模拟；
	5. 新蛋白质元件人工设计合成及应用

三、国际分子细胞科学领域发展现状

下文将一方面以分子细胞科学领域的重要基础分子细胞生物学作为对象，从定量的角度分析其论文产出情况，另一方面从定性的角度对现阶段分子细胞科学领域的最新进展进行分析。两部分相结合，以期展现国际分子细胞科学领域的基础和前沿两方面的现状。

在科研论文的检索方面，采用美国科学信息研究所（Institute for Scientific Information，ISI）的科学引文索引（Science Citation Index Expanded，SCI-Expanded）数据库作为数据源，论文统计类型限定为“article”和“review”，即仅限定在纯科研型论文范畴。检索方法是选取 Web of Science 平台期刊分类中细胞生物学学科期刊，检索其中发表的分子细胞生物学研究的论文。

论文分析结果显示，作为分子细胞科学基础和重要组成部分的分子细胞生物学领域始终维持着较快的发展速度。从论文发表情况可以看到，1980～2020年，全球共发表分子细胞生物学研究论文680 055篇（图2-2）。其中，论文数量在1990～1995年和2013～2017年两个区间分别有一个明显的快速增长期，论文年均增长率分别达到19.77%和8.35%。在此期间，基因组、蛋白质组研究理念开始建立，使得科学家在分子细胞结构和功能研究领域取得了重要进展，并成功解析细胞分裂、生长、运动、兴奋性、分化、衰老与病变过程中的多种机制。与此同时，细胞工程技术逐渐成熟，伴随着人们对分子细胞功能和调控机制的认识愈发深刻，该技术在生物医药合成、基因治疗、动植物育种等领域的应用被不断推广。

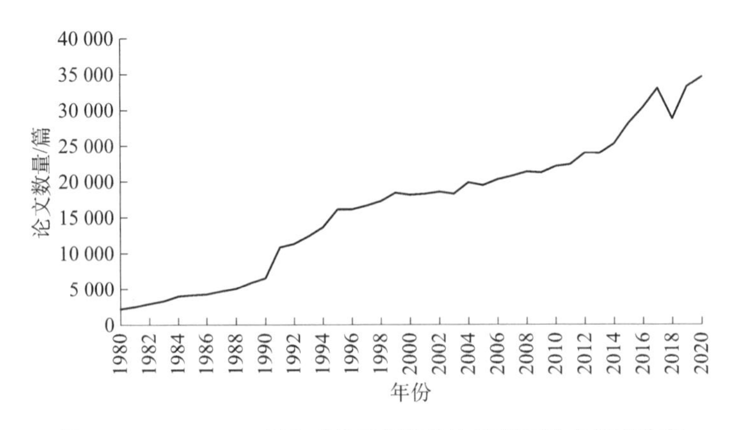

图2-2 1980～2020年全球分子细胞生物学研究论文年度分布

多学科交叉技术的发展促进了分子细胞生物学的进一步优化发展，使得分子细胞研究朝着看得更清、算得更快、通量更高、改得更好的方向不断迈进，人们对细胞生命活动规律及其分子机制的理解得到进一步升华，从而进入分子细胞科学发展新阶段。

例如，显微技术的不断进步提高了生物分子成像的分辨率和数据质量，为了解细胞内生物大分子提供了新机遇。一方面，基于对荧光分子光转化能力和单分子定位技术的探索，超高分辨率荧光显微技术得以发明。该技术突破了“阿贝衍射极限”，使光学显微镜步入纳米时代，为病毒和亚细胞结构等精细结构的研究提供了强有力的支撑。另一方面，生物学与物理、数学、信息科学交叉融合，推动了冷冻电镜技术的成熟。其中，算法的优化为高效、高精度地完成生物大分子三维重构奠定了基础，而直接电子探测器的发明和应用更是使冷冻电镜技术得以突破分辨率的障碍，迎来了“分辨率革命”，其解析精度达到了近原子水平，进而开启了冷冻电镜时代。

人工智能技术的发展引领了新一轮科技革命，而基于图像识别、神经网络等关键技术的进步，人工智能在医学分子影像诊断、活性分子发现与筛选中也得到了广泛应用，并为破解生物大分子结构开辟了全新的发展道路。2020年，谷歌公司旗下的英国人工智能技术公司 DeepMind 研发出新一代算法 AlphaFold2。借助该算法，研究人员能够在几天内基于氨基酸序列精确预测蛋白质的 3D 结构，准确性可以与使用冷冻电子显微镜、核磁共振或 X 射线晶体学等实验手段解析的 3D 结构相媲美。应用人工智能技术解析蛋白质结构，大幅降低了研究成本、提高了结构解析速度，并扩大了蛋白质解析的范围。

微流控技术作为操纵微量液体的新兴技术，尤其适合单细胞这样的微观物体操纵，其在单细胞测序中的应用开启了单细胞测序的规模化时代。与此同时，随着单细胞技术和高通量组学方法的进步，在单细胞水平同时整合基因组、转录组、表观基因组、蛋白质组或代谢组等两种或多种组学信息的单细胞多组学技术应运而生，为深入分析和理解细胞状态和命运决定的内在机理铺平了道路，该技术也被《自然－方法》评选为 2019 年的年度技术。

在分子细胞改造方面，在深入认识分子细胞信号调控网络及调控机制的基础上，科学家进一步融入工程化的思想，对生物体进行有目标的设计、改造乃至重新合成，实现基于空间结构的催化元件序列数字化设计、精确控制细胞行为的基因回路设计等，突破了生物自然进化的局限。

四、我国分子细胞科学领域发展现状

1980～2020年，我国共发表分子细胞生物学研究论文 80 904 篇（图 2-3）。1981 年，我国在该领域论文数量为 1 篇；之后的 10 年中，论文数量整体增长较慢，每年论文数量维持在 10 篇左右；1991 年后，我国分子细胞生物学研究领域论文数量开始呈现增长趋势，于 1999 年首次突破 100 篇，到 2007 年突破 1000 篇；之后，我国分子细胞生物学研究领域进入快速发展阶段，从 2011 年的 2441 篇增长至 2020 年的 12 560 篇，论文数量年均增长率达到 19.96%。

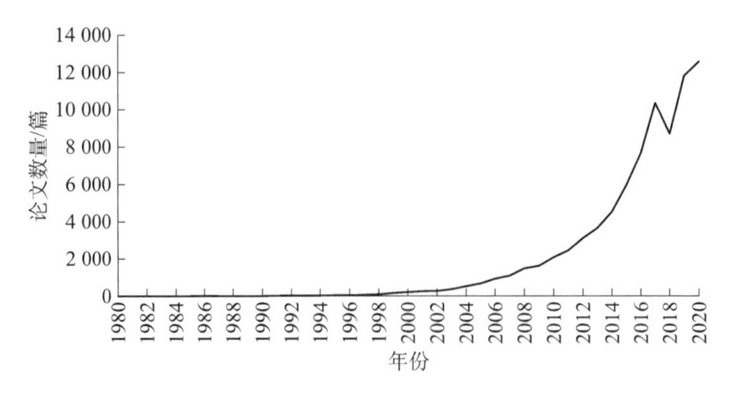

图 2-3　1980～2020 年我国分子细胞生物学研究论文年度分布

我国在分子细胞科学领域的发展紧跟全球领先水平，在单细胞测序技术、冷冻电镜技术、合成生物学等分子细胞研究关键交叉技术及其应用中取得了国际领先成果。

例如，在单细胞测序方面，北京大学汤富酬团队相继开发了单细胞转录组测序技术、单细胞 DNA 甲基化组测序技术——单细胞简化代表性重亚硫酸盐测序技术（single-cell reduced representation bisulfite sequencing，scRRBS）、单细胞表观多组学测序技术——单细胞染色质整体组尺度景观测序技术（single-cell chromatin overall omic-scale landscape sequencing，scCOOL-seq）（Guo et al.，2017）等高通量单细胞测序技术；北京大学谢晓亮团队于 2017 年开发了一种新型单细胞全基因组线性扩增方法——转座子插入线性扩增（linear amplification via transposon insertion，LIANTI）（Chen et al.，2017）；浙江大学郭国骥团队自主研发了 Microwell-seq 高通量单细胞分析平台，并于 2018 年和 2020 年分别发布首个小鼠细胞图谱（Han et al.，2018）和人类细胞图谱（Han et al.，2020）。

在冷冻电镜技术方面，我国多个高校和科研院所已经建立了冷冻电镜分析平台，目前全球大约有 100 个冷冻电镜研究机构，而我国目前正在运行的冷冻电镜中心有 9 个，约占世界总量的 10%；全球正在使用的 300 kV 高端冷冻电镜总量约为 200 台，我国有 20 台，约占 10%（郭振玺等，2020）。基于冷冻电镜技术，我国科学家做出了多项重大性突破，包括解析了分子生物学领域困扰人们 30 多年的 30 nm 染色质纤维高级结构（Song et al.，2014）。

在合成生物学方面，中国科学院分子植物科学卓越创新中心覃重军团队将单细胞真核生物酿酒酵母天然的 16 条染色体人工创建为具有完整功能的单条染色体，在国际上首次人工创建了单条染色体的真核细胞（Shao et al.，2018），获得了合成生物学领域具有里程碑意义的突破——创造出了一个简约化的生命体。

第三节　分子细胞科学领域竞争力分析

一、学科地位分析

选取 2020 年生命科学领域的高被引论文作为分析对象，以关键词聚类[①]的方法来反映当年生命科学不同领域受关注的情况。

从图 2-4 可以看出，2020 年生命科学领域高被引论文聚焦在细胞分子结构功能解析、新型冠状病毒（SARs-Cov-2，简称新冠病毒）相关研究、临床试验三大主题，可见分子细胞科学研究在整个生命科学领域处于极其重要的位置，旨在综合多学科力量利用前沿技术揭示细胞命运、细胞表型、细胞功能分子表达调控机制、细胞信号转导调控、细胞分子结构及互作等关键问题。

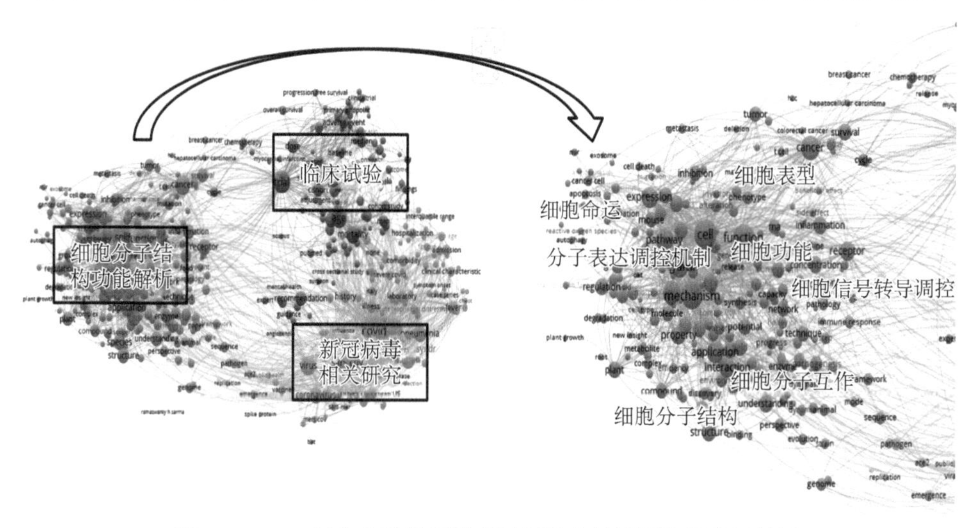

图 2-4　2020 年生命科学领域高被引论文关键词聚类分析结果

① 分析工具为 VOSviewer.

二、国家竞争力分析

分子细胞生物学是分子细胞科学的基础和重要组成部分，从分子细胞生物学科研产出角度进行分析，总结梳理部分国家在分子细胞领域的研究水平。

2011～2020年，分子细胞生物学领域论文数量Top10国家如图2-5所示，美国发表论文97 794篇，占全球分子细胞生物学论文总量的34.46%，位居全球首位。中国发表论文70 756篇，占全球分子细胞生物学论文总量的24.93%，排在第2位。德国、英国、日本分别发表论文24 742篇、21 684篇和18 621篇，分别占全球分子细胞生物学论文总量的8.72%、7.64%和6.56%，位列第3至第5位。

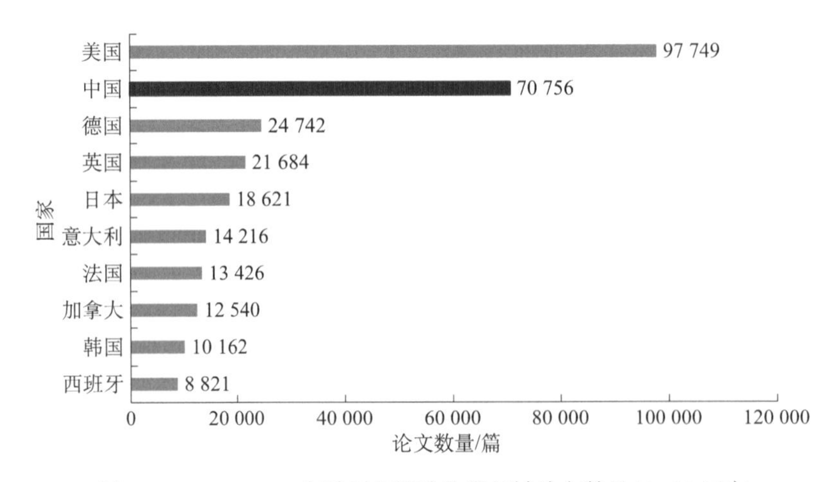

图2-5　2011～2020年分子细胞生物学领域论文数量Top10国家

从分子细胞生物学领域论文数量Top5国家发文年度分布来看，2011～2020年，中国分子细胞生物学领域论文数量迅速增长，年均增长率达到19.96%，其他4个国家该领域论文数量在这10年变化不大，年均增长率为1%～2%（图2-6）。

对分子细胞生物学论文数量Top10国家的论文被引用情况进行分析，美国论文的总被引频次和篇均被引频次均居首位，分别为3 991 780次和40.82次。中国分子细胞生物学论文的总被引频次为1 218 012次，居10个国家的第2位，而篇均被引频次为17.21次，居10个国家的第10位（表2-4）。

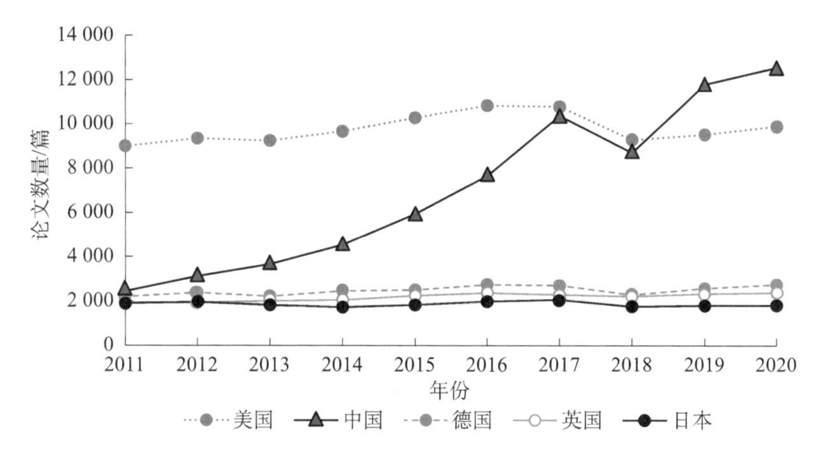

图 2-6　2011～2020 年分子细胞生物学领域论文数量 Top5 国家发文年度分布

表 2-4　2011～2020 年分子细胞生物学领域论文数量 Top10 国家论文被引用情况

排名	国家	论文数量 / 篇	总被引频 / 次	篇均被引频次 / 次
1	美国	97 794	3 991 780	40.82
2	中国	70 756	1 218 012	17.21
3	德国	24 742	854 699	34.54
4	英国	21 684	861 650	39.74
5	日本	18 621	484 565	26.02
6	意大利	14 216	421 354	29.64
7	法国	13 426	489 611	36.47
8	加拿大	12 540	412 393	32.89
9	韩国	10 162	219 316	21.58
10	西班牙	8 821	299 016	33.90

二、机构竞争力分析

从全球研究机构排名来看，2011～2020 年，美国哈佛大学共发表分子细胞生物学论文 9 225 篇，占该领域论文总数的 3.25%，居全球首位（图 2-7）。

中国共有3家机构进入论文数量Top10，中国科学院、上海交通大学和中山大学分别发表分子细胞生物学论文5803篇、4376篇和3231篇，分居第5位、第7位和第10位。

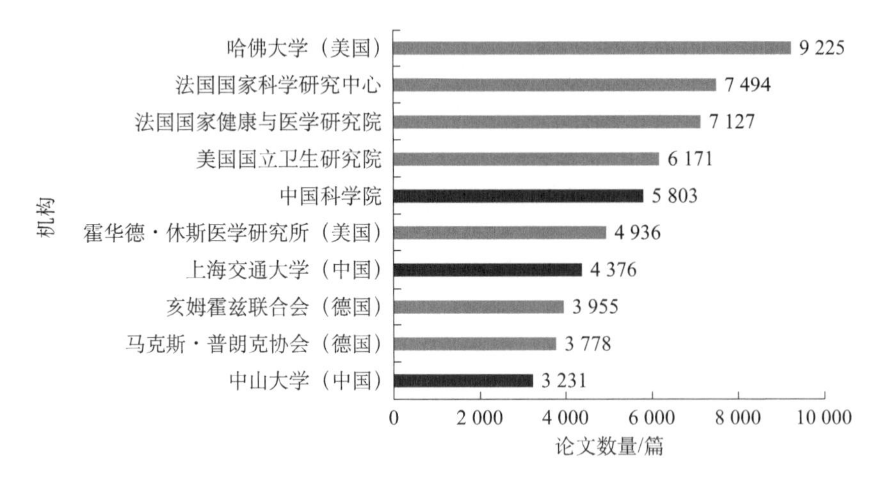

图2-7　2011～2020年全球分子细胞生物学领域论文数量Top10研究机构

对分子细胞生物学论文数量Top10机构的论文被引用情况进行分析，美国哈佛大学论文总被引频次为626 968次，位于首位，其篇均被引频次为67.96次。霍华德·休斯医学研究所的篇均被引频次居全球首位，达95.29次。中国科学院、上海交通大学和中山大学的篇均被引频次分别为30.81次、20.00次和21.54次（表2-5）。

表2-5　2011～2020年分子细胞生物学领域论文数量Top10研究机构论文被引用情况

排名	机构	论文数量/篇	总被引频次/次	篇均被引频次/次
1	哈佛大学（美国）	9 225	626 968	67.96
2	法国国家科学研究中心	7 494	275 463	36.76
3	法国国家健康与医学研究院	7 127	275 169	38.61
4	美国国立卫生研究院	6 171	305 486	49.50
5	中国科学院	5 803	178 762	30.81
6	霍华德·休斯医学研究所（美国）	4 936	470 356	95.29
7	上海交通大学（中国）	4 376	87 533	20.00
8	亥姆霍兹联合会（德国）	3 955	166 578	42.12

续表

排名	机构	论文数量 / 篇	总被引频次 / 次	篇均被引频次 / 次
9	马克斯・普朗克协会（德国）	3 778	192 646	50.99
10	中山大学（中国）	3 231	69 610	21.54

从中国研究机构排名来看，中国科学院共发表分子细胞生物学论文 5803 篇，占该领域中国论文总数的 8.20%，居中国首位（图 2-8）。上海交通大学和中山大学发表分子细胞生物学论文数量分别居第 2 位和第 3 位，分别占该领域中国论文总数的 6.18% 和 4.57%。

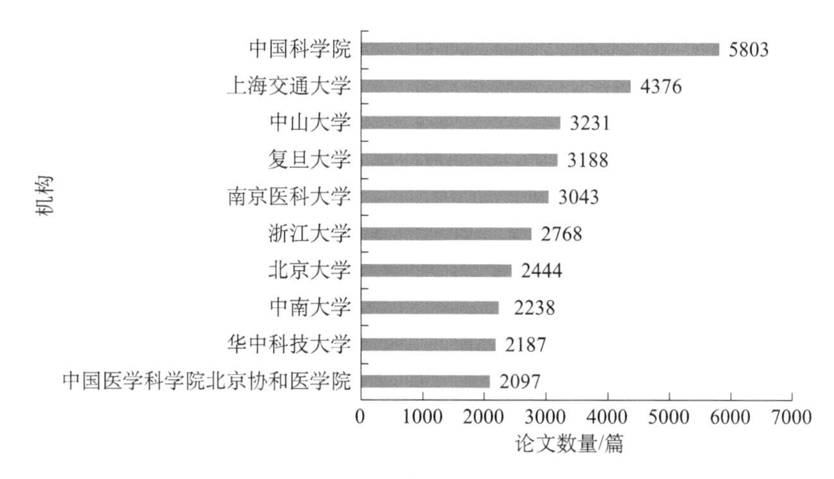

图 2-8　2011～2020 年中国分子细胞生物学领域论文数量 Top10 研究机构

对中国分子细胞生物学论文数量 Top10 机构的论文被引用情况进行分析，中国科学院论文总被引频次为 178 762 次，篇均被引频次为 30.81，均居中国首位（表 2-6）。

表 2-6　2011～2020 年中国分子细胞生物学领域论文数量 Top10 研究机构论文被引用情况

排名	机构	论文数量 / 篇	总被引频次 / 次	篇均被引频次 / 次
1	中国科学院	5 803	178 762	30.81
2	上海交通大学	4 376	87 533	20.00
3	中山大学	3 231	69 610	21.54
4	复旦大学	3 188	71 257	22.35
5	南京医科大学	3 043	54 299	17.84
6	浙江大学	2 768	61 163	22.10

续表

排名	机构	论文数量 / 篇	总被引频次 / 次	篇均被引频次 / 次
7	北京大学	2 444	65 724	26.89
8	中南大学	2 238	39 712	17.74
9	华中科技大学	2 187	41 248	18.86
10	中国医学科学院北京协和医学院	2 097	45 654	21.77

四、高水平论文情况

ESI 高水平论文包括 ESI 高被引论文和 ESI 热点论文，分别对应近 10 年内发表且被引频次排在相应学科领域全球前 1% 以内的论文、近 2 年内发表且在近 2 个月内被引频次排在相应学科领域全球前 1‰以内的论文。

2011～2020 年，全球分子细胞生物学领域 ESI 高水平论文共 3833 篇。从国家排名来看，美国发表 ESI 高水平论文 2473 篇，居全球首位。英国、中国、德国和法国分别发表 ESI 高水平论文 571 篇、541 篇、487 篇和 273 篇，分居第 2 位至第 5 位（图 2-9）。

从研究机构排名来看，在分子细胞生物学领域 ESI 高水平论文数量 Top10 的研究机构中，除了中国科学院和法国国家健康与医学研究院，其余都是美国的研究机构。其中，哈佛大学发表 ESI 高水平论文 519 篇，居全球首位。中国科学院发表 ESI 高水平论文 159 篇，居全球第 8 位（图 2-10）。

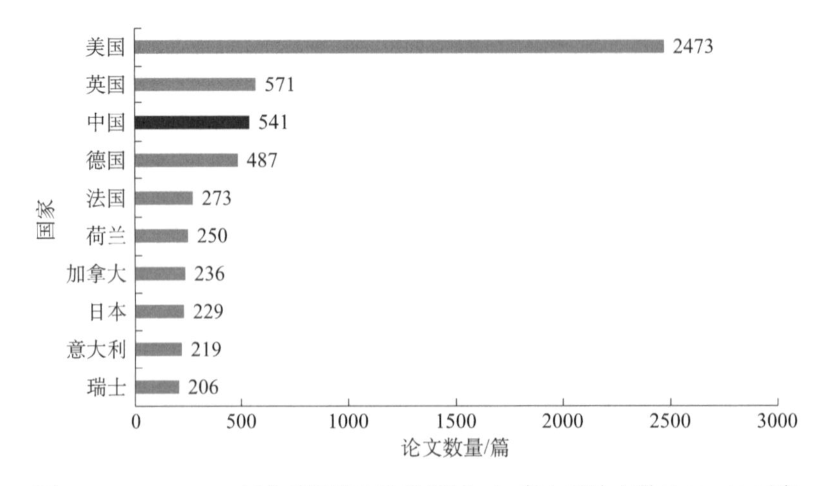

图 2-9　2011～2020 年分子细胞生物学领域 ESI 高水平论文数量 Top10 国家

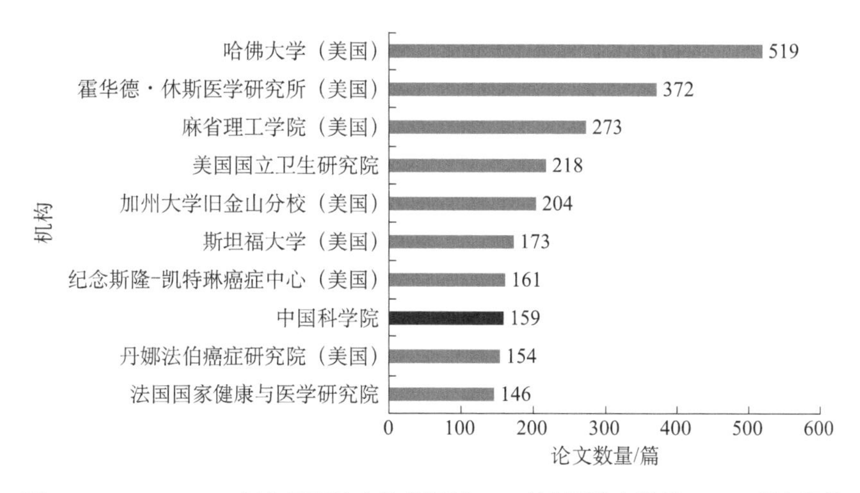

图 2-10　2011～2020 年分子细胞生物学领域 ESI 高水平论文数量 Top10 研究机构

第四节　分子细胞科学领域发展政策建议

随着各国不断推出分子细胞科学相关的重大专项与科技规划，这一前沿领域逐渐成为全球生命科学研究中的热点。在此，对我国分子细胞科学领域发展提出如下政策建议。

一、增加财政投入提升研发水平

我国在生命科学基础研究和应用研究的投入相比美国仍然有巨大差距，建议增加科研经费投入，以提升我国在生命科学基础研究和应用研究的整体水平。在国家科技发展的中长期规划中纳入分子细胞科学研究领域，通过科学的规划和系统的激励，把我国的分子细胞科学研究提升至国际引领水平，以自主源头创新把握发展主动权。

二、设立重大项目引导前沿研究

设立分子细胞科学领域的重大科技项目。美国、欧盟、日本等发达国家/组织通过设立大型科技专项有力地支持了地区的分子细胞科学研究的发展。在国家重点研发计划、国际自然科学基金重大项目、中国科学院战略先导科技专项等重大科技规划中长期设立分子细胞科学专项，做好前瞻性战略部署，通过系统地规划和跟踪，鼓励前沿探索，同时滚动资助优秀项目，把学问做深，把领域做强。

三、建立研究基地打造学术高地

建立分子细胞科学领域的国家重点实验室，实现人才队伍建设和研究能力建设。通过稳定支持科学家工作室模式，鼓励原始创新“无人区”式的基础研究；通过多元化研究资助模式，鼓励目标导向、需求导向的应用基础研究；通过支撑平台的服务能力建设，缩短科研人员与前沿技术的距离，实现研究水平跨代提升；通过青年研究人员培养，为领域的可持续发展不断输送创新性优秀人才。以国家重点实验室为实体，打造分子细胞科学的前沿研究、应用转化、技术服务和人才培养的学术高地。

四、组织品牌会议促进学科交叉

多学科交叉应用是提升分子细胞科学研究维度的关键。我国学科体系较为完备，科研水平整体处于高速发展时期，各学科的最新成果和技术不断迭代。分子细胞科学的发展正处于这样良好的发展环境中。建议通过定期组织召开主题清晰、规模适中的高水平学术会议，通过不同领域科学家们的交流，提升与其他学科领域特别是与信息工程、人工智能、临床医学等方向的交叉融合，促进知识的联系，拓展领域的外延，突破学科的边界，实现分子细胞科学的跨越式发展。

五、扩大国际交流引进国际人才

近30年的国际化人才培养是我国生命科学领域高速发展的原动力之一。目前，我国在分子细胞科学相关领域和国际领先水平仍然存在较大差距。建议进一步加强国际人才交流，培养具有国际化视野的高端创新人才，同时继续支持和加强同领先国家及机构的科技合作和交流，吸纳国内外优秀人才以多种方式参与分子细胞科学研究。

本章参考文献

郭燕丽，范校周. 2014. 超声分子影像学：现状与将来. 第三军医大学学报，36（1）：6-10.

郭振玺，王晋，张丽娜，等. 2020. 我国冷冻电镜平台建设现状及其发展. 中国科技资源导刊，52（6）：52-62.

贺玖明，李铁钢，何菁菁. 2012. 常压敞开式离子化质谱技术研究进展. 分析测试学报，31（9）：1151-1160.

姜鹏. 2021. 基因编辑与细胞农业. 北京规划建设，(06)：165-169.

时盈晨，刘海广. 2018. X射线自由电子激光的原理和在生物分子结构测定研究中的应用. 物理，47（7）：426-436.

Dance A, 李楠. 2019. 活细胞成像：更深、更快、更宽. 科学新闻，556（1）：59-62.

Annaluru N, Muller H, Mitchell L A, et al. 2014. Total synthesis of a functional designer eukaryotic chromosome. Science, 344（6179）: 55-58.

Center for Research and Development Strategy, Japan Science and Technology Agency. 2019. Live Cell Atlas Deciphering Dynamics of Biological Systems via Multi-Dimensional Analysis. https://www.jst.go.jp/crds/pdf/2018/SP/CRDS-FY2018-SP-09.pdf[2021-03-05].

Center for Research and Development Strategy, Japan Science and Technology Agency. 2020. 4D Cellome: Toward a new phase of research into cell structure, dynamics and function. https://www.jst.go.jp/crds/pdf/2019/SP/CRDS-FY2019-SP-05.pdf [2021-03-05].

Chen C Y, Xing D, Tan L Z, et al. 2017. Single-cell whole-genome analyses by Linear Amplification via Transposon Insertion（LIANTI）. Science, 356（6334）: 189-194.

Cookson B T, Brennan M A. 2001. Pro-inflammatory programmed cell death. Trends in

Microbiology, 9（3）: 113-114.

European Commission. 2019. Launch of six European initiatives with potential for transformational impact on society and the economy. https://ec.europa.eu/digital-single-market/en/news/launch-six-european-initiatives-potential-transformational-impact-society-and-economy[2020-11-08].

Gibson D G, Glass J I, Lartigue C, et al. 2010. Creation of a bacterial cell controlled by a chemically synthesized genome. Science, 329（5987）: 52-56.

Guo F, Li L, Li J Y, et al. 2017. Single-cell multi-omics sequencing of mouse early embryos and embryonic stem cells. Cell Research, 27（8）: 967-988.

Guo H S, Zhu P, Wu X L, et al. 2013. Single-cell methylome landscapes of mouse embryonic stem cells and early embryos analyzed using reduced representation bisulfite sequencing. Genome Research, 23（12）: 2126-2135.

Han X P, Wang R Y, Zhou Y C, et al. 2018. Mapping the mouse cell atlas by microwell-seq. Cell, 172（5）: 1091-1107.

Han X P, Zhou Z M, Fei L J, et al. 2020. Construction of a human cell landscape at single-cell level. Nature, 581（7808）: 303-309.

HuBMAP Consortium. 2019. The human body at cellular resolution: the NIH Human Biomolecular Atlas Program. Nature, 574（7777）: 187-192.

HUMAN CELL ATLAS. 2016. More about the human cell atlas. https://www.humancellatlas.org/learn-more/[2021-03-05].

Larsson C, Grundberg I, Söderberg O, et al. 2010. *In situ* detection and genotyping of individual mRNA molecules. Nature Methods, 7（5）: 395-397.

Lee J H, Daugharthy E R, Scheiman J, et al. 2014. Highly multiplexed subcellular RNA sequencing *in situ*. Science, 343（6177）: 1360-1363.

Liao M F, Cao E H, Julius D. 2013. Structure of the TRPV1 ion channel determined by electron cryo-microscopy. Nature, 504（7478）: 107-112.

LifeTime Initiative. 2020. LifeTime Strategic Research Agenda. https://lifetime-initiative.eu/wp-content/uploads/2020/08/LifeTime-Strategic-Research-Agenda.pdf[2021-03-05].

Liu Z, Cai Y J, Wang Y, et al. 2018. Cloning of macaque monkeys by somatic cell nuclear transfer. Cell, 172（4）: 881-887.

Medical Research Council. 2017. Molecular and cellular medicine Board strategy. https://mrc.ukri.org/funding/science-areas/molecular-cellular/board-strategy-mcmb/[2021-03-05].

National Institutes of Health. 2015. Program Snapshot. https://commonfund.nih.gov/4Dnucleome[2021-03-05].

National Science Foundation. 2021. Programs: Division of Molecular and Cellular Biosciences. https://www.nsf.gov/funding/programs.jsp?org=MCB[2021-03-05].

Navin N, Kendall J, Troge J, et al. 2011. Tumour evolution inferred by single-cell sequencing. Nature, 472（7341）: 90-94.

Service R F. 2020. 'The game has changed.' AI triumphs at solving protein structures. https://www.sciencemag.org/news/2020/11/game-has-changed-ai-triumphs-solving-protein-structures[2021-03-05].

Shao Y Y, Lu N, Wu Z F, et al. 2018. Creating a functional single-chromosome yeast. Nature, 560（7718）: 331-335.

Shi J J, Zhao Y, Wang K, et al. 2015. Cleavage of GSDMD by inflammatory caspases determines pyroptotic cell death. Nature, 526（7575）: 660-665.

Song F, Chen P, Sun D P, et al. 2014. Cryo-EM study of the chromatin fiber reveals a double helix twisted by tetranucleosomal units. Science, 344（6182）: 376-380.

Takahashi K, Yamanaka S. 2006. Induction of pluripotent stem cells from mouse embryonic and adult fibroblast cultures by defined factors. Cell, 126（4）: 663-676.

Tang F C, Barbacioru C, Wang Y Z, et al. 2009. mRNA-seq whole-transcriptome analysis of a single cell. Nature Methods, 6（5）: 377-382.

第二篇

分子细胞科学
关键科学与技术
问题及其发展方向

第三章

基因组的结构与演化

第一节 概　述

基因组包含的遗传信息是生命活动的基础，而染色体是真核细胞遗传信息储存和传递的结构载体。染色体由单条线性 DNA 双链及与其相互作用的蛋白质组成。这种由 DNA、组蛋白及其他蛋白组成的复合物即染色质，其在细胞分裂期可以进一步组装成为分裂中期染色体。不同物种中染色体的数目、大小虽各不相同，但其基本结构有序而又保守。根据不同的凝聚程度，染色质可分为高度凝聚、转录沉默的异染色质和相对松散、转录活跃的常染色质。端粒、着丝粒等组成性异染色质是染色体性状与功能完整性维系的结构元件。在核内三维空间内，线性染色质的结构和组分在高度动态调控中形成了表观基因组，继而调控细胞时空特异的遗传信息储存、解读和传递。基因组所包含的遗传信息在生物发育过程中执行高度组织细胞特异性的基因表达程序，进而形成复杂的生命体。基因组的解码将线性编码信息转换为多彩的细胞生命活动，这是基因组结构研究的重点和前沿。

基因组包含了个体发育所需的绝大部分关键信息。在多细胞生物中，同一个体不同细胞的基因组基本相同。但是，细胞生命活动过程中往往伴随着各种外源或内源性基因组 DNA 损伤的发生。这些 DNA 损伤可能影响到重要基因的功能，甚至会进一步通过复制将错误信息遗传给子代细胞，引起灾难性的后果。因此，生物进化出多套损伤修复通路来维持基因组的稳定性。但“祸兮福所倚，福兮祸所伏”，基因组信息又非绝对一成不变，遗传信息的变异是生命进化的分子基础，特定的细胞类型会通过发生程序性基因组遗传信息变异，从而改变编码信息赋予细胞新的功能，如免疫细胞、神经细胞、生殖细胞中的 DNA 序列多样化过程。近年，单细胞基因组测序技术进一步揭示了多细胞生物中不同细胞单体的基因组差异。在漫长的时间尺度中积累的“随机”基因组信息改变是否遵循一定的生物学规律、遗传信息的突变与稳定如何在进化中平衡等已成为研究基因组结构和演化的重要科学问题。

2018 年，我国科学家成功地将具有 16 条染色体的酿酒酵母改造为单条染色体酵母（Shao et al.，2018），为探索基因组演化和生命进化开辟了一条新的路径，领跑了这一领域的发展。这种人造的单条线性染色体以及后续构建的单条环形染色体（Shao et al.，2019）具有重构的染色体三维结构，但仍具有正常的细胞功能，彰显了基因组结构的可塑性。开展基于单细胞真核模式生物的人工改造研究，有助于探索染色体数目、大小乃至端粒等功能元件在进化中的功能和意义。多细胞生物的生命调控更为复杂，染色质表观调控机制更为繁复。通过“人造生命”等合成生物学方法可以进一步解答基因组信息储存、解读和传递过程中的重大基础科学问题，同时也可以启示我们生命进化的未来。

第二节　关键科学问题

一、基因组结构与遗传信息编解码的分子基础

基因组信息被编码至染色体载体中，而同一套信息又在不同时空中被有

序解读为复杂而又程序化的生命活动。基因组如何层层折叠形成有序的结构，机体发育过程中如何协调基因组遗传信息使之产生高度组织化的细胞特异的基因表达程序进而形成多细胞生物体，仍是基因组结构研究的谜题。遗传信息编解码依赖于基因组结构上多个层次的调控，包括染色体元件的区室化、三维基因组塑造以及染色质表观调控等。

（1）染色质结构元件如端粒和着丝粒等，处于组成性异染色质状态，在细胞核内被隔绝在无明确边界的区室内发挥其结构功能。动态变化的染色质区域亦可根据其凝聚状态形成结构和功能的区室。对于这种染色质区室化调控的分子基础，我们目前仍知之甚少。

（2）基因组 DNA 通过不同尺度上的染色质折叠而在核内形成多个层次的三维结构。这种高级结构的形成与塑造，既是高效存储遗传信息所必需的，又是遗传信息解码的重要调控机制。三维基因组塑造中顺式元件的作用、折叠的动力和功能等仍是未来急需解决的难题。

（3）后基因组时代多个大科学项目的开展为全景式理解基因组表观动态调控提供了可能，并促进了多种表观遗传学概念的提出。近年来，新的表观修饰不断涌现，关键作用酶被发现，基于结构的机制解析逐渐清晰，但基因组表观动态调控的分子基础仍有许多未知。

二、基因组完整性维持与序列变异的分子机制

在漫长的进化过程中，生物体获得了多套 DNA 损伤修复通路，从而维持基因组结构的稳定性和完整性。然而，DNA 损伤的发生往往是“福祸相依”的，程序性 DNA 损伤在免疫系统、生殖系统等中起始了众多生物学过程。“变”与“不变”的抉择是基因组演化、生物进化的分子基础。近年来的研究发现，多细胞生物中体细胞基因组存在大量的序列变异，变异的累积可能会导致基因组的不稳定，进而导致肿瘤的发生。基因组结构完整性的维持和 DNA 序列突变或主动改变的分子机制仍是基因组结构与演化研究的聚焦点。

（1）DNA 修复途径的知识拼图仍有多个明显空白。例如，真核生物错配修复的链识别的分子机制是什么？在染色质结构中 DNA 结合蛋白、表观修

饰、非编码 RNA 等如何协同修复三维基因组中的损伤?

（2）程序性 DNA 损伤在免疫系统、生殖系统中起始了不同的生物学过程，可以被“解读”为不同的结果并影响细胞的命运。程序性 DNA 损伤与一般性 DNA 损伤的发生和修复机制有何异同?细胞如何选择保真和易错修复通路来得到预期序列多样化结果，从而影响特定的生物学过程?遗传信息的突变与跨代稳定传递如何达到微妙的平衡?

（3）癌症基因组中碱基突变与结构变异产生的驱动因素是什么?是否还存在尚未描述的结构变异?在漫长的时间尺度中积累的“随机”突变是否遵循一定的生物学规律，这些突变是不是某种生理功能所必需的?基因组变异中 DNA 损伤的来源解析急需新的检测技术和功能研究方法。

三、基因组的演化与人工合成

基因组测序技术和分子方法的突飞猛进大大促进了基因组演化的研究。近年来，这一领域取得了令人目不暇接的成果。我国科学家在家养动物起源与驯化、经济作物基因组进化、脊椎动物演化、鸟类起源、人群演化等领域进行了开创性的研究。分子细胞科学是进化生物学等学科研究和发展的基石。分子细胞科学理论与技术如何继续驱动基因组演化的研究、如何促进生命的人工合成都是分子细胞科学领域未来需要解决的问题。反之，人工合成生命的设计、验证也可以进一步解答基因组信息储存、解读和传递过程中的重大基础科学问题，启迪我们生命进化的未来。

2010 年，科学家利用人工合成基因组构建了存活的支原体，打开了人工合成生命的大门。如何在分子细胞基础研究的基础上，将基因回路等元件标准化并在系统生物学层面进行整合，实现基因组的重编程，是这一方向内大科学计划与课题组研究面临的共同问题。

基因组的人工合成依赖于分子细胞科学产生的各种技术方法。目前，技术限制仍是阻碍这一学科发展的主要因素之一。例如，目前人工合成基因组仍聚焦于原核生物和简单的真核生物细胞，从头构建哺乳动物人工染色体仍是一个难题。人工合成基因组急需在 DNA 合成、组装、构建、转移等各个层面发展新的突破技术。如何设计非天然碱基、非天然氨基酸乃至镜像生物大分子等并将其引入人工合成生命，都是分子细胞科学有待解决的重大科学技术问题。

第三节　重要研究方向的发展现状和趋势分析

一、基因组高级结构与遗传信息编解码

1. 染色体元件区室化调控

作为真核细胞遗传信息储存的载体，染色体的不同区域具有不同功能。其中，端粒和着丝粒是染色体结构与功能完整性维系的重要元件（Liu et al., 2020b）。例如，染色体通过着丝粒与动态纺锤体微管结合决定染色体运动的时空动态性及姐妹染色单体分离的时间与质量；端粒是染色体末端的核蛋白复合物，在保持线性染色体的完整性和控制细胞周期分裂潜能等过程中发挥着至关重要的作用。染色体结构元件处于组成性异染色质内，在整个细胞周期中都处于压缩状态。这些结构元件与基因编码区具有截然不同的功能，而又能影响遗传信息的解码（Gottschling et al., 1990；Robin et al., 2014）。在细胞核内，染色体的不同区域或不同元件被进一步区室化，从而在遗传信息解码或结构维持等方面行使其独特的生物学功能。

利用生物化学和遗传学方法对染色体元件组分进行鉴定、在体外重组结构元件的复合物，以及基于结构生物学的关键蛋白局部结构解析大大促进了对端粒和着丝粒等结构元件的理解（de Lange, 2018；Holland and Cleveland, 2009；McKinley and Cheeseman, 2016）。与此同时，有关核内染色体元件的原位动态组装机制的研究也逐渐展开，并对这一领域的研究技术提出了新的挑战。染色体元件在有限的区室中聚集了多种核蛋白复合物，其中单分子的动态观察对光学成像技术提出了更高的要求。利用超分辨光学显微技术，科学家在细胞中观测到了端粒 T 环的结构（Doksani et al., 2013），并且精细描

述了细胞核内端粒酶组装和招募的过程（Schmidt et al.，2016）。这一系列工作展示了开发与运用高时空分辨新技术与新方法对染色体元件原位动态组装研究的重要性。细胞核内染色体元件高度区室化的特征也提示对于这一过程的理解急需新的视角和理论。基于大分子相变凝聚体形成理论，可能是理解染色体元件区室化调控的关键。

自 2009 年 Brangwynne 等研究者发现在细胞内存在大分子相变凝聚体的证据后，生物大分子在生理条件下发生相变和相分离的相关理论已用于解释多种生物学过程。例如，我国研究者发现纺锤体基质蛋白 BuGZ 有自身“液-液相分离”的能力，而且这个特性对于微管活性的调控来说是必需的（Jiang et al.，2015）。染色质也可以通过相分离方式构筑其区室化的元件。研究发现，异染色质区域表现出“液-液相分离”的特性（Strom et al.，2017）。人类异染色质蛋白 1（Heterochromatin Protein 1，HP1）中的一个亚型 HP1α，具有自身“液-液相分离”的能力（Larson et al.，2017）。染色质在生理条件下具备相分离的潜力（Gibson et al.，2019；Sanulli et al.，2019；Shakya et al.，2020）。虽然亦有证据表明染色质在生理条件下呈现固态的特征（Strickfaden et al.，2020），但是液-液相分离已成为解释染色体元件区室化的重要视角。相分离在着丝粒组装（Trivedi et al.，2019）、端粒动态调控（Min et al.，2019；Zhang et al.，2020a，2020b）中可能发挥重要作用。我国科学家在这一领域率先解析了 MeCP2 蛋白介导异染色质相变的机制（Wang et al.，2020a）。在前期研究基础上，我国科学家还提出了基于相分离的染色体元件区室化调控的理论和这一领域面临的挑战（Liu et al.，2020b）。目前，活细胞内相分离的研究方法仍然很不完善，急需发展能够进行细胞内大分子凝聚体功能检测的技术和分析手段。例如，光电关联等高时空分辨成像技术的开发和应用、大分子相分离的理论计算和模拟等。当前，相关理论处于初步探索阶段，存在着意见争鸣，急需大量后继研究，夯实相变在染色体元件区室化运作中的功能机制，并提出新的理论模型。

2. 三维基因组的塑造

人类基因组 DNA 长度约 3G bp（Lander et al.，2001；Venter et al.，2001），如以 B 型双链 DNA 计算单一细胞基因组 DNA 的长度近 2 m。在细

胞核内，多层次不同尺度的染色质折叠可使基因组 DNA 和其结合蛋白形成三维空间结构，高效存储遗传信息。同时，基因组的这种三维空间结构也在基因组遗传信息的解码过程中发挥重要的调节作用。近年来，基于高通量测序、超高分辨成像等三维基因组技术呈现了染色质的高级结构，一系列遗传学操作也证明了染色质高级结构对基因表达的影响。在癌细胞基因组中，碱基突变或结构变异能够引起染色质高级结构的变化，进而影响多种顺式作用元件与基因编码区域的调控，最终引起肿瘤的发生与发展。

染色体构象捕获（chromosome conformation capture，3C）技术（Dekker et al.，2002）引领了三维基因组研究领域的发展。在 3C 技术的基础上，研究者进一步发展出了定点构象捕获 4C 技术、定范围构象捕获 5C 技术、全基因组构象捕获 Hi-C 等技术、特定蛋白质介导构象捕获的 ChIA-PET 等技术，以及单细胞全基因组构象捕获（single-cell Hi-C）技术（Kempfer and Pombo，2020）。DNA-FISH 及活细胞成像技术进一步在视觉上呈现了三维基因组的构象（Kempfer and Pombo，2020）。目前，科学家利用序列荧光原位杂交（sequential fluorescence *in situ* hybridization，seqFISH）（Takei et al.，2021）或多重抗误差矫正荧光原位杂交技术（multiplexed error-robust fluorescence *in situ* hybridization，MERFISH）（Su et al.，2020）等技术，已经能够观测到单细胞内数千个基因组位点的空间定位。另外，细胞核碎片或切片的测序技术，如基因组结构绘制（genome architecture mapping，GAM）、利用标签隔离池识别相互作用（split-pool recognition of interactions by tag extension，SPRITE）及基于微流控技术的染色质相互作用分析技术（chromatin interaction analysis，ChIA-Drop）等，也可用于呈现三维基因组内的空间相互作用（Kempfer and Pombo，2020）。三维基因组技术的发展实现了从千碱基对量级到兆碱基对量级不同尺度上的对三维基因组的多层结构解析。基因调控顺式作用元件增强子能跨越数十万个碱基与靶标启动子相互作用，激活特定基因转录表达。超级增强子簇作为调控中枢，能够在一定的空间内调控多个目的基因的表达（Dowen et al.，2014）。在更高层次上，三维基因组被分隔成大小约 100 万 bp 的染色质结构域，这些区域内部 DNA 间相互作用的概率显著提高，被称作拓扑结构域（topological associated domain，TAD）（Dixon et al.，2012；Nora et al.，2012）。近年来，科学家发现黏连蛋白（cohesin）介导的环挤压是形成拓扑结构域的关键（Rao et al.，2017；Schwarzer et al.，2017）。研究发现，在拓扑结构域的边界往往富集对向的CTCF蛋白结合序列（Dixon et al.，2012；Rao et al.，2014）；黏连蛋白通过其依赖ATP的分子马达推动染色质挤压成环（Davidson et al.，2019；Kim et al.，2019），形成拓扑结构域（Vian et al.，2018）；而CTCF蛋白结合的区域能阻止染色质穿过，形成拓扑结构域的边界（Guo et al.，2015；Nora et al.，2017）。多个拓扑结构域又可以组装形成A和B两个染色质区室，分别对应常染色质和异染色质（Lieberman-Aiden et al.，2009）。

三维基因组的空间结构在不同的细胞类型中具有共性，在一定程度上也具有细胞类型特异性。染色质区室的空间位置在不同细胞类型之间有所不同，一部分的拓扑结构域边界是细胞类型特异性的。这种三维基因组的可塑性调控了细胞类型特异的转录活性。在这一系列前沿研究中，我国科学家利用遗传学方法揭示了CTCF蛋白结合区域在三维基因组中的功能（Guo et al.，2015），利用动物模型获得了胚胎早期发育（Du et al.，2017；Ke et al.，2017；Zheng and Xie，2019）、组织发育（Luo et al.，2021），以及人类胚胎早期发育（Chen et al.，2019）中的三维基因组可塑性调控图谱。三维基因组研究领域的迅猛发展得益于多种技术手段的发展。基于染色体构象捕获的技术目前应用最为广泛，而成像技术以及不依赖于染色体构象捕获的测序技术仍在继续发展中。我们期待新的技术能够呈现出新的基因组结构层次和揭示新的机制。单细胞染色体构象捕获技术的进一步普及，为揭示生物体发育、复杂疾病中三维基因组的功能机制提供了可能。在三维基因组结构塑造中，基因组重复序列、顺式元件、染色体折叠的动力、环挤压在各种生物学功能中的作用等仍是未来急需解决的难题。黏连蛋白介导的环挤压模型已经成为解释基因组上DNA损伤应答（Arnould et al.，2021）、免疫多样性产生等多种生物学过程的新理论（Ba et al.，2020；Dai et al.，2021；Hill et al.，2020；Zhang et al.，2019a，2019b），其在多种生物学过程中的作用机制将是未来研究的重点之一。

3. 基因组结构的表观调控

基因组遗传信息的解读受到表观遗传的直接调控，如DNA以及组蛋白的表观修饰。在人类基因组计划完成后，多个聚焦于基因组功能调控机制和表观遗传图谱的国际大科学项目开始启动，如DNA元件百科全书计划（Encyclopedia of DNA Elements，ENCODE）、表观遗传组学蓝图计划（Roadmap Epigenomics）等（Consortium，2012；Roadmap Epigenomics Consortium et al.，2015）。这类大科学项目的开展为全景式理解基因组表观动态调控提供了可能，促进了多种表观遗传学概念的提出。基于高通量测序的各种基因组结构和表观修饰分析方法的出现大大促进了这一领域的发展。同时，生化分析和结构生物学等领域的研究还在不断地揭示新型表观修饰及其分子基础和机制。近年来，真核生物DNA胞嘧啶去甲基化的分子基础和机制得到了明确的阐述，多种关键作用蛋白或酶的结构得到了清晰的解析，各种令人出乎意料的新型表观修饰方式不断被发现。这些研究前沿发现提示我们，对基因组表观动态调控的理解才刚刚开始，这一动态调控的分子基础仍存在许多的未知。

我国科学家在基因组表观调控领域做出了卓越的贡献，在结构解析与功能机制、发育与细胞命运重编程中表观遗传动态和新表观修饰的发现等方面都领跑了这一领域的发展。在此，我们以DNA表观修饰为例，管中窥豹，简要介绍这一领域的发展及我国科学家的贡献。

真核生物DNA胞嘧啶去甲基化的发生机制曾是表观遗传领域苦思求解的难题。2009年美国科学家发现了TET1可能的作用（Tahiliani et al.，2009）之后，我国科学家（He et al.，2011）和国际同行一道阐释了这一生化通路的分子机制，引领了这一研究领域的发展。在随后一系列的工作中，我国科学家领跑了DNA去甲基化在胚胎发育、重编程等过程中的分子机制研究（Dai et al.，2016；Gu et al.，2011；Guo et al.，2014；Hu et al.，2014；Zhang et al.，2013），对于这一通路中关键酶结构进行了测定（Hu et al.，2013，2015b），进一步明确了这一反应的生化机制。

真核生物DNA上的新修饰也在这一过程中被逐渐发现。我国科学家在模式生物果蝇中报道了在真核生物DNA上的N6-甲基腺嘌呤（6mA）修饰（Zhang et al.，2015）。这一研究成果，以及国际同行在线虫中发现的6mA修饰，首次揭示了真核生物DNA上的新修饰（Greer et al.，2015；Zhang et al.，2015）。随后，国际上多项工作进一步发现6mA修饰在其他真核生物特别是哺乳动物细胞中同样存在（Koziol et al.，2016；Mondo et al.，2017；Wu et al.，2016）。这种原本发现于原核生物基因组的DNA修饰方式，可能也在真核生物基因组结构等方面发挥重要的功能。这类稀有修饰是否受到了实验条件的影响仍具有一定的争议，围绕这类稀有修饰的学术争鸣促进了这一领域的发展（Liu et al.，2017，2021b；O'Brown et al.，2019；Schiffers et al.，2017）。化学生物学家在DNA修饰的鉴定和分析方面做了很多研究工作，例如，利用化学生物学和高通量测序方法对基因组DNA上的修饰或稀有碱基进行鉴定（Shu et al.，2018；Xia et al.，2015；Zhu et al.，2017），为研究这类修饰的生物学功能提供修饰的全景图谱。模式生物的选择和应用为这一领域拓展了新的方向。2019年，我国科学家在莱茵衣藻（*Chlamydomonas reinhardtii*）中发现了一种全新的DNA修饰（Xue et al.，2019）。研究者从DNA双加氧酶TET蛋白的同源蛋白CMD1出发，利用一系列生化、分子生物学方法发现，CMD1能够以维生素C为底物，在DNA的5-甲基胞嘧啶（5-methylcytosine）上催化产生一种全新的DNA修饰——5-甘油基-甲基胞嘧啶（5gmC），这种独特的DNA修饰参与了衣藻中光合作用的调控。之后，我国科学家又进一步从结构生物学的角度解析了这一修饰的发生机制（Li et al.，2021）。

我国科学家在DNA甲基化的建立与维持、跨代传递和功能机制等方面做出了多项原创、系统性的研究工作。例如，揭示了卵子发生中DNA甲基化模式建立的机制（Li et al.，2018），发现了DNA甲基图谱遗传规律（Jiang et al.，2013；Wang et al.，2014），解析了DNA甲基化稳态维持机制（Zhang et al.，2011）等。同样，围绕基因组结构中从DNA到染色体的过程，我国科学家首次解析了30 nm染色质纤维的结构及其动态变化（Song et al.，2014），为研究基因组结构形成提供了多层次的信息。

综上，我国科学家在新修饰、新机制和新功能等多方面都做出了原创性重要工作。基因组结构的表观调控仍存在许多的未知，新模型生物的引入和新方法的建立都有可能拓展我们的认知，解决悬而未决的问题并提出新的理论。

二、基因组完整性维持与序列变异

1. DNA 复制与损伤修复知识拼图中的空白

细胞中的基因组 DNA 每天都会遭受成百上千次的外源性或内源性损伤。如果这些 DNA 损伤没有得到及时修复，就可能影响到重要基因（如抑癌基因）的功能，甚至会进一步通过复制而遗传给子代细胞，引起灾难性的后果。因此，细胞发展出多种类型的修复途径来应对各种 DNA 损伤，包括碱基切除修复、错配修复、核苷酸切除修复以及双链断裂修复等。在过去 50 年的研究历史中，细胞内 DNA 损伤应答和修复的“工具箱”得到了详细的阐明。2015 年度的诺贝尔化学奖被授予了分别开创了碱基切除修复、核苷酸切除修复和错配修复研究的三位科学家。相关分子机制的研究已经勾勒出细胞内 DNA 损伤修复分子网络的轮廓。但是，在目前的知识拼图中，仍然存在很多未知的空白。近年来，新的因子和新的机制陆续被发现，例如 RNA 等“意想不到的”分子直接参与了 DNA 的损伤修复。损伤修复途径的功能丧失会导致 DNA 损伤积累，最终引起细胞功能缺陷、细胞死亡和癌变等后果。因此，DNA 损伤相关小分子也是肿瘤治疗的重要潜在药物。在临床上，合成致死等概念被用于指导肿瘤的用药和克服单一药物的耐药性。这些临床应用都依赖于基础研究的深入开展。例如，多个多腺苷二磷酸核糖聚合酶［poly（ADP-ribose）polymerase，PARP］制剂已经被广泛应用于同源重组相关基因突变的肿瘤治疗中，而其耐药性也是目前研究的热点。

在近年的研究中，RNA 在 DNA 双链损伤修复中的功能愈发清晰。2012 年，我国科学家在植物细胞中发现双链 DNA 损伤处存在特殊的小 RNA（Wei et al.，2012），并在随后的一系列工作中研究了损伤相关 RNA 产生的分子机制（Ba and Qi，2013）。同年，国际同行发现了类似的非编码 RNA 在 DNA 双链断裂响应中的功能（Francia et al.，2012）。这些研究提示了非编码 RNA 在 DNA 损伤响应中的可能作用，也掀开了 RNA 在 DNA 损伤响应修复中的神秘面纱。随后的研究发现，RNA 可以在同源重组中作为重组的模板（Keskin et al.，2014），在非同源末端连接中 RNA 可以被整合入 DNA 的连接点（Pryor et al.，2018）。这些研究揭示了 RNA 在 DNA 损伤修复中长期被忽略的直接作用，但其功能意义尚不清晰。2021年，我国科学家发现，在双链DNA损伤处由RNA聚合酶Ⅲ转录产生的RNA能够与DNA单链产生RNA-DNA杂交结构，这种结构维持了同源重组中长单链DNA的稳定性（Liu et al.，2021a）。这一工作解决了非同源末端连接中的长单链DNA维持的问题，提出了同源重组修正的工作模型，为研究RNA参与的DNA损伤修复开辟了新的领域。

在新机制不断被诠释的同时，新的DNA损伤修复因子也在近年陆续被发现。例如，HMCES蛋白最初被鉴定为5-羟甲基胞嘧啶（5hmC）结合蛋白，但后续研究证明HMCES蛋白能够通过共价交联结合DNA上的脱碱基位点，从而启动下游一系列的跨损伤DNA复制修复（Mohni et al.，2019）。这种针对脱碱基位点的“自杀式”修复机制（Halabelian et al.，2019；Mohni et al.，2019；Thompson et al.，2019）丰富了人们对于DNA损伤修复的认知。同样，Shieldin蛋白复合物的发现和功能解析也是DNA损伤修复研究领域的热点。2015年，两项研究结果（Boersma et al.，2015；Xu et al.，2015）表明*REV7*基因突变是PARP抑制剂耐药性产生的原因之一。之后，科学家发现REV7蛋白与细胞内尚未有功能注释的三个蛋白形成了一个新的复合物（Gupta et al.，2018），这一复合物随后被命名为Shieldin蛋白复合物，并作为53BP1下游的效应因子抑制DNA末端的切除（Dev et al.，2018；Ghezraoui et al.，2018；Mirman et al.，2018；Noordermeer et al.，2018）。Shieldin蛋白复合物的发现，拓展了我们对于同源重组和非同源末端连接通路抉择机制以及PARP抑制剂耐药性的理解。我国科学家发现和报道了DNA损伤修复网络各条通路中的多个新的因子，如同源重组中的AUNIP蛋白（Lou et al.，2017）、非同源末端连接锚定蛋白IFFO1（Li et al.，2019），以及非同源末端连接装载蛋白ERCC6L2（Liu et al.，2020a）等。同样，在翻译后修饰调控的DNA损伤修复机制[如MRE11的UFMylation修饰、ATM去乙酰化与RNF168磷酸化修饰（Tang et al.，2019；Wang et al.，2019；Xie et al.，2018）]，以及表观遗传修饰与DNA损伤修复的交互关系（Yang et al.，2017）等方面，我国科学家都做出了系统性的原创工作。我国与国际同行齐头并进（Francica et al.，2020；Olivieri et al.，2020；Qin et al.，2019），共同促进了这一领域的发展。

新因子的发现拓展了我们对DNA损伤修复网络的认识，也促进了临床肿瘤治疗的发展。2020年，我国科学家发现CXorf67蛋白通过与乳腺癌易感基因编码产物 BRCA2 蛋白竞争性结合 PALB2 蛋白，抑制同源重组修复途径（Han et al.，2020）。在临床上，这一基因可以用来作为后颅窝 A 组（posterior fossa group A，PFA）型室管膜瘤的生物标志物指导 PARP 抑制剂等疗法的应用（Han et al.，2020）。对于 DNA 复制的分子机制，我国科学家也做出了多项创造性的工作。例如，我国科学家利用设计巧妙的质谱方法，发现了 DNA 复制过程中染色质核小体组装的规律（Xu et al.，2010）；揭示了组蛋白变体 H2A.Z 在真核生物 DNA 复制起点中的作用（Long et al.，2020）。在 DNA 复制与损伤修复领域，科学家通过新因子的发现、新机制的解析逐步绘制了细胞内 DNA 代谢网络的图谱，但在这一看似成熟的领域仍存在许多明显的知识空白和缺失的环节。例如，错配修复在 DNA 复制中执行校正的功能，这一修复机制能够特异性识别碱基错误的 DNA 子链而非复制母链模板。在大肠杆菌中，错配修复可以通过模板母链上高度甲基化的 GATC 序列与甲基化程度较低的新合成子链区分开来，从而借由半甲基化的双链 DNA 来实现错配链识别（Li，2008）。然而，在真核细胞中错配修复如何辨别复制子链和母链仍是不解之谜。目前提出的多种科学假设（Ghodgaonkar et al.，2013；Kadyrov et al.，2006；Lujan et al.，2013；Pluciennik et al.，2010）并不能完美地解释这一现象。在染色质高级结构中，DNA 结合蛋白、表观修饰、非编码 RNA 等如何协同工作从而高效修复 DNA 损伤等是这一领域亟须解决的科学问题。

2. 免疫系统、生殖系统等系统中程序性 DNA 损伤的发生与功能解读

细胞基因组 DNA 序列并非一成不变，程序性 DNA 损伤是免疫系统、生殖系统中众多重要生命活动起始所必需的，而神经细胞分化、激活等过程也伴随着 DNA 损伤的产生。在适应性免疫系统中，免疫细胞受体分子的多样化是其识别各类病原体的分子基础。20 世纪 80 年代初，抗体基因 V（D）J 重组的发现首次证实了体细胞基因组中基因片段重组造成 DNA 序列多样化的过程。这一发现开启了人们对体细胞基因组序列变化的认识。进一步，免疫多样化中程序性 DNA 损伤发生的关键性的内切酶［重组激活基因（recombination activating gene, RAG）编码的核酸内切酶］和脱氨酶［活化诱导胞嘧啶核苷脱氨酶（activation-induced cytidine deaminase, AID）］得到陆续鉴定，程序性 DNA 损伤发生和处理的一系列分子机制慢慢得到揭示。不同于

一般的 DNA 损伤的无错修复，免疫受体基因上的损伤容易被错误解读为多样化的 DNA 序列。类似的程序性 DNA 损伤也发生于生殖系统减数分裂过程中。有性生殖生物体通过减数分裂形成配子时，Ⅱ型拓扑异构酶 SPO11 能够造成 DNA 双链断裂，从而容许控制不同性状的基因重新组合即基因重组的发生。减数分裂中的基因重组是形成生物遗传多样性的重要基础。在程序性 DNA 损伤发生和修复中，细胞应用了不同来源的 DNA 代谢酶起始损伤，并利用了普遍存在的 DNA 损伤修复机制进行高效修复。这一过程具有自身的特性，又与其他的损伤修复具有共性。这一领域不断涌现的研究成果，逐渐加深了我们对基因组结构和序列进化的认识。

程序性 DNA 损伤是在生理条件下染色质高级结构内发生的损伤，并同时需要高效的修复以维持免疫细胞或生殖细胞的生物学功能。近年的研究发现，程序性 DNA 损伤往往被限制在三维基因组的拓扑结构域中。例如，RAG 核酸内切酶的切割作用只能发生在一定的线性区域内，而 CTCF 结合位点定义了 RAG 核酸内切酶切割的边界（Hu et al.，2015a）。又如，AID 脱氨作用在不同的拓扑结构域能够高效发生（Senigl et al.，2019）。三维基因组拓扑结构域将这种程序性 DNA 损伤限制在一定的空间，从而有效地避免了各种附带损伤。在拓扑结构域内黏连蛋白介导的环挤压模型使得 RAG 核酸内切酶能够有序地切割 D 基因和 J 基因，进而切割 V 基因，完成程序性的 V（D）J 重组而非杂乱的重组（Zhang et al.，2019b）。在人或小鼠的基因组中，抗体重链基因簇涵盖了约 2M bp 的距离，而数百个 V 基因散布在这一范围。RAG 核酸内切酶如何以近似的概率靶向每一个 V 基因，从而使 B 细胞群体表达多样化的 V 基因，一直是这一领域悬而未决的谜题。2020 年和 2021 年的三项工作完美地解答了这一难题（Ba et al.，2020；Dai et al.，2021；Hill et al.，2020）。在距离重组中心较远的 V 基因附近含有 CTCF 结合位点并发生活跃的转录，这些 CTCF 结合位点或活跃的转录机器能够阻滞黏连蛋白介导的环挤压过程，从而使得 RAG 核酸内切酶有机会发生切割（Ba et al.，2020；Dai et al.，2021；Hill et al.，2020）。利用遗传学操作降解 CTCF 蛋白（Ba et al.，2020）或环挤压负调控因子（Dai et al.，2021），能够在体外培养细胞中重现环挤压过程。这一系列的发现揭示了程序性 DNA 损伤在三维基因组高级结构中的发生机制。免疫多样化中程序性 DNA 损伤的修复具有其自身的特性。例如，在抗体类型转换中不同 DNA 末端的连接受到方向性的调控，即端粒一侧的断裂总是和中心粒一侧的断裂相连接（Dong et al.，2015）。这种方向性的调控与三维基因组的高级结构密切相关（Zhang et al.，2019a）。我国科学家利用遗传学筛选方法找到了调控 DNA 重组方向性的关键因子 ERCC6L2（Liu et al.，2020a），这为研究三维基因组高级结构中 DNA 末端连接提供了新的思路（Wu et al.，2020）。

程序性 DNA 损伤在染色质高级结构中的发生机制研究，同样是配子发生领域研究的热点。在减数分裂过程中程序性 DNA 损伤起始于 SPO11 和 GM960 复合物所介导的 DNA 切割（Hunter，2015）。SPO11 在基因组上的切割位点并非随机分布，而是更偏好性地作用于一些热点（Hunter，2015）。在哺乳动物细胞中热点的选择上，PRDM9 及其介导的 H3K4me3 修饰发挥着关键作用（Baudat et al.，2010；Myers et al.，2010；Parvanov et al.，2010）。2020 年，我国科学家提出了 PRDM9 通过影响局部染色质环境从而影响 SPO11 切割和下游损伤修复的新理论（Chen et al.，2020）。这一工作揭示了染色质表观修饰与 DNA 损伤修复通路抉择的全新关系。免疫系统和生殖系统中基因组序列多样化过程是已知特定细胞类型中程序性 DNA 损伤发生和修复的范例。DNA 损伤修复在其他系统中可能也存在类似的作用。例如，非同源末端连接因子的缺陷在动物模型中往往会导致神经系统发育的异常，提示神经系统发育过程中可能会产生大量的 DNA 双链断裂。利用体外培养的神经干细胞，科学家发现许多超长的神经发育相关基因上发生了 DNA 复制和转录的冲撞，从而产生了大量的双链断裂（Wei et al.，2016）。而孤独症谱系障碍来源的神经干细胞更容易发生这种高频的 DNA 双链断裂（Wang et al.，2020b）。我国科学家进一步利用动物模型揭示了神经干细胞中 DNA 双链断裂发生的分子机制（Li et al.，2020a）。研究发现，在成熟的神经细胞中，由拓扑异构酶造成的 DNA 损伤能够促进神经元激活基因的表达（Madabhushi et al.，2015）。在其他系统中是否会有新型程序性 DNA 损伤的发现、程序性 DNA 损伤与普通损伤的修复过程有何特异性、程序性 DNA 损伤的失调如何导致各种系统疾病的发生等问题都是未来研究的热点。

3.“随机”突变的生物学规律

在肿瘤的发生和发展中，基因组变异是其重要特征和驱动因素之一。癌症基因组内的DNA变异遵循一定的规律，DNA损伤修复信号通路突变会导致特定基因组变异的发生。例如，错配修复相关基因的突变，往往会导致基因组内微卫星序列不稳定；DNA双链断裂响应基因的突变在动物模型中会导致基因转位高频发生。DNA损伤修复通路基因在细胞中的结构变异和碱基突变，为理解癌症基因组变异提供了重要的遗传学证据。过去十余年中，得益于测序技术和方法的进步，癌症基因组分析为我们揭示了更具细节的突变规律。例如，目前科学家在癌症基因组分析中已经发现50种以上不同的单碱基突变特征。但是大部分碱基突变特征的发生机制尚不明确，已发现的可能的分子机制也缺乏直接的遗传学证据。近年的研究揭示，长期被忽略的染色体结构变异（如染色体外环状DNA等）在肿瘤发生和发展中可能发挥的作用。这些染色体结构变异如何发生、如何在肿瘤细胞中维持、如何影响肿瘤细胞的生长等仍需要进一步的研究。基因组变异并不是肿瘤细胞所独有的，近年多项研究证实多细胞生物的体细胞基因组中存在各种程度的变异。这些低频的基因组变异来源于细胞何种内源或外源损伤、是否具有一定的生理功能等都是这一研究方向急需解决的科学问题，如何研究这些低频的基因组变异所需时间也对现有的技术方法提出了挑战。

体细胞发生基因突变可能来源于细胞内源的生命活动或外源的DNA损伤（Wu et al.，2018）。在癌症基因组中，发生于原癌基因或抑癌基因上的突变，其频率往往显著高于背景突变频率，这可能对肿瘤发生和发展起关键作用，这种高突变频率的基因也被定义为肿瘤驱动基因（Lawrence et al.，2013）。肿瘤驱动基因突变可以发生在编码区和启动子、增强子等调控区（Li et al.，2020b；Weinhold et al.，2014）。得益于多组学数据的综合运用和机器学习等方法的应用（Han et al.，2019；Martini，1986），通过基于编码区突变的驱动基因分析目前已经揭示了多种肿瘤中的驱动基因。然而，基于调控区突变的驱动基因鉴定仍然是一个重大的技术挑战。大量实验和临床数据的积累以及基于机器学习等人工智能方法的应用，可能是解决这一问题的可行途径。癌症基因组中基因变异可以粗略地分为单核苷酸位点变异、小片段插入和缺失、拷贝数变异和结构变异等。近年癌症基因组的研究揭示了更多种类的基因变异，如染色体外环状DNA。染色体外环状DNA早在1964年就被报道（Hotta and Bassel，1965），并存在于多种真核细胞中。近年来，研究发现，染色体外环状 DNA 能够支持癌基因的扩增，从而促进肿瘤的发展（Turner et al.，2017）。后续的研究进一步探索了染色体外环状 DNA 通过染色质高级结构和调控元件促进癌基因的过量表达（Morton et al.，2019；Wu et al.，2019），以及对基因组完整性的进一步影响（Koche et al.，2020）。

目前多种体细胞突变的来源仍未得到解析。利用突变特征分析，癌症基因组科学家将碱基突变的特征分为几十种（Alexandrov et al.，2020）。大部分突变产生的分子机制仍不明确，APOBEC 脱氨酶造成的突变谱式机制是其中目前了解最为详细的。2012 年，科学家对 21 个乳腺癌样本的基因组突变进行分析发现，其中一种突变特征与 APOBEC 在基因组上造成的突变特征高度相似（Nik-Zainal et al.，2012）；随后发现 APOBEC3B 是乳腺癌基因组中产生这种突变特征的主要脱氨酶（Burns et al.，2013a）；继而科学家发现这种 APOBEC 依赖的突变特征同样存在于其他种类的癌症基因组中（Burns et al.，2013b；Nik-Zainal et al.，2016；Roberts et al.，2013）。人 *APOBEC3* 基因簇中有 7 种 *APOBEC3* 同源基因，科学家发现 APOBEC3A 同样可能在乳腺癌基因组中造成基因组突变（Chan et al.，2015；Nik-Zainal et al.，2014）。APOBEC3A/B 如何在基因组上产生突变，仍是一个未解之谜。利用酵母模型或多组学数据分析（Haradhvala et al.，2016；Hoopes et al.，2016；Seplyarskiy et al.，2016），科学家发现 APOBEC3A/B 能够靶向 DNA 复制过程中后随链上的单链区域产生突变。利用单细胞来源克隆的基因组分析，科学家进一步发现在肿瘤细胞传代中 APOBEC 依赖的突变呈现出偶尔爆发的方式（Petljak et al.，2019）。突变来源的研究面临着很多技术挑战，如上述大量证据来源于关联分析，缺乏遗传学因果关系等。科学家分析了几十种外源因素与基因突变特征的关系（Kucab et al.，2019），为突变特征的分子机制解析提供了丰富的信息。如何建立细胞内源 DNA 损伤与癌症基因组变异之间的关联是这一领域的重要科学问题。

按照 DNA 复制酶效率计算，人类基因组每个细胞周期单个碱基的突变率约为 10^{-8}，基因组上突变的分布与多个因素密切相关，包括基因表达量（Pleasance et al.，2010）、DNA 复制时间（Stamatoyannopoulos et al.，2009）、组织类型（Garcia-Nieto et al.，2019）、染色质状态（Schuster-Bockler and Lehner，2012）等。科学家对 GTEx（Genotype-Tissue Expression）项目中正

常人群的组织进行了体细胞突变的分析，发现大量的肿瘤相关基因发生了突变，其中突变频率最高的是 *TP53* 和 *NOTCH1*，且突变细胞的丰度随年龄增大而升高（Yizhak et al.，2019）。另一个类似的研究也发现，正常食管组织中存在大量的癌基因突变，其中 *NOTCH1* 的突变频率甚至比食管癌组织中的还高（Martincorena et al.，2018）。我国科学家（Li et al.，2020c）与国际同行（Lawson et al.，2020）共同解析了肿瘤细胞和正常细胞中基因组 DNA 变异的图谱，呈现了基因组变异的可能机制。正常细胞中低频基因组变异与癌症基因组突变机制有何异同、在细胞功能方面体细胞克隆突变是否具有生理意义等问题急需解答。单细胞 RNA 和 DNA 测序技术的大量应用为呈现正常细胞中基因组变异提供了新的手段。这一方向的研究也需要进一步开发合适的研究体系和新型技术方法。

三、分子细胞科学驱动的人工合成基因组

1. 人工合成基因组

基因组的人工合成是建立在分子细胞科学、进化生物学等多个学科基础上的一个研究方向。核酸的化学合成、大片段拼装和转移技术的出现直接促进了这一方向的发展。在研究策略上，早期研究中合成基因组局限于天然基因组的从头复制合成。之后，多项工作通过精简基因数目、改变密码子数量、减少基因组非必需片段等方法，完成了人工设计和改造合成基因组。在研究对象上，合成基因组研究从原核生物，如支原体和大肠杆菌，逐渐拓展到单细胞真核生物酿酒酵母。在研究规模上，从早期的单个研究小组或单个公司团队的研究，逐步发展到国际人类基因组编写计划的提出。在学科方向交叉上，合成基因组的最初设计策略建立于分子细胞学科对生命本质探索的基础上，而之后对“人工生命”不断地测试为理解生命本质提供了独特的材料和视角。

在过去的 20 年内，科学家在合成基因组领域进行了各种尝试，例如，2002年脊髓灰质炎病毒基因组cDNA的化学合成（Cello et al.，2002），2003年 phiX174 噬菌体基因组 DNA 的合成（Smith et al.，2003）等。2008 年，

美国克雷格 • 文特尔研究所（J. Craig Venter Institute，JCVI）用一系列方法合成了约 582 kb 的生殖支原体 *Mycoplasma genitalium* 基因组 DNA（Gibson et al.，2008）。这一工作形成了从头合成基因组 DNA 的一系列技术雏形，如片段组装技术以及基于细菌和酵母的人工染色体拼装技术等，为人工合成基因组乃至“合成生命”提供了可能。两年后，JCVI 的科学家团队设计合成了基于丝状支原体 *Mycoplasma mycoides* 的第一个人工合成基因组细胞 JCVI-syn1.0（Gibson et al.，2010）。科学家首先合成并拼装了约 1.08 Mb 的基因组 DNA，并利用了 JCVI 发明的基于酵母质粒合成细菌基因组的方法（Lartigue et al.，2009）实现了 JCVI-syn1.0 的构建。JCVI-syn1.0 首次展示了人造生命的可能，开辟了一个全新的领域，引起了全世界的轰动。但是，JCVI-syn1.0 基因组的结构和大部分的遗传信息并没有改变，合成基因组还只是已知天然序列的复制。6 年后，JCVI 的科学家团队构建了 JCVI-syn3.0（Hutchison et al.，2016），再一次刷新了我们的认知。在构建 JCVI-syn3.0 过程中，科学家做了大量的减法，删去了 JCVI-syn1.0 基因组中多个非必需基因。利用转座子突变方法，科学家找到了支原体基因组内的必需基因、非必需基因和影响生长的基因。通过不同基因片段的排列组合，科学家得到了可以支持细胞存活的 JCVI-syn2.0 基因组。在 JCVI-syn2.0 细胞中，科学家进一步利用多轮转座子突变方法，最终得到了包含 438 个编码蛋白基因和 35 个编码 RNA 基因，基因组长度为 531 kb 的 JCVI-syn3.0（Hutchison et al.，2016）。在目前已知的自主存活生物中，JCVI-syn3.0 基因组是最小的。在最小生命体的人工构建上，JCVI 的科学家不断地刷新我们对生命的认知，开启了人类对于合成生命的无限想象。

从病毒到原核生物基因组的合成促使科学家开始了真核生物基因组的合成工作，相对简单且研究深入的模式生物酿酒酵母成为人工合成的主要对象。2011 年，科学家提出了人工合成酵母基因组 Sc2.0 计划以及初步的实验合成项目（Dymond et al.，2011）。其中，合成酵母基因组包含了多个人为设计的特征，如基因组重排元件 SCRaMbLE 的引入等。三年后，科学家合成了改造的酵母 3 号染色体（Annaluru et al.，2014）。在人工改造的染色体上，科学家尝试了密码子优化、重复区域敲除、SCRaMbLE 元件引入等多种操作。多国科学家的参与进一步加速了人工合成酵母基因组 Sc2.0 计划的进行，这一合成基因组采用了统一的设计（Richardson et al.，2017），并在2017年完成了近一半基因组的合成和组装（Mitchell et al.，2017；Shen et al.，2017；Wu et al.，2017；Xie et al.，2017；Zhang et al.，2017a）。我国科学家在其中发挥了主要的作用（Shen et al.，2017；Wu et al.，2017；Xie et al.，2017；Zhang et al.，2017a）。人工合成Sc2.0的三维基因组空间结构发生了显著变化（Mercy et al.，2017），同时体现了这一人工合成基因组在自主进化等方面的潜能。在此基础上，科学家提出了“人类基因组编写计划”（HGP-Write）（Boeke et al.，2016）。目前，这一计划已经吸引了全球16个国家200多位科学家的加入。

从头合成基因组造价不菲，同时需要大量工程化操作，对生命科学领域独立研究小组的工作模式提出了挑战。但同时，独立研究小组往往可以通过巧妙的构思，在合成基因组的设计、改造中有原创发现。例如，美国科学家在2016年成功构建了只有57个密码子的大肠杆菌（Ostrov et al.，2016）。在大部分的生命体中，多个冗余的遗传密码子对应一种氨基酸或提供翻译终止信号。在这项工作中，科学家选择了大肠杆菌基因组中7个冗余的密码子，在基因组水平替换掉这7个密码子，得到了*rE.coli-57*的菌株（Ostrov et al.，2016），这一菌株为未来非天然氨基酸的引入等提供了可能。合成基因组的设计和改造仍然面临很多的难题。在基因组的演化中，真核生物染色体的数目和大小的变化也有着重要影响，但传统分子细胞生物学方法很难对染色体数目与大小的进化展开研究。2018年，我国科学家成功地将具有16条染色体的酿酒酵母改造为单条染色体酵母（Shao et al.，2018），为染色体数目与大小的研究开辟了新的方向。在这项工作中，研究团队利用CRISPR/Cas9技术进行了染色体的融合实验，通过15轮的融合人工构建了单条线性染色体酵母。单条线性染色体酵母虽然在三维基因组层面单条线性染色体有了巨大的改变，但是其基因转录和细胞生长并未受到显著影响。同年，美国科学家利用类似的思路构建了双条线性染色体酵母（Luo et al.，2018）。在后继工作中，我国科学家进一步构建了单环形染色体（Shao et al.，2019），打破了真核生物线性染色体对于端粒结构的依赖。这种染色体数目简约版本的合成基因组显示了基因组结构的可塑性，为人工合成基因组提供了新的思路。

2. 从头构建哺乳动物人工染色体

目前人工合成基因组仍聚焦于原核生物和简单的真核生物细胞，哺乳动物合成基因组的研究仍进展缓慢。哺乳动物人工染色体在基因治疗、大片段转基因动物模型、人类基因组写入等方面都具有显而易见的应用价值。但是，哺乳动物染色体具有复杂的着丝粒和端粒区域，且其复制原点又没有明显的序列特征，因此，科学家最初采用了改造天然染色体得到微型染色体的方法。早在20世纪90年代初，科学家就开发了利用端粒序列的随机或定点插入，不断地缩短染色体从而得到微型染色体的方法（Farr et al.，1992；Itzhaki et al.，1992）。利用这种方法，科学家基于不同的染色体可以得到多种数兆碱基对长度的微型染色体（Farr et al.，1995；Heller et al.，1996；Mills et al.，1999）。利用同样的染色体克隆策略，数兆碱基对级别的大片段操作成为可能（Kuroiwa et al.，2000）。结合微细胞介导的染色体转移（microcell mediated chromosome transfer，MMCT）技术（Fournier and Ruddle，1977；Tomizuka et al.，1997），科学家利用染色体克隆策略成功制作了抗体人源化小鼠模型（Tomizuka et al.，2000）。但是，微型染色体并不能满足人类基因组编写的需要，从头构建哺乳动物人工染色体仍是一个技术和理论难题。1997年，科学家首次报道了在HT1080细胞系中从头构建人类人工染色体的方法（Harrington et al.，1997）。这种方法以及随后的大量改进方法都依赖于着丝粒高度重复的α卫星重复序列。这种着丝粒高度重复序列的合成以及相对不稳定的人工染色体为人类人工染色体的普遍应用设置了难以逾越的障碍。20余年后，科学家终于发明了不依赖于α卫星序列的人类人工染色体构建方法。基于前期对于着丝粒的分子细胞基础研究，科学家在2019年利用定点招募着丝粒组分蛋白HJURP的方法，成功实现了着丝粒上CENP-A核小体的组装（Logsdon et al.，2019）。这种不依赖于重复序列的人类人工染色体构建方法，为人类基因组的人工合成提供了重要的工具，也为其他哺乳动物或植物的人工染色体构建提供了借鉴。在哺乳动物DNA复制机制方面，我国已经有了多项创造性的发现，如染色质核小体组装的规律（Xu et al.，2010）和真核生物DNA复制起点的表观调控（Long et al.，2020）等。这些分子细胞基础研究的成果都可能应用于人工染色体的构建中。在不久的将来，哺乳动物人工染色体技术可能会引发一系列的生物技术变革。

3. 非天然碱基、氨基酸或镜像生物大分子在人工合成生命中的引入

人工合成基因组不仅是天然遗传信息的重新写入、合并简约或排列组合，更应引入非天然的因素从而构建更为复杂或独特的生命体。在人工合成生命中引入非天然碱基对是这一类研究的范例。天然核苷酸含有腺嘌呤（A）、鸟嘌呤（G）、胸腺嘧啶（T）和胞嘧啶（C）4种碱基，其中在DNA双螺旋结构中A—T、C—G分别配对。化学生物学家根据天然碱基对的特征，在体外合成和设计了多套非天然碱基对，并利用蛋白质进化的方法对DNA复制酶进行了优化。2014年，科学家终于将非天然碱基对dNaM—d5SICS（研究人员将其命名为X—Y碱基对）写入细菌基因组中，构建了含有三种碱基对的合成生命（Malyshev et al.，2014）。在这一研究中，科学家通过引入核苷三磷酸转运蛋白成功实现了非天然碱基原料的供给，实现了X—Y碱基对的相对稳定遗传（Malyshev et al.，2014）。通过对dNaM:d5SICS碱基对的进一步优化，科学家还开发了更有利于稳定遗传的dNaM—dTPT3、dPTMO—dTPT3等碱基对（Dien et al.，2018；Zhang et al.，2017b）。这一系列的工作解决了非天然碱基对遗传信息存储、传代的功能问题。2017年，同一组科学家利用人工改造的转录和翻译系统，实现了非天然碱基对遗传信息的解读（Zhang et al.，2017c）。在这一工作中，科学家引入了人工设计的密码子、tRNA分子和非天然氨基酸，使dNaM—dTPT3碱基对能够被转录成为特殊的密码子并被特殊的tRNA识别，从而将非天然氨基酸插入目的蛋白中（Zhang et al.，2017c）。至此，科学家成功地将非天然碱基融入遗传信息储存、解读、传代的功能中，将密码子从61种扩充到67种（Fischer et al.，2020）。随着更多非天然碱基对的设计，如Hachimoji核酸中dZ—dP、dS—dB碱基对（Hoshika et al.，2019），我们相信未来合成基因组将能容纳更复杂、更独特的信息。自然界中存在的天然生物大分子都是单一手性的分子，而我国科学家利用全化学方法合成了一种DNA聚合酶的镜像蛋白，实现了镜像DNA的体外合成与转录（Wang et al.，2016）。未来，利用基于镜像异构体的人工合成基因组有可能实现“镜像生命”的人工合成。

第四节　我国发展战略与重点方向

一、总体思路和发展目标

1. 总体思路

整合分子细胞生物学研究方法，结合多学科理论和技术，发展基因组结构和演化研究的新方法；聚焦基因组结构和演化研究中的关键科学问题，发现新因子、解析新机制、提出新理论；解决基因组演化中分子细胞生物学中的重要问题，“机制解析”与“人工合成”并进。

2. 发展目标

在“基因组结构和演化”研究的重要方向取得突破性成果；为疾病诊疗提供新的思路；进一步提升我国科学家在该研究领域的国际影响力。

二、优先发展领域或重点研究方向

1. 多学科交叉的基因组结构研究

发展分子细胞生物学技术，结合物理、化学、计算科学等发展基因组结构研究的方法技术；聚焦染色体元件区室化的分子基础、三维基因组的动态塑造、表观基因组的协同调控等关键科学问题；发展新模式生物，发现新因子、新机制和新功能。

2. 基因组完整性和序列变异研究

解决 DNA 复制、损伤修复、重组等机制中的关键科学问题；探寻遗传信息“变”与“不变”中的通路抉择和细胞命运调控机制；建立研究基因组中低频、“随机”的碱基突变和结构变异的方法体系。

3. 基因组演化与人工合成研究

充分发挥大科学计划与独立研究小组模式，实现合成基因组更高层次的重编，探索哺乳动物细胞合成基因组；发展分子细胞生物学方法，建立更为复杂、独特的人造生命；解决人造生命中若干技术壁垒，开展相关伦理研究。

本章参考文献

Alexandrov L B, Kim J, Haradhvala N J, et al. 2020. The repertoire of mutational signatures in human cancer. Nature, 578（7793）: 94-101.

Annaluru N, Muller H, Mitchell L A, et al. 2014. Total synthesis of a functional designer eukaryotic chromosome. Science, 344（6179）: 55-58.

Arnould C, Rocher V, Finoux A L, et al. 2021. Loop extrusion as a mechanism for formation of DNA damage repair foci. Nature, 590（7847）: 660-665.

Ba Z, Lou J, Ye A Y, et al. 2020. CTCF orchestrates long-range cohesin-driven V（D）J recombinational scanning. Nature, 586（7828）: 305-310.

Ba Z, Qi Y. 2013. Small RNAs: emerging key players in DNA double-strand break repair. Sci China Life Sci, 56（10）: 933-936.

Baudat F, Buard J, Grey C, et al. 2010. PRDM9 is a major determinant of meiotic recombination hotspots in humans and mice. Science, 327（5967）: 836-840.

Boeke J D, Church G, Hessel A, et al. 2016. GENOME ENGINEERING. The Genome Project-Write. Science, 353（6295）: 126-127.

Boersma V, Moatti N, Segura-Bayona S, et al. 2015. MAD2L2 controls DNA repair at telomeres and DNA breaks by inhibiting 5' end resection. Nature, 521（7553）: 537-540.

Burns M B, Lackey L, Carpenter M A, et al. 2013a. APOBEC3B is an enzymatic source of mutation in breast cancer. Nature, 494（7437）: 366-370.

Burns M B, Temiz N A, Harris R S. 2013b. Evidence for APOBEC3B mutagenesis in multiple human cancers. Nat Genet, 45（9）: 977-983.

Cello J, Paul A V, Wimmer E. 2002. Chemical synthesis of poliovirus cDNA: generation of infectious virus in the absence of natural template. Science, 297（5583）: 1016-1018.

Chan K, Roberts S A, Klimczak L J, et al. 2015. An APOBEC3A hypermutation signature is distinguishable from the signature of background mutagenesis by APOBEC3B in human cancers. Nat Genet, 47（9）: 1067-1072.

Chen X, Ke Y, Wu K, et al. 2019. Key role for CTCF in establishing chromatin structure in human embryos. Nature, 576（7786）: 306-310.

Chen Y, Lyu R, Rong B, et al. 2020. Refined spatial temporal epigenomic profiling reveals intrinsic connection between PRDM9-mediated H3K4me3 and the fate of double-stranded breaks. Cell Res, 30（3）: 256-268.

Consortium E P. 2012. An integrated encyclopedia of DNA elements in the human genome. Nature, 489（7414）: 57-74.

Dai H Q, Hu H, Lou J, et al. 2021. Loop extrusion mediates physiological *Igh* locus contraction for RAG scanning. Nature, 590（7845）: 338-343.

Dai H Q, Wang B A, Yang L, et al. 2016. TET-mediated DNA demethylation controls gastrulation by regulating Lefty-Nodal signalling. Nature, 538（7626）: 528-532.

Davidson I F, Bauer B, Goetz D, et al. 2019. DNA loop extrusion by human cohesin. Science, 366（6471）: 1338-1345.

de Lange T. 2018. Shelterin-mediated telomere protection. Annu Rev Genet, 52: 223-247.

Dekker J, Rippe K, Dekker M, et al. 2002. Capturing chromosome conformation. Science, 295（5558）: 1306-1311.

Dev H, Chiang T W, Lescale C, et al. 2018. Shieldin complex promotes DNA end-joining and counters homologous recombination in BRCA1-null cells. Nat Cell Biol, 20（8）: 954-965.

Dien V T, Holcomb M, Feldman A W, et al. 2018. Progress toward a semi-synthetic organism with an unrestricted expanded genetic alphabet. J Am Chem Soc, 140（47）: 16115-16123.

Dixon J R, Selvaraj S, Yue F, et al. 2012. Topological domains in mammalian genomes identified by analysis of chromatin interactions. Nature, 485（7398）: 376-380.

Doksani Y, Wu J Y, de Lange T, et al. 2013. Super-resolution fluorescence imaging of telomeres reveals TRF2-dependent T-loop formation. Cell, 155（2）: 345-356.

Dong J, Panchakshari R A, Zhang T, et al. 2015. Orientation-specific joining of AID-initiated DNA breaks promotes antibody class switching. Nature, 525（7567）: 134-139.

Dowen J M, Fan Z P, Hnisz D, et al. 2014. Control of cell identity genes occurs in insulated neighborhoods in mammalian chromosomes. Cell, 159（2）: 374-387.

Du Z, Zheng H, Huang B, et al. 2017. Allelic reprogramming of 3D chromatin architecture during early mammalian development. Nature, 547（7662）: 232-235.

Dymond J S, Richardson S M, Coombes C E, et al. 2011. Synthetic chromosome arms function in yeast and generate phenotypic diversity by design. Nature, 477（7365）: 471-476.

Farr C J, Bayne R A, Kipling D, et al. 1995. Generation of a human X-derived minichromosome using telomere-associated chromosome fragmentation. EMBO J, 14（21）: 5444-5454.

Farr C J, Stevanovic M, Thomson E J, et al. 1992. Telomere-associated chromosome fragmentation: applications in genome manipulation and analysis. Nat Genet, 2（4）: 275-282.

Fischer E C, Hashimoto K, Zhang Y, et al. 2020. New codons for efficient production of unnatural proteins in a semisynthetic organism. Nat Chem Biol, 16（5）: 570-576.

Fournier R E, Ruddle F H. 1977. Microcell-mediated transfer of murine chromosomes into mouse, Chinese hamster, and human somatic cells. Proc Natl Acad Sci U S A, 74（1）: 319-323.

Francia S, Michelini F, Saxena A, et al. 2012. Site-specific DICER and DROSHA RNA products control the DNA-damage response. Nature, 488（7410）: 231-235.

Francica P, Mutlu M, Blomen V A, et al. 2020. Functional radiogenetic profiling implicates ERCC6L2 in non-homologous end joining. Cell Rep, 32（8）: 108068.

Garcia-Nieto P E, Morrison A J, Fraser H B. 2019. The somatic mutation landscape of the human body. Genome Biol, 20（1）: 298.

Ghezraoui H, Oliveira C, Becker J R, et al. 2018. 53BP1 cooperation with the REV7-shieldin complex underpins DNA structure-specific NHEJ. Nature, 560（7716）: 122-127.

Ghodgaonkar M M, Lazzaro F, Olivera-Pimentel M, et al. 2013. Ribonucleotides misincorporated into DNA act as strand-discrimination signals in eukaryotic mismatch repair. Mol Cell, 50（3）: 323-332.

Gibson B A, Doolittle L K, Schneider M W G, et al. 2019. Organization of chromatin by intrinsic and regulated phase separation. Cell, 179（2）: 470-484, e421.

Gibson D G, Benders G A, Andrews-Pfannkoch C, et al. 2008. Complete chemical synthesis, assembly, and cloning of a *Mycoplasma genitalium* genome. Science, 319（5867）: 1215-1220.

Gibson D G, Glass J I, Lartigue C, et al. 2010. Creation of a bacterial cell controlled by a chemically synthesized genome. Science, 329（5987）: 52-56.

Gottschling D E, Aparicio O M, Billington B L, et al. 1990. Position effect at *S. cerevisiae* telomeres: reversible repression of Pol Ⅱ transcription. Cell, 63（4）: 751-762.

Greer E L, Blanco M A, Gu L, et al. 2015. DNA methylation on N6-adenine in *C. elegans*. Cell, 161（4）: 868-878.

Gu T P, Guo F, Yang H, et al. 2011. The role of Tet3 DNA dioxygenase in epigenetic reprogramming by oocytes. Nature, 477（7366）: 606-610.

Guo F, Li X, Liang D, et al. 2014. Active and passive demethylation of male and female pronuclear DNA in the mammalian zygote. Cell Stem Cell, 15（4）: 447-459.

Guo Y, Xu Q, Canzio D, et al. 2015. CRISPR inversion of CTCF sites alters genome topology and enhancer/promoter function. Cell, 162（4）: 900-910.

Gupta R, Somyajit K, Narita T, et al. 2018. DNA repair network analysis reveals shieldin as a key regulator of NHEJ and PARP inhibitor sensitivity. Cell, 173（4）: 972-988, e923.

Halabelian L, Ravichandran M, Li Y, et al. 2019. Structural basis of HMCES interactions with abasic DNA and multivalent substrate recognition. Nat Struct Mol Biol, 26（7）: 607-612.

Han J, Yu M, Bai Y, et al. 2020. Elevated CXorf67 expression in PFA ependymomas suppresses DNA repair and sensitizes to PARP inhibitors. Cancer Cell, 38（6）: 844-856, e847.

Han Y, Yang J, Qian X, et al. 2019. DriverML: a machine learning algorithm for identifying driver genes in cancer sequencing studies. Nucleic Acids Res, 47（8）: e45.

Haradhvala N J, Polak P, Stojanov P, et al. 2016. Mutational strand asymmetries in cancer genomes reveal mechanisms of DNA damage and repair. Cell, 164（3）: 538-549.

Harrington J J, Van Bokkelen G, Mays R W, et al. 1997. Formation of *de novo* centromeres and construction of first-generation human artificial microchromosomes. Nat Genet, 15（4）: 345-355.

He Y F, Li B Z, Li Z, et al. 2011. Tet-mediated formation of 5-carboxylcytosine and its excision by TDG in mammalian DNA. Science, 333（6047）: 1303-1307.

Heller R, Brown K E, Burgtorf C, et al. 1996. Mini-chromosomes derived from the human Y chromosome by telomere directed chromosome breakage. Proc Natl Acad Sci U S A, 93（14）: 7125-7130.

Hill L, Ebert A, Jaritz M, et al. 2020. *Wapl* repression by Pax5 promotes *V* gene recombination by *Igh* loop extrusion. Nature, 584（7819）: 142-147.

Holland A J, Cleveland D W. 2009. Boveri revisited: chromosomal instability, aneuploidy and tumorigenesis. Nat Rev Mol Cell Biol, 10（7）: 478-487.

Hoopes J I, Cortez L M, Mertz T M, et al. 2016. APOBEC3A and APOBEC3B preferentially deaminate the lagging strand template during DNA replication. Cell Rep, 14（6）: 1273-1282.

Hoshika S, Leal N A, Kim M J, et al. 2019. Hachimoji DNA and RNA: A genetic system with eight building blocks. Science, 363（6429）: 884-887.

Hotta Y, Bassel A. 1965. Molecular size and circularity of DNA in cells of mammals and higher plants. Proc Natl Acad Sci U S A, 53（2）: 356-362.

Hu J, Zhang Y, Zhao L, et al. 2015a. Chromosomal loop domains direct the recombination of antigen receptor genes. Cell, 163（4）: 947-959.

Hu L, Li Z, Cheng J, et al. 2013. Crystal structure of TET2-DNA complex: insight into TET-mediated 5mC oxidation. Cell, 155（7）: 1545-1555.

Hu L, Lu J, Cheng J, et al. 2015b. Structural insight into substrate preference for TET-mediated oxidation. Nature, 527（7576）: 118-122.

Hu X, Zhang L, Mao S Q, et al. 2014. Tet and TDG mediate DNA demethylation essential for mesenchymal-to-epithelial transition in somatic cell reprogramming. Cell Stem Cell, 14（4）: 512-522.

Hunter N. 2015. Meiotic recombination: The essence of heredity. Cold Spring Harb Perspect Biol, 7（12）: a016618.

Hutchison C A, 3rd, Chuang R Y, Noskov V N, et al. 2016. Design and synthesis of a minimal bacterial genome. Science, 351（6280）: aad6253.

Itzhaki J E, Barnett M A, MacCarthy A B, et al. 1992. Targeted breakage of a human chromosome mediated by cloned human telomeric DNA. Nat Genet, 2（4）: 283-287.

Jiang H, Wang S, Huang Y, et al. 2015. Phase transition of spindle-associated protein regulate spindle apparatus assembly. Cell, 163（1）: 108-122.

Jiang L, Zhang J, Wang J J, et al. 2013. Sperm, but not oocyte, DNA methylome is inherited by zebrafish early embryos. Cell, 153（4）: 773-784.

Kadyrov F A, Dzantiev L, Constantin N, et al. 2006. Endonucleolytic function of MutLα in human mismatch repair. Cell, 126（2）: 297-308.

Ke Y, Xu Y, Chen X, et al. 2017. 3D chromatin structures of mature gametes and structural reprogramming during mammalian embryogenesis. Cell, 170（2）: 367-381 e320.

Kempfer R, Pombo A. 2020. Methods for mapping 3D chromosome architecture. Nat Rev Genet, 21（4）: 207-226.

Keskin H, Shen Y, Huang F, et al. 2014. Transcript-RNA-templated DNA recombination and repair. Nature, 515（7527）: 436-439.

Kim Y, Shi Z, Zhang H, et al. 2019. Human cohesin compacts DNA by loop extrusion. Science,

366（6471）: 1345-1349.

Koche R P, Rodriguez-Fos E, Helmsauer K, et al. 2020. Extrachromosomal circular DNA drives oncogenic genome remodeling in neuroblastoma. Nat Genet, 52（1）: 29-34.

Koziol M J, Bradshaw C R, Allen G E, et al. 2016. Identification of methylated deoxyadenosines in vertebrates reveals diversity in DNA modifications. Nat Struct Mol Biol, 23（1）: 24-30.

Kucab J E, Zou X, Morganella S, et al. 2019. A compendium of mutational signatures of environmental agents. Cell, 177（4）: 821-836, e816.

Kuroiwa Y, Tomizuka K, Shinohara T, et al. 2000. Manipulation of human minichromosomes to carry greater than megabase-sized chromosome inserts. Nat Biotechnol, 18（10）: 1086-1090.

Lander E S, Linton L M, Birren B, et al. 2001. Initial sequencing and analysis of the human genome. Nature, 409（6822）: 860-921.

Larson A G, Elnatan D, Keenen M M, et al. 2017. Liquid droplet formation by HP1αsuggests a role for phase separation in heterochromatin. Nature, 547（7662）: 236-240.

Lartigue C, Vashee S, Algire M A, et al. 2009. Creating bacterial strains from genomes that have been cloned and engineered in yeast. Science, 325（5948）: 1693-1696.

Lawrence M S, Stojanov P, Polak P, et al. 2013. Mutational heterogeneity in cancer and the search for new cancer-associated genes. Nature, 499（7457）: 214-218.

Lawson A R J, Abascal F, Coorens T H H, et al. 2020. Extensive heterogeneity in somatic mutation and selection in the human bladder. Science, 370（6512）: 75-82.

Li G M. 2008. Mechanisms and functions of DNA mismatch repair. Cell Res, 18（1）: 85-98.

Li J, Shang Y, Wang L, et al. 2020a. Genome integrity and neurogenesis of postnatal hippocampal neural stem/progenitor cells require a unique regulator Filia. Sci Adv, 6（44）: eaba0682.

Li K, Zhang Y, Liu X, et al. 2020b. Noncoding variants connect enhancer dysregulation with nuclear receptor signaling in hematopoietic malignancies. Cancer Discov, 10（5）: 724-745.

Li R, Du Y, Chen Z, et al. 2020c. Macroscopic somatic clonal expansion in morphologically normal human urothelium. Science, 370（6512）: 82-89.

Li W, Bai X, Li J, et al. 2019. The nucleoskeleton protein IFFO1 immobilizes broken DNA and suppresses chromosome translocation during tumorigenesis. Nat Cell Biol, 21（10）: 1273-1285.

Li W, Zhang T, Sun M, et al. 2021. Molecular mechanism for vitamin C-derived C（5）-glyceryl-methylcytosine DNA modification catalyzed by algal TET homologue CMD1. Nat Commun, 12（1）: 744.

Li Y, Zhang Z, Chen J, et al. 2018. Stella safeguards the oocyte methylome by preventing de novo methylation mediated by DNMT1. Nature, 564（7734）: 136-140.

Lieberman-Aiden E, van Berkum N L, Williams L, et al. 2009. Comprehensive mapping of long-range interactions reveals folding principles of the human genome. Science, 326（5950）: 289-293.

Liu B, Liu X, Lai W, et al. 2017. Metabolically generated stable isotope-labeled deoxynucleoside code for tracing DNA N（6）-methyladenine in human cells. Anal Chem, 89（11）: 6202-6209.

Liu S, Hua Y, Wang J, et al. 2021a. RNA polymerase III is required for the repair of DNA double-strand breaks by homologous recombination. Cell, 184（5）: 1314-1329, e1310.

Liu X, Lai W, Li Y, et al. 2021b. N（6）-methyladenine is incorporated into mammalian genome by DNA polymerase. Cell Res, 31（1）: 94-97.

Liu X, Liu T, Shang Y, et al. 2020a. ERCC6L2 promotes DNA orientation-specific recombination in mammalian cells. Cell Res, 30（9）: 732-744.

Liu X, Liu X, Wang H, et al. 2020b. Phase separation drives decision making in cell division. J Biol Chem, 295（39）: 13419-13431.

Logsdon G A, Gambogi C W, Liskovykh M A, et al. 2019. Human artificial chromosomes that bypass centromeric DNA. Cell, 178（3）: 624-639, e619.

Long H, Zhang L, Lv M, et al. 2020. H2A.Z facilitates licensing and activation of early replication origins. Nature, 577（7791）: 576-581.

Lou J, Chen H, Han J, et al. 2017. AUNIP/C1orf135 directs DNA double-strand breaks towards the homologous recombination repair pathway. Nat Commun, 8（1）: 985.

Lujan S A, Williams J S, Clausen A R, et al. 2013. Ribonucleotides are signals for mismatch repair of leading-strand replication errors. Mol Cell, 50（3）: 437-443.

Luo J, Sun X, Cormack B P, et al. 2018. Karyotype engineering by chromosome fusion leads to reproductive isolation in yeast. Nature, 560（7718）: 392-396.

Luo X, Liu Y, Dang D, et al. 2021. 3D Genome of macaque fetal brain reveals evolutionary innovations during primate corticogenesis. Cell, 184（3）: 723-740, e721.

Madabhushi R, Gao F, Pfenning A R, et al. 2015. Activity-induced DNA breaks govern the expression of neuronal early-response genes. Cell, 161（7）: 1592-1605.

Malyshev D A, Dhami K, Lavergne T, et al. 2014. A semi-synthetic organism with an expanded genetic alphabet. Nature, 509（7500）: 385-388.

Martincorena I, Fowler J C, Wabik A, et al. 2018. Somatic mutant clones colonize the human esophagus with age. Science, 362（6417）: 911-917.

Martini A K. 1986. Reconstruction of pronation of the forearm（experiences using the Zancolli operation）. Handchir Mikrochir Plast Chir, 18（5）: 271-274.

McKinley K L, Cheeseman I M. 2016. The molecular basis for centromere identity and function. Nat Rev Mol Cell Biol, 17（1）: 16-29.

Mercy G, Mozziconacci J, Scolari V F, et al. 2017. 3D organization of synthetic and scrambled chromosomes. Science, 355（6329）: eaaf4597.

Mills W, Critcher R, Lee C, et al. 1999. Generation of an approximately 2.4 Mb human X centromere-based minichromosome by targeted telomere-associated chromosome fragmentation in DT40. Hum Mol Genet, 8（5）: 751-761.

Min J, Wright W E, Shay J W. 2019. Clustered telomeres in phase-separated nuclear condensates engage mitotic DNA synthesis through BLM and RAD52. Genes Dev, 33（13-14）: 814-827.

Mirman Z, Lottersberger F, Takai H, et al. 2018. 53BP1-RIF1-shieldin counteracts DSB resection through CST- and Pol alpha-dependent fill-in. Nature, 560（7716）: 112-116.

Mitchell L A, Wang A, Stracquadanio G, et al. 2017. Synthesis, debugging, and effects of synthetic chromosome consolidation: syn Ⅵ and beyond. Science, 355（6329）: eaaf4831.

Mohni K N, Wessel S R, Zhao R, et al. 2019. HMCES maintains genome integrity by shielding abasic sites in single-strand DNA. Cell, 176（1-2）: 144-153, e113.

Mondo S J, Dannebaum R O, Kuo R C, et al. 2017. Widespread adenine N6-methylation of active genes in fungi. Nat Genet, 49（6）: 964-968.

Morton A R, Dogan-Artun N, Faber Z J, et al. 2019. Functional enhancers shape extrachromosomal oncogene amplifications. Cell, 179（6）: 1330-1341 e1313.

Myers S, Bowden R, Tumian A, et al. 2010. Drive against hotspot motifs in primates implicates the PRDM9 gene in meiotic recombination. Science, 327（5967）: 876-879.

Nik-Zainal S, Alexandrov L B, Wedge D C, et al. 2012. Mutational processes molding the genomes of 21 breast cancers. Cell, 149（5）: 979-993.

Nik-Zainal S, Davies H, Staaf J, et al. 2016. Landscape of somatic mutations in 560 breast cancer whole-genome sequences. Nature, 534（7605）: 47-54.

Nik-Zainal S, Wedge D C, Alexandrov L B, et al. 2014. Association of a germline copy number polymorphism of APOBEC3A and APOBEC3B with burden of putative APOBEC-dependent mutations in breast cancer. Nat Genet, 46（5）: 487-491.

Noordermeer S M, Adam S, Setiaputra D, et al. 2018. The shieldin complex mediates 53BP1-dependent DNA repair. Nature, 560（7716）: 117-121.

Nora E P, Goloborodko A, Valton A L, et al. 2017. Targeted degradation of CTCF decouples local insulation of chromosome domains from genomic compartmentalization. Cell, 169（5）: 930-944, e922.

Nora E P, Lajoie B R, Schulz E G, et al. 2012. Spatial partitioning of the regulatory landscape of the X-inactivation centre. Nature, 485（7398）: 381-385.

O'Brown Z K, Boulias K, Wang J, et al. 2019. Sources of artifact in measurements of 6mA and 4mC abundance in eukaryotic genomic DNA. BMC Genomics, 20（1）: 445.

Olivieri M, Cho T, Alvarez-Quilon A, et al. 2020. A genetic map of the response to DNA damage in human cells. Cell, 182（2）: 481-496, e421.

Ostrov N, Landon M, Guell M, et al. 2016. Design, synthesis, and testing toward a 57-codon genome. Science, 353（6301）: 819-822.

Parvanov E D, Petkov P M, Paigen K. 2010. Prdm9 controls activation of mammalian recombination hotspots. Science, 327（5967）: 835.

Petljak M, Alexandrov L B, Brammeld J S, et al. 2019. Characterizing mutational signatures in human cancer cell lines reveals episodic APOBEC mutagenesis. Cell, 176（6）: 1282-1294, e1220.

Pleasance E D, Cheetham R K, Stephens P J, et al. 2010. A comprehensive catalogue of somatic mutations from a human cancer genome. Nature, 463（7278）: 191-196.

Pluciennik A, Dzantiev L, Iyer R R, et al. 2010. PCNA function in the activation and strand direction of MutLα endonuclease in mismatch repair. Proc Natl Acad Sci U S A, 107（37）: 16066-16071.

Pryor J M, Conlin M P, Carvajal-Garcia J, et al. 2018. Ribonucleotide incorporation enables repair of chromosome breaks by nonhomologous end joining. Science, 361（6407）: 1126-1129.

Qin B, Yu J, Nowsheen S, et al. 2019. UFL1 promotes histone H4 ufmylation and ATM activation. Nat Commun, 10（1）: 1242.

Rao S S, Huntley M H, Durand N C, et al. 2014. A 3D map of the human genome at kilobase resolution reveals principles of chromatin looping. Cell, 159（7）: 1665-1680.

Rao S S P, Huang S C, Glenn St Hilaire B, et al. 2017. Cohesin loss eliminates all loop domains. Cell, 171（2）: 305-320, e324.

Richardson S M, Mitchell L A, Stracquadanio G, et al. 2017. Design of a synthetic yeast genome.

Science, 355（6329）: 1040-1044.

Roadmap Epigenomics Consortium, Kundaje A, Meuleman W, et al. 2015. Integrative analysis of 111 reference human epigenomes. Nature, 518（7539）: 317-330.

Roberts S A, Lawrence M S, Klimczak L J, et al. 2013. An APOBEC cytidine deaminase mutagenesis pattern is widespread in human cancers. Nat Genet, 45（9）: 970-976.

Robin J D, Ludlow A T, Batten K, et al. 2014. Telomere position effect: regulation of gene expression with progressive telomere shortening over long distances. Genes Dev, 28（22）: 2464-2476.

Sanulli S, Trnka M J, Dharmarajan V, et al. 2019. HP1 reshapes nucleosome core to promote phase separation of heterochromatin. Nature, 575（7782）: 390-394.

Schiffers S, Ebert C, Rahimoff R, et al. 2017. Quantitative LC-MS provides no evidence for m（6）dA or m（4）dC in the genome of mouse embryonic stem cells and tissues. Angew Chem Int Ed Engl, 56（37）: 11268-11271.

Schmidt J C, Zaug A J, Cech T R. 2016. Live cell imaging reveals the dynamics of telomerase recruitment to telomeres. Cell, 166（5）: 1188-1197, e1189.

Schuster-Bockler B, Lehner B. 2012. Chromatin organization is a major influence on regional mutation rates in human cancer cells. Nature, 488（7412）: 504-507.

Schwarzer W, Abdennur N, Goloborodko A, et al. 2017. Two independent modes of chromatin organization revealed by cohesin removal. Nature, 551（7678）: 51-56.

Senigl F, Maman Y, Dinesh R K, et al. 2019. Topologically associated domains delineate susceptibility to somatic hypermutation. Cell Rep, 29（12）: 3902-3915, e3908.

Seplyarskiy V B, Soldatov R A, Popadin K Y, et al. 2016. APOBEC-induced mutations in human cancers are strongly enriched on the lagging DNA strand during replication. Genome Res, 26（2）: 174-182.

Shakya A, Park S, Rana N, et al. 2020. Liquid-liquid phase separation of histone proteins in cells: role in chromatin organization. Biophys J, 118（3）: 753-764.

Shao Y, Lu N, Cai C, et al. 2019. A single circular chromosome yeast. Cell Res, 29（1）: 87-89.

Shao Y, Lu N, Wu Z, et al. 2018. Creating a functional single-chromosome yeast. Nature, 560（7718）: 331-335.

Shen Y, Wang Y, Chen T, et al. 2017. Deep functional analysis of synII, a 770-kilobase synthetic yeast chromosome. Science, 355（6329）: eaaf4791.

Shu X, Liu M, Lu Z, et al. 2018. Genome-wide mapping reveals that deoxyuridine is enriched in the human centromeric DNA. Nat Chem Biol, 14（7）: 680-687.

Smith H O, Hutchison III C A, Pfannkoch C, et al. 2003. Generating a synthetic genome by whole genome assembly: ΦX174 bacteriophage from synthetic oligonucleotides. Proc Natl Acad Sci U S A, 100（26）: 15440-15445.

Song F, Chen P, Sun D, et al. 2014. Cryo-EM study of the chromatin fiber reveals a double helix twisted by tetranucleosomal units. Science, 344（6182）: 376-380.

Stamatoyannopoulos J A, Adzhubei I, Thurman R E, et al. 2009. Human mutation rate associated with DNA replication timing. Nat Genet, 41（4）: 393-395.

Strickfaden H, Tolsma T O, Sharma A, et al. 2020. Condensed chromatin behaves like a solid on the mesoscale *in vitro* and in living cells. Cell, 183（7）: 1772-1784 e1713.

Strom A R, Emelyanov A V, Mir M, et al. 2017. Phase separation drives heterochromatin domain formation. Nature, 547（7662）: 241-245.

Su J H, Zheng P, Kinrot S S, et al. 2020. Genome-scale imaging of the 3D organization and transcriptional activity of chromatin. Cell, 182（6）: 1641-1659, e1626.

Tahiliani M, Koh K P, Shen Y, et al. 2009. Conversion of 5-methylcytosine to 5-hydroxymethylcytosine in mammalian DNA by MLL partner TET1. Science, 324（5929）: 930-935.

Takei Y, Yun J, Zheng S, et al. 2021. Integrated spatial genomics reveals global architecture of single nuclei. Nature, 590（7845）: 344-350.

Tang M, Li Z, Zhang C, et al. 2019. SIRT7-mediated ATM deacetylation is essential for its deactivation and DNA damage repair. Sci Adv, 5（3）: eaav1118.

Thompson P S, Amidon K M, Mohni K N, et al. 2019. Protection of abasic sites during DNA replication by a stable thiazolidine protein-DNA cross-link. Nat Struct Mol Biol, 26（7）: 613-618.

Tomizuka K, Shinohara T, Yoshida H, et al. 2000. Double trans-chromosomic mice: maintenance of two individual human chromosome fragments containing Ig heavy and kappa loci and expression of fully human antibodies. Proc Natl Acad Sci U S A, 97（2）: 722-727.

Tomizuka K, Yoshida H, Uejima H, et al. 1997. Functional expression and germline transmission of a human chromosome fragment in chimaeric mice. Nat Genet, 16（2）: 133-143.

Trivedi P, Palomba F, Niedzialkowska E, et al. 2019. The inner centromere is a biomolecular condensate scaffolded by the chromosomal passenger complex. Nat Cell Biol, 21（9）: 1127-1137.

Turner K M, Deshpande V, Beyter D, et al. 2017. Extrachromosomal oncogene amplification drives tumour evolution and genetic heterogeneity. Nature, 543（7643）: 122-125.

Venter J C, Adams M D, Myers E W, et al. 2001. The sequence of the human genome. Science, 291（5507）: 1304-1351.

Vian L, Pekowska A, Rao S S P, et al. 2018. The energetics and physiological impact of cohesin extrusion. Cell, 175（1）: 292-294.

Wang L, Hu M, Zuo M Q, et al. 2020a. Rett syndrome-causing mutations compromise MeCP2-mediated liquid-liquid phase separation of chromatin. Cell Res, 30（5）: 393-407.

Wang L, Zhang J, Duan J, et al. 2014. Programming and inheritance of parental DNA methylomes in mammals. Cell, 157（4）: 979-991.

Wang M, Wei P C, Lim C K, et al. 2020b. Increased neural progenitor proliferation in a hiPSC model of autism induces replication stress-associated genome instability. Cell Stem Cell, 26（2）: 221-233, e226.

Wang Z, Gong Y, Peng B, et al. 2019. MRE11 UFMylation promotes ATM activation. Nucleic Acids Res, 47（8）: 4124-4135.

Wang Z, Xu W, Liu L, et al. 2016. A synthetic molecular system capable of mirror-image genetic replication and transcription. Nat Chem, 8（7）: 698-704.

Wei P C, Chang A N, Kao J, et al. 2016. Long neural genes harbor recurrent DNA break clusters in neural stem/progenitor cells. Cell, 164（4）: 644-655.

Wei W, Ba Z, Gao M, et al. 2012. A role for small RNAs in DNA double-strand break repair. Cell, 149（1）: 101-112.

Weinhold N, Jacobsen A, Schultz N, et al. 2014. Genome-wide analysis of noncoding regulatory mutations in cancer. Nat Genet, 46（11）: 1160-1165.

Wu L, Schatz D G. 2020. Making ends meet in class switch recombination. Cell Res, 30（9）: 711-712.

Wu S, Turner K M, Nguyen N, et al. 2019. Circular ecDNA promotes accessible chromatin and high oncogene expression. Nature, 575（7784）: 699-703.

Wu S, Zhu W, Thompson P, et al. 2018. Evaluating intrinsic and non-intrinsic cancer risk factors. Nat Commun, 9（1）: 3490.

Wu T P, Wang T, Seetin M G, et al. 2016. DNA methylation on N（6）-adenine in mammalian embryonic stem cells. Nature, 532（7599）: 329-333.

Wu Y, Li B Z, Zhao M, et al. 2017. Bug mapping and fitness testing of chemically synthesized chromosome X. Science, 355（6329）: eaaf4706.

Xia B, Han D, Lu X, et al. 2015. Bisulfite-free, base-resolution analysis of 5-formylcytosine at the genome scale. Nat Methods, 12（11）: 1047-1050.

Xie X, Hu H, Tong X, et al. 2018. The mTOR-S6K pathway links growth signalling to DNA damage response by targeting RNF168. Nat Cell Biol, 20（3）: 320-331.

Xie Z X, Li B Z, Mitchell L A, et al. 2017. "Perfect" designer chromosome V and behavior of a ring derivative. Science, 355（6329）: eaaf4704.

Xu G, Chapman J R, Brandsma I, et al. 2015. REV7 counteracts DNA double-strand break resection and affects PARP inhibition. Nature, 521（7553）: 541-544.

Xu M, Long C, Chen X, et al. 2010. Partitioning of histone H3-H4 tetramers during DNA replication-dependent chromatin assembly. Science, 328（5974）: 94-98.

Xue J H, Chen G D, Hao F, et al. 2019. A vitamin-C-derived DNA modification catalysed by an algal TET homologue. Nature, 569（7757）: 581-585.

Yang Q, Zhu Q, Lu X, et al. 2017. G9a coordinates with the RPA complex to promote DNA damage repair and cell survival. Proc Natl Acad Sci U S A, 114（30）: E6054-E6063.

Yizhak K, Aguet F, Kim J, et al. 2019. RNA sequence analysis reveals macroscopic somatic clonal expansion across normal tissues. Science, 364（6444）: eaaw0726.

Zhang G, Huang H, Liu D, et al. 2015. N6-methyladenine DNA modification in *Drosophila*. Cell, 161（4）: 893-906.

Zhang H, Zhao R, Tones J, et al. 2020a. Nuclear body phase separation drives telomere clustering in ALT cancer cells. Mol Biol Cell, 31（18）: 2048-2056.

Zhang J, Gao Q, Li P, et al. 2011. S phase-dependent interaction with DNMT1 dictates the role of UHRF1 but not UHRF2 in DNA methylation maintenance. Cell Res, 21（12）: 1723-1739.

Zhang J M, Zou L. 2020b. Alternative lengthening of telomeres: from molecular mechanisms to therapeutic outlooks. Cell Biosci, 10（1）: 30.

Zhang R R, Cui Q Y, Murai K, et al. 2013. Tet1 regulates adult hippocampal neurogenesis and cognition. Cell Stem Cell, 13（2）: 237-245.

Zhang W, Zhao G, Luo Z, et al. 2017a. Engineering the ribosomal DNA in a megabase synthetic chromosome. Science, 355（6329）: eaaf3981.

Zhang X, Zhang Y, Ba Z, et al. 2019a. Fundamental roles of chromatin loop extrusion in antibody class switching. Nature, 575（7782）: 385-389.

Zhang Y, Lamb B M, Feldman A W, et al. 2017b. A semisynthetic organism engineered for the stable expansion of the genetic alphabet. Proc Natl Acad Sci U S A, 114（6）: 1317-1322.

Zhang Y, Ptacin J L, Fischer E C, et al. 2017c. A semi-synthetic organism that stores and retrieves increased genetic information. Nature, 551（7682）: 644-647.

Zhang Y, Zhang X, Ba Z, et al. 2019b. The fundamental role of chromatin loop extrusion in physiological V（D）J recombination. Nature, 573（7775）: 600-604.

Zheng H, Xie W. 2019. The role of 3D genome organization in development and cell differentiation. Nat Rev Mol Cell Biol, 20（9）: 535-550.

Zhu C, Gao Y, Guo H, et al. 2017. Single-cell 5-formylcytosine landscapes of mammalian early embryos and ESCs at single-base resolution. Cell Stem Cell, 20（5）: 720-731, e725.

第四章

生物分子的结构、功能与设计

第一节 概　　述

生物分子（biomolecule）是自然存在于生物体内的分子的总称，是生命的承载体和生命活动的执行者。生物分子的内涵非常丰富，包括核酸、蛋白质、碳水化合物（糖类）和脂质等大分子，以及代谢物、次级代谢物和天然产物等小分子。生物学经历了演化生物学、遗传学到现代分子生物学的发展进程。人们对生命体的认识由表及里、由浅入深，从宏观现象归纳到微观层面认识构成生命体基本分子的结构与性质。DNA、RNA 等核酸分子的发现阐明了遗传的物质基础，DNA 双螺旋结构的解析首开了分子生物学发展之先河，中心法则（DNA⇨RNA⇨ 蛋白质）的发现与完善则奠定了现代生物学研究的基础。

遗传信息表达流由 DNA、RNA 和蛋白质构成，其中 RNA 是生命起源的最初分子形式，是所有生物遗传信息表达的传递者和调控者。核酸研究不仅能够带来生命科学概念和理论的系列重大突破，而且能够颠覆传统的生物医学，开拓核酸的前沿技术产业。蛋白质是执行生命活动的一类重要生物大分子。蛋白质基本的构成单元是氨基酸，自然存在的常见氨基酸有20种。翻译后修饰是蛋白质在翻译后的化学修饰过程，糖基化是其中广泛存在且结构多变的一个类型。糖链是继核酸、蛋白质后被发现具有强大功能的第三条重要的生物分子链，其潜在信息编码容量远远大于核酸与蛋白质。修饰的糖链结构复杂，糖型丰富，不同细胞中存在的各种聚糖结构可以组成庞大的糖组信息。糖链通过与凝集素相互作用参与细胞识别、细胞分化、发育、信号转导、免疫应答等多种重要生命活动。脂质分子是构成细胞和生命体的重要组成成分之一，也是储存能量的主要分子。脂类物质包括脂肪酸、胆固醇、胆固醇酯、磷脂、甘油三酯、鞘磷脂、醚脂等。脂质分子还被用于合成胆汁酸、甾醇类激素、信号分子等，在生命活动中发挥重要作用。

在生命活动的过程中，不同类别的生物分子存在互作，从而协同发挥作用。蛋白质与核酸相互作用形成的高分子复合物在细胞内几乎无处不在。基因组高级结构的形成、核仁的形成、转录起始复合物的组装与动态变化、核糖体的组装、某些特定类型的RNA与蛋白质组装形成的无膜亚细胞结构等，均是蛋白质与核酸相互作用调控细胞重大基础功能的例证。蛋白质存在多种翻译后修饰，其中分泌蛋白和膜蛋白自内质网上附着的核糖体合成后被糖基化，糖基化修饰是蛋白质加工成熟的标记。高度糖基化的蛋白聚糖是细胞外基质的重要组成成分。Wnt、Hedgehog、BMP等形态发生素存在特定的脂酰化修饰，脂酰化修饰促进形态发生素向细胞外分泌，同时在发育过程形成浓度梯度，促使细胞分化与体轴的建立。除此以外，糖类和脂类作为细胞活动的供能物质，也参与细胞基本结构的形成，如构成细胞膜的多种脂质分子，负责细胞粘连与识别的糖蛋白。革兰氏阴性细菌的脂多糖是细胞壁外膜的重要成分，同时也是诱发炎症与先天性免疫的主要原因。

因此，进一步加深对生物分子结构和功能的认识，有助于全面揭示生命本质以及肿瘤、内分泌疾病、免疫性疾病、感染性疾病等重大疾病的发生、发展机制，在寻找疾病的诊治靶点和创新药物研究等方面具有重大科学意义。

第二节 关键科学问题

一、生物分子的结构与识别

生物分子的特异性识别是其发挥一切功能的基本前提，而其三维结构往往在识别过程中发挥决定性作用。DNA与RNA分别承载着遗传信息的储存与传递功能，是遗传的物质基础。这两种分子的区别仅限于一个羟基，细胞是如何来识别这种相对微小的区别的至今尚不清楚。RNA一般以前体的形式被转录，经加工与修饰等步骤成熟以后，被转运至相应的位点发挥作用；而一些转录和加工的副产物及一些异常的RNA则需要被快速识别并降解。细胞如何识别前体与成熟体、正常与异常RNA，这些同样是关乎遗传信息正确传递的重要问题，目前我们知之甚少。另外，近年来的研究发现了多种全新的、不同于经典蛋白质编码基因的非编码RNA，它们在多种重要的生命过程中也发挥关键调控作用。编码与非编码RNA的特异性识别机制同样是亟待研究的关键科学问题。同样，具有相似氨基酸序列的不同蛋白质及同种蛋白质的不同翻译后修饰的特异性识别，是细胞执行各种复杂功能的基础。不同糖链结构可能影响所修饰的蛋白质的空间结构，并在被识别之后改变蛋白质的稳定性和功能。非常有趣的是，同样的生物分子在不同的组织和细胞中，甚至在同种细胞中不同条件下，经常会发挥不同的功能，这种功能区别背后的一种重要机制就是生物分子的识别受到严密的调控。因此，基于结构等的生物分子的特异性识别，可以说是细胞内生化反应和细胞维持正常功能的基础。

二、生物分子的功能与调控

生物分子的功能研究是剖析生命奥秘的根本。对生物分子及其调控网络在细胞命运决定、个体发育和应激、生殖与代际遗传和群体进化等重要生命现象中的功能进行研究，有助于阐明各种复杂生命现象背后的基础分子机制，推动生命科学理论的突破性发展。研究生物分子代谢中的各个过程及其相互之间步骤的耦联机制，以全新的角度与方式诠释包括人类在内的各生物基因组，可为重大疾病的干预、防治及药物靶点研究等提供全新的思路与技术。发展和应用多组学、生物分子高灵敏实时示踪和分析技术，从亚细胞到宏观个体尺度探索生物分子代谢的发生、代谢物的运输和转化，可阐明各组织器官的通信、协调和互作，进而揭示机体在生理与病理情况下代谢改变与重塑的规律，以及代谢与机体健康和疾病发展的关系。同时，阐明个体代谢差异的环境与遗传基础，发展精准药物与健康生活策略，可实现从分子到群体及不同时间阶段的对代谢的全面认识，研发新型药物并发展治疗相关疾病的新策略。此外，从生物分子组学层面阐明疾病发生发展的分子机制，构建多种生物分子－疾病整合数据库，可为疾病治疗提供新的靶点和分子标志物。随着基因组学、转录组学、蛋白质组学、代谢组学等组学技术的快速发展，以及生物信息与大数据科学的交叉应用，对疾病和特定患者进行个性化精准诊断与防治成为可能。

三、生物分子的改造与设计

随着对生物分子的结构与功能的理解，根据需求改造和设计生物分子已经成为现实，并且取得了鼓舞人心的阶段性进展。蛋白质设计已被应用于包裹 RNA 和纳米颗粒疫苗。靶向目标蛋白界面的从头设计可用于调节与疾病有关的信号蛋白，从而达到药物干预的效果。另外，通过设计目标蛋白分子内和分子间不同的结合力，可以实现制作蛋白质开关，蛋白质开关被应用于合成生物学和细胞疗法。从头设计结合复杂的、高度功能化的小分子蛋白质可以生产基于蛋白质的小分子载体和开发抑制致病蛋白功能的小分子药物。近来，国内外研究者发现，血清 / 血浆以及外泌体中的小 RNA 可作为癌症及心血管系统等疾病早期诊断及预后评价的分子指标。国际上有 40 多项结合 RNA 设计、RNA 干扰和 CRISPR 等技术的核酸药物处于Ⅱ～Ⅲ期临床试验中，涉及癌症、糖尿病、呼吸系统疾病、眼部疾病和神经系统疾病治疗药物。

2016年，有2项通过操纵RNA来治疗遗传病的重大新药在欧美上市，取得了极好的疗效。如何根据需求理性改造和设计生物分子正在成为人类探索生命奥秘和提升健康的关键科学问题。

第三节　重要研究方向的发展现状与趋势分析

一、RNA的代谢、功能与调控

RNA生物学正在成为现代生命科学中最有影响力的领域。大量新型RNA的发现以及RNA在各类生命活动中的生理功能，不断刷新着人们对传统生物学的认知。正因为如此，目前国际上对RNA的研究十分重视，欧美各个国家都相继启动了以RNA为主体的重大研究计划。在人类基因组计划完成后，欧盟提出了“RNA调控网络与健康和疾病”计划，通过“地平线欧洲”框架计划展开对该领域的资助，旨在确立其在RNA领域的领导地位。随后，美国牵头启动了ENCODE计划，并以美国国家科学基金会和国立卫生研究院为主导机构进行资助。日本也同样启动了功能RNA研究项目、哺乳动物基因组功能注释计划等重要研究。美国国立卫生研究院于2013年8月6日设立了以人类疾病RNA为主题的“胞外RNA（exRNA）通信”重大研究计划。在RNA化学修饰研究方面，美国国立卫生研究院投入了千万美元量级的经费，资助了首个以“动态RNA表观转录组”为研究重点的基因组科学研究中心。2016年开始的人类细胞图谱（Human Cell Atlas）计划，在很大程度上致力于从单细胞水平弄清人类各种细胞类型中RNA的表达情况及特征。这些重大计划的实施，目标是将癌症中RNA相关的最新研究成果进行临床转化应用，开发其作为疾病生物标记的潜力。另外，RNA编辑正在成为治疗遗传疾病最有前景的治疗手段之一。国际上针对RNA前沿技术的投入在不断加大，继美国ENCODE计划后，美国国立卫生研究院共同基金2015年启动了4D细胞核组计划，涵盖拟开发的创新性技术，研究的重点也关注到RNA形成的核仁结构、亚核小体、胞质小体以及相变等对基因表达调控的影响。

近十年来RNA研究一直保持快速发展，研究人员发现了微RNA（microRNA，miRNA）、长链非编码RNA等一大批在生命过程中发挥关键调控作用的重要细胞因子，解析了RNA剪接体、核糖体等一批重要RNA-蛋白质复合体的结构和功能，发明和发展了基因沉默（如RNAi）、基因编辑（如CRISPR）和基因测序等具有划时代意义的新技术，核酸研究已成为生命科学领域发展最快的研究方向之一。基础科学的发展推动了产业技术突破，基因沉默、基因检测、核酸疫苗、mRNA表达等技术正在发展成为疾病诊断和治疗等领域的前沿方向。特别是以RNA为靶标的创新药物有望成为继抗体类药物后的新一代药物，因此得到了国际大型制药企业的广泛重视。国际上以RNA研究和临床应用为核心，围绕生命科学试剂和服务、临床检测、生物医药三大领域布局，形成了万亿美元规模的产业。可以预见，未来10年，核酸研究依然是生命科学领域的最前沿，非编码RNA、表观遗传、转录组学等核酸研究领域将迎来新发展，更多的新核酸分子、生物学新规律将被发现和阐明，数以千计的核酸分子有望成为疾病诊断和治疗的新靶标，核酸治疗药物、核酸检测等新型医药产品将在临床上广泛应用，核酸新兴产业将成为生物医药产业的重要组成部分，RNA产业必将迎来新一轮“井喷”式全面发展。

我国RNA研究开展得并不晚。研究初期，我国科学家对RNA的研究主要集中在tRNA和rRNA的结构与功能方面，其中在tRNA的人工合成方面曾有过辉煌的成就（人工合成酵母丙氨酸tRNA）。在理论方面，早在20世纪80年代初，我国科学家就已经开始采用信息论、几何学、密码学等方法对核酸序列的特征进行理论分析。在RNA代谢与调控研究领域，我国科学家进行tRNA参与蛋白质合成质量控制的经典功能和蛋白质合成以外的非经典功能研究近30年，这些工作在国际RNA研究领域受到广泛认可（Hilander et al.，2018；Ji et al.，2017；Liu et al.，2015）。研究人员在秀丽隐杆线虫中率先鉴定了约100个新型的中等长度非编码RNA，并预测了线虫中存在约2700个小RNA（Deng et al.，2006；He et al.，2006，2007；Yuan et al.，2014；Zhao et al.，2016）。20世纪90年代，Ma等（2013）在新的核仁小RNA（small nucleolar RNA，snoRNA）结构与功能研究方面取得了一系列进展。Xue等（2013）发现剪接因子多聚嘧啶序列结合蛋白（polypyrimidine tract binding protein, PTBP）在各种细胞中广泛调节 miRNA 和 mRNA 的相互作用，并且发现了一条全新的 miRNA 调控的细胞重编程途径，其可以使各种非神经细胞向功能神经细胞转分化。Zhou 等（2019）在 Piwi 蛋白相互作用 RNA（piRNA）研究方面获得了一系列重要成果，他们发现了 piRNA 在精子发生后期触发小鼠 PiWi 蛋白（小鼠的 PiWi 蛋白称作 MiWi 蛋白）降解的现象，又发现无精子症患者中有一类拮抗人 PiWi 蛋白泛素化修饰和降解突变可导致雄性不育，并探究了其发生机制。Wei 等（2012）以拟南芥为模式生物，发现了一类在 DNA 损伤修复中起重要作用的新型小分子 RNA。Li 等（2017）和 Zhang 等（2014，2016）发现了若干新型长线状和环状非编码 RNA，同时揭示了这些 RNA 加工和功能发挥的分子机制。Fan 等（2017）、Chen 等（2019b）、Wang 等（2019a）发现了新生 RNA 的出核转运或核内降解的命运决定的全新机制与调控模式。Wang 等（2012）、Yue 等（2016）发现了一个与 Dscam 互斥剪切有关 RNA 结构性基因座控制区域，还发现了互斥剪切模式涉及 RNA 二级结构的竞争。Zhao 等（2014）、Xiao 等（2016）揭示了非编码 miRNA 介导的 m6A 甲基化位点形成的选择性机制，鉴定了 m6A 去甲基化酶及 RNA 甲基化修饰具有可逆性并调控 RNA 加工代谢。Li 等（2016a，2016b）开发了一系列新型 RNA 甲基化、假尿嘧啶化修饰的测序新技术，绘制了转录组中这些新型 RNA 修饰的高清图谱。Bai 等（2017）、Zhang 等（2017b）、Zhan 等（2018）在剪接体的高分辨率结构研究方面取得了一系列突破性进展，揭示了剪接体的组装和工作机理，为理解 RNA 剪接过程奠定了重要基础。Liu 等（2017b）解析了 1.9Å 分辨率的 Gemin5 蛋白 N 端与核小 RNA（snRNA）底物的复合物晶体结构；阐明了包括核糖 2′-*O*- 甲基化修饰和 m5C 在内的一系列 tRNA 修饰酶的底物结合机制和催化机理。

在 RNA 生理和疾病研究领域，研究人员发现了调控免疫系统中树突状细胞分化和病毒感染的长链非编码 RNA（Wang et al.，2014）；解析了在癌症发生过程中具有重要作用的长链非编码 RNA 和小 RNA 的功能（Yu et al.，2007）；发现了 RNA 修饰在生命信息传递的剪切过程、诱导多能干细胞重编程、精子发生和造血干细胞命运决定过程中具有重要作用（Xu et al.，2017）；报道了调控胚胎干细胞自我更新的长链非编码 RNA，以及由 tRNA 产生的小 RNA 片段在代谢性疾病隔代遗传中的关键作用（Luo et al.，2016；Yin et al.，

2015）。另外，Peng 等（2012）还发现细胞外的小 RNA 在多种生命活动过程中可能具有重要作用。Zhu 等（2017）发现食物中的 miRNA 可以影响蜜蜂和果蝇的机体发育，并且可进一步影响物种的进化。这些 RNA 生理功能相关研究成果的涌现，极大地提高了我国 RNA 的研究水平，使得我国在 RNA 研究领域的国际地位逐渐上升。我国在 RNA 生理功能研究领域涌现出的优秀科研团队和卓越科研成果，为整个 RNA 研究领域的进一步发展打下了良好的基础。

在 RNA 与农学领域，近十年来，研究人员建立了小 RNA 通过表观遗传参与作物株型调控的分子理论，揭示了小 RNA 进入不同效应复合物的分拣机制，发现了参与 DNA 损伤修复的新型小 RNA，阐述了 miRNA 通过 DNA 甲基化参与基因转录水平调控的作用方式，解读了 miRNA 决定植物年龄路径的分子模式，诠释了 RNA 指导的 DNA 甲基化（RNA directed DNA methylation，RdDM）在维持基因组稳定性中的重要意义，解析了植物小 RNA 抗病毒的分子机制等。此外，我国科学家证明，非编码 RNA 在水稻、棉花等农作物的抗虫、理想株型建成、籽粒增大增产、激素信号转导、光敏雄性不育等重要农艺性状方面发挥着重大作用，并探索了将非编码 RNA 应用于农作物重要农艺性状育种改良工作中的可能性。目前，我国在植物 RNA 领域已积累了丰富的研究经验和相当的科研实力，各 RNA 研究团队特色鲜明、专长互补，并产生了协同效应。这为下一步研究计划的开展奠定了扎实的工作基础，今后我国植物 RNA 研究领域将继续保持蓬勃发展的良好态势。

在 RNA 前沿技术领域，我国科学家在单细胞 RNA 测序（RNA-seq）方面做了大量开创性工作，系统分析了早期胚胎发育过程中的转录组动态变化特点（Fan et al.，2020b）。此外，我国在单细胞液滴测序（Drop-seq）技术研发方面也取得了较好的进展。在 RNA 结构研究方面，Li 等（2020）在转录组范围内系统解析了 RNA 的二级结构。Zhou 等（2019）在利用 CLIP-seq 及 GRID-seq 技术检测 RNA 在染色质上的靶标方面取得了突破性进展。在 RNA 的体内转导方面，Zhao 等（2013）、Gou 等（2017）独创了睾丸原位转导技术，并首次阐释了 Piwi 蛋白突变与男性不育的关系及详细作用机制。在长链非编码 RNA 的系统发现方面，Chen（2016）、Wu 等（2017）通过对不含多腺苷酸［poly（A）］的 RNA 测序发现并鉴定了大量的环形 RNA 和包含完整的 snoRNA 序列的非编码 RNA（long noncoding RNAs with snoRNA Ends，sno-lncRNA），并利用超高分辨率技术系统地研究了 sno-lncRNA 的功能和代谢。Li 等（2015）研发了系列单位点 RNA 化学修饰检测和高通量测序技术。

在核酸信息学分析技术利用领域，我国科学家起步早、基础好并取得了丰硕的成果。我国科学家开发了非编码 RNA 综合数据平台，提出了一套以非编码 RNA 所参与的细胞生化过程和在此过程中发挥的作用为标准的统一分类体系（Liu et al.，2005；Zhao et al.，2016）；此外还提出了“双色网络”“核酸计算机”等重大前沿科学概念和设想。在 RNA 产业化领域，特别是在 miRNA 领域，我国发表的论文在世界上名列前茅，这得益于我国的 miRNA 科学试剂和服务产业已初具规模，几十家中小企业提供技术服务和销售 RNA 试剂产品。我国 RNA 临床检测技术虽然起步晚，但已有数十家企业开展了 RNA 测序、外周血中核酸检测、肿瘤核酸检测等前沿临床检测试验，未来 5 年内将有一批核酸诊断试剂应用于临床。

二、蛋白质的结构、功能与设计

蛋白质的结构是其特异性识别和功能发挥的基础。以 G 蛋白耦联受体（GPCR）为例，当前市面上 1/3 的药物以其作为靶点，GPCR 结构与功能的研究加速了相关药物研发的进程（Rask-Andersen et al.，2011）。所有的 GPCR 都有 7 次跨膜结构，但它们的结构、动力学特征、配体选择性、调控因子以及下游信号分子却千差万别（Hilger et al.，2018；Latorraca et al.，2017；Wacker et al.，2017）。不同类别和亚家族的 GPCR 如此，即使是同一亚家族序列高度相似的 GPCR，其结构也有所不同（Chien et al.，2010；Fan et al.，2020a；Im et al.，2020；Wang et al.，2018，2017；Xiao et al.，2021；Zhuang et al.，2021）。不同活性状态的 GPCR 结构为生物化学、生物物理及生物信息学研究 GPCR 功能和动力学提供了原子水平的 3D 模型（Cherezov et al.，2007；Rasmussen et al.，2011；Staus et al.，2016）。GPCR 能以同源二聚体、异源二聚体和寡聚体多种形式发挥功能，晶体结构揭示了同源二聚体相互作用的细节（Gonzalez-Maeso，2011；Pin et al.，2004；Terrillon and

Bouvier，2004）。近年来，单颗粒冷冻电镜技术的快速发展推进了对 GPCR 与下游效应蛋白复合体的结构研究，加深了对 GPCR 活化过程中构象变化的认识，同时揭示了 GPCR 激活下游效应蛋白的分子机制（Garcia-Nafria et al.，2018；Hilger et al.，2020；Koehl et al.，2018；Kuhlbrandt，2014；Lin et al.，2020；Maeda et al.，2019；Zhang et al.，2017c）。此外，GPCR 的精细结构还揭示了离子、脂质、多肽以及合成分子对 GPCR 的别构调控（Conn et al.，2009；Liu et al.，2012，2019；Wu et al.，2014；Yang et al.，2020a；Yu et al.，2020），这对加深对 GPCR 功能和药物应答机制的认识大有裨益，同时也说明 GPCR 是一套复杂的别构机器，不仅仅受控于药物配体。此外，利用计算机模拟 GPCR 结构与功能的关系是目前研究 GPCR 的重要手段（Hollingsworth and Dror，2018；Kato et al.，2019；Latorraca et al.，2020），利用计算模拟的研究策略可以填补 GPCR 超家族和分子水平相互作用研究的空白区域，并为药物研发提供了平台（McCorvy et al.，2018）。

抗生素是近代医学史上的伟大发现（Gould，2016），自 20 世纪 40 年代以来，它拯救了数亿人的生命（Hutchings et al.，2019）。然而，抗生素的滥用加速了细菌的进化，导致抗多种药物的“超级细菌”（Patini et al.，2020）不断出现，以致细菌仍然是感染致死疾病中的首要病原体（Naylor et al.，2018；Rather et al.，2017；Zaman et al.，2017）。跨膜蛋白因承担了物质转运、信号转导、生化反应催化等重要功能，是生物体的必需组分，因而是重要的抗生素作用靶点（Choi and Lee，2019；Epand et al.，2016；Hart et al.，2019；Imai et al.，2019；Luther et al.，2019；Nie et al.，2020；Urfer et al.，2016）。但对其结构进行研究仍然富有挑战性，需要在技术上进行创新与突破。一个重要的研究思路是，对重要病原微生物生长、侵染所必需的跨膜蛋白进行三维结构解析，并根据结构设计对其活性有抑制作用的小分子，从而获得新型抗生素的先导分子。

病毒，特别是呼吸道病毒，传播速度快、范围广，急需有效的检测方法和治疗药物（Perlman，2020）。小分子药物因为潜在毒性及脱靶效应等而发展周期长、成功率低；而生物大分子抗体药物的评价期相对更短，安全性更高（Imai and Takaoka，2006；Lu et al.，2020）。目前，病毒抗体主要来自康复患者（Venkat Kumar et al.，2020），由于研发周期长，且前期需要 P3/P4 实验室，普通研究机构难以进行抗体研发。采用病毒蛋白进行动物免疫和单抗筛选前期无生物安全风险，但周期较长，急需建立对P3/P4实验室依赖程度低、快速、普适性的抗体筛选方法，以应对突发的由新型病毒引起的人群感染。在发展检测试剂和治疗药物时，除有效性外，时效性也非常关键。无论是通过康复患者还是动物免疫途径，获得抗体的速度均较慢。可以考虑采用人工抗体筛选（Sall et al.，2016）的创新思路，其主要优点包括周期短、无生物安全问题，并且方便进行亲和力改造。因此，建立人工抗体筛选体系可望加快应对新型病毒流行的速度。

蛋白质设计通常被认为是蛋白质折叠的反问题，需要确定能稳定蛋白质所需构象的氨基酸序列。蛋白质设计大致分为两类：一类是基于天然蛋白质改造的蛋白质工程；另一类是生成新蛋白质骨架和序列的蛋白质从头设计。其中，蛋白质从头设计为理解蛋白质的结构和功能提供了重要的检验工具，并允许创建自然界不存在的蛋白质。当前，蛋白质从头设计的方法也发展迅速。最早的蛋白质是使用简单的物理化学原理和分子力学力场设计的。新近开发的方法依赖于骨架片段和结构生物信息学，采用可折叠的蛋白质结构空间，并且在设计工作流程中使用分子动力学计算方法进行测试和验证（Korendovych and DeGrado ，2020）。目前，机器学习已经应用于蛋白质从头设计来生成三级结构和氨基酸序列（Gao et al.，2020）。

蛋白质从头设计的应用十分广泛，在多个领域取得了重要进展（Kuhlman and Bradley，2019）。在基于蛋白质结合的改造方面，蛋白质自组装设计产生了具有对称性超结构的自组装抗体纳米笼，可用来包裹RNA的蛋白质组合体，可应用于药物递送和疫苗研发（Butterfield et al.，2017；He et al.，2020）；基于目标蛋白界面的靶向设计的人造蛋白可以结合并调节与疾病有关的信号蛋白从而达到药物干预的效果（Silva et al.，2019）；蛋白质开关被应用于合成生物学和细胞疗法（Lajoie et al.，2020；Ng et al.，2019）；从头设计结构复杂的、高度功能化的小分子蛋白质可以产生基于蛋白质的小分子载体和开发抑制致病蛋白功能的小分子药物（Polizzi and DeGrado，2020）。

不仅如此，蛋白质从头设计在探索和扩展化学催化过程中也起到了重要的作用（Korendovych and DeGrado，2020）。从头设计金属蛋白为多种应用设计含辅因子蛋白质创造了更多可能性，并将传统的过渡金属催化剂的优势与蛋白质的多功能性结合了起来（Nastri et al.，2019）。蛋白质从头设计在非金属蛋白的化学催化研究中也有可喜的成就（Kries et al.，2013）。通过定向进化具有微弱催化功能的设计蛋白可以得到具有天然酶催化效率的构建体（Bunzel et al.，2021）。蛋白质从头设计的另一个重大进展是实现了跨膜运输的生物功能（Korendovych and DeGrado，2020）。定制设计的跨膜转运蛋白在单分子传感中的应用可能将随之而来，例如用于 RNA 和 DNA 测序。总而言之，蛋白质从头设计已发展成为一种用于创造拥有全新结构和功能的蛋白质的重要工具。鉴于设计方法的逐渐成熟，可以预期蛋白质从头设计将在医学、工业等多领域研究和应用中发挥巨大的作用。

三、糖的识别、结构与功能

蛋白质或脂质分子糖基修饰包含重要生物信息，这种生物信息的突出特征之一即是由复杂多变的糖链结构所带来的高度“异质性”。糖链可与不同的聚糖结合蛋白相结合，通过体内的识别系统呈递信息。由于聚糖存在多种构象异构体，因此“糖密码”呈现出显著的三维特异性。在体内，凝集素是对特定聚糖结构具有高度选择性的聚糖结合蛋白，在聚糖变异研究中具有非常重要作用。基于凝集素的糖组学正是利用凝集素能够解密“糖密码”的生物特性发展而来的一种综合研究策略，利用聚糖和凝集素的相互作用，分析聚糖轮廓差异。尽管凝集素不能提供连接位点的确定结构，但它揭示了有关糖复合物特征的末端修饰的差异。聚糖与凝集素糖基化位点的连接方式是其构象开关，通过构象变化调节其对特异凝集素的亲和力。凝集素和同源聚糖可以相互作用每个聚糖因其空间结构的差异而具有特殊含义，即可被相应的受体“读取”。而当这种糖基化修饰发生在脂质分子头部基团时，其与蛋白质糖基化具有相似的特征，即聚糖的结构变化可通过凝集素特异性识别。

糖脂分子通常分布在生物膜结构中，酰基链、蛋白质以及胆固醇之间的疏水性接触有助于促使糖脂以聚集体（微区）的形式组织，形成局部高浓度，成为病原体和细胞表面相互作用的主要部位。例如，霍乱弧菌的 AB5 毒素的凝集素部分就是通过选择独特的构象异构体，与细胞表面神经节苷脂 GM1 的五糖结合。同时，糖脂还能调节蛋白质构象，该活性与病毒感染、传播和阿尔茨海默病和克－雅病的淀粉样聚集等致病过程相关。细胞膜结构中含量和种类丰富的鞘糖脂，可为分子识别过程提供多样性的标准。此外，鉴定参与糖脂识别的蛋白质结构域至关重要，近年来，除凝集素外的糖脂结合蛋白也越来越受到关注，它们需要至少一个完全暴露于溶剂环境的芳香族残基。除芳香族残基外，这一类糖脂结合域还包含引起碳链转向的特定残基和碱性残基，增加该结构域的整体亲水性，使位于蛋白质表面的芳香族残基完全暴露于溶剂中。由于糖脂结构复杂，且不同糖脂结合蛋白间结构同源性差，蛋白质分子糖脂结合结构域的研究仍未有重大突破。

糖链介导的生物分子相互作用为其在肿瘤诊断、肿瘤治疗、炎症治疗、抗病毒感染和糖尿病治疗等众多领域的应用提供了广阔的前景。近年来，国际上关于糖链在肿瘤免疫逃逸等重大疾病发病机制中的功能研究取得了一系列突破。例如，唾液酸识别蛋白 Siglec-15 可抑制 T 细胞的增殖和活化，其被认为是免疫治疗领域的又一重磅靶点（Wang et al.，2019b）；肿瘤细胞表面 CD24 蛋白通过和巨噬细胞表面 Sigle-10 结合来向巨噬细胞传递“别吃我”的信号（Barkal et al.，2019）；免疫检查点分子 PD-L1 的 N- 糖基化阻碍了 PD-L1 单抗的识别（Lee et al.，2019）。由此可见，研究糖链在重大疾病发生、发展中的变化规律、功能和生物学意义已是当今国际生命科学研究中最受关注的领域之一。

目前，我国对糖链在发育、癌症发生等过程中的作用做了一些具有国际前沿水平的研究和尝试，取得了多项成果，部分领域保持了与国际发展水平同步。例如，在蛋白质糖基化肿瘤诊断方面，发现 IgG 半乳糖基化程度可以作为潜在的肿瘤标志物用于肿瘤早期诊断（Ren et al.，2016）；发现肿瘤干细胞表面的高甘露糖型 N- 糖链介导了其与淋巴管内皮细胞表面高甘露糖受体间的相互作用，并促进了肿瘤干细胞免疫逃逸和自我更新（Wei et al.，2019）；设计出将点击化学与蛋白质芯片技术相结合的方法，可实现快速高通量发现 O-GalNAc 糖基转移酶修饰蛋白质底物，为系统研究 O-GalNAc 糖基化修饰的调控机制提供了一种新手段；揭示了 ppGalNAc-T2 酶中一个关键结构域的构象变化过程及其对底物识别和酶催化反应的分子作用机制；发现蛋白质翻译起始阶段的蛋白质因子（eIF4A 和 eIF4G）O-GlcNAc 糖基化修饰，揭示了重要的蛋白质翻译调控模式（Duan et al.，2018）。

近年来，在糖蛋白富集材料和方法、位点特异性糖基化的样品预处理方法及糖链分析的深度覆盖方面取得了一些突出成果（Suttapitugsakul et al.，2020）。研究人员开发了结合酶促反应和化学反应的靶向富集带有 Tn 抗原的糖蛋白（Yang et al.，2020b）以及高灵敏硼酸亲和富集 N- 糖蛋白组的新方法（Xiao et al.，2018）；利用多种化学固相富集糖蛋白，结合糖链库和蛋白质库的分步鉴定和整合，提供了规模化解析位点特异性糖型的方法（Chen et al.，2020；Sun et al.，2016）。在糖链定性、定量分析方法的开发方面，研究人员开发了一种高灵敏度唾液酸连接特异性化学衍生结合毛细管电泳 - 电喷雾 - 质谱联用（CE-ESI-MS）的方法，用于从复杂生物样品中深度鉴定 N- 聚糖，鉴定水平达到了埃摩尔级（Lageveen-Kammeijer et al.，2019）。

蛋白质糖链的结构分析是功能研究的必要前提。我国在位点特异性蛋白质糖基化的组学分析和糖链精准定量方法开发方面处于国际优势地位。在位点特异性蛋白质糖基化的组学分析方面，研究人员开发出阶梯能量高能碰撞裂解（higher-energy collisional dissociation，HCD）完整糖肽串级质谱图的采集方法及检索算法，并利用该方法获得了小鼠组织中最大的完整糖肽数据集（Liu et al.，2017a）；开发了完整糖肽电荷转移高效碎裂方法，利用该方法显著提高了血清完整糖肽的肽段鉴定率（Brauns and Cramer，2019）；开发了完整糖肽分步富集方法并通过自行构建糖链数据库，鉴定并获得了人血清中最大规模的数据集（Shu et al.，2020）。此外，在糖基化的解析软件、糖链的标记方法及糖蛋白的富集方法和材料方面也都开展了创新工作。化学标记方法进一步为糖链检测和解析提供了有力手段（Jin et al.，2019；Peng et al.，2019）。小分子抑制剂则可调控对应糖基化修饰的水平，开发具有细胞选择性和蛋白质特异性的糖基化标记方法与小分子抑制剂及激动剂，将成为研究糖链生物学功能的有力工具。

在糖链结构和功能研究方面，我国正迎头赶上国际发展潮流。充分利用我国和世界各国糖生物学研究的进展，开展后续的疾病相关糖蛋白及明确起关键作用的糖链研究，是当前我国生命科学面临的新课题。

四、脂代谢网络的运行机制与功能

近年来，脂代谢领域的研究进展日益迅猛。例如，在生物化学方面，不断发掘传统代谢酶的新功能，深入挖掘不同信号转导途径对代谢的调控作用，揭示多种代谢物的感应机制，鉴定并阐释新的脂类分子及其与疾病的关系。在细胞生物学方面，揭示新型细胞器及经典细胞器在代谢营养感知、代谢物质运输中的关键作用。在机体层面，发现并阐释代谢性器官间的物质与信息交流方式、新型代谢性器官的生理与病理意义。在遗传方面，基于人群或特色家系的研究，发现了许多影响代谢的新基因或新变异，以及表观遗传修饰对脂代谢的调控作用。同时，在不同的内外因素刺激及生理病理条件下，组织或机体的脂代谢又会进行动态的重构或重编程，以适应不同营养条件或应激状态，进而促进或抑制疾病的发生发展。

对近 5 年受同行高度关注的重要领域（高被引论文，即被引频次在同领域排名前 1% 的论文）进行分析，总结出全球代谢调控研究的若干重点子领域。其中，肥胖、糖尿病及其并发症的发病机制和治疗研究是该领域的重要热点，随着 2 型糖尿病的发病率呈爆发式增长，探索胰岛素分泌和胰岛素敏感性的调控机制成为该领域的重大问题。如二甲双胍对胰岛素敏感性调节的作用机制；mTORC、AMP 活化蛋白激酶（AMP-activated protein kinase，AMPK）、核受体和 GPCR 等感受营养与能量状态的机制研究；各种代谢物中间体和其衍生物的信号功能、感知机制、信号转导和生物学效应；线粒体等细胞器在胰岛素抵抗与 2 型糖尿病中的作用；等等。在糖尿病治疗药物的研发方面，研究人员尤为关注肠道菌群稳态、胰高血糖素样肽 -1（glucagon-like peptide-1，GLP-1）和钠－葡萄糖协同转运蛋白 2（sodium-dependent glucose transporter 2，SLGT2）抑制剂的研究；脂肪、肌肉等组织的新型分泌因子不断涌现，在能量平衡调控和肥胖的发生发展研究中发挥重要作用。除了糖尿病与肥胖，代谢紊乱所导致的其他相关疾病（如阿尔茨海默病等神经退行性疾病、肿瘤、心血管疾病、衰老等）的机制研究也在代谢调控领域有着重要的位置。先前研究报道了线粒体缺陷和氧化应激在阿尔茨海默病和帕金森病中的表现情况。1931 年，在肿瘤研究遇到瓶颈时，诺贝尔奖得主德国生化学家奥托·瓦尔堡（Otto Warburg）发现恶性肿瘤细胞在有氧条件下仍以糖酵解方式获取能量（即瓦氏效应），引发了近年来肿瘤代谢研究的复兴。除此之外，大量研究表明肥胖是癌症的重要危险因素。

近年来，我国在脂代谢领域的前沿研究取得了许多重大突破。例如，发现了心脏和血管等组织的新的前体细胞；发现了相关激素、细胞因子、代谢物和非编码 RNA 等代谢调控分子对代谢的调控作用，揭示了胆固醇和脂肪代谢的稳态调控机制；阐明了生理状态下细胞感知葡萄糖水平的 AMPK 激活通路、细胞感知氨基酸缺乏的机制（Zhang et al.，2017a）；揭示了肠道菌群代谢物是影响胰岛素敏感性的重要因素（Zhao et al.，2018）；发现并解析了亲代代谢异常引起的疾病跨代遗传现象和机制（Chen et al.，2016）；完成了中国人 2 型糖尿病和极端肥胖个体的全基因组关联分析（genome wide association study，GWAS），发现并确定了中国人特有的 2 型糖尿病易感基因谱（Yan et al.，2021）；从人群中鉴定出了新的胆固醇吸收基因（Zhang et al.，2018）；建立了亚洲最大的血液脂肪酸谱、国际最大的维生素 B_1 谱、尿液离子谱等组学数据库；阐释了三羧酸循环（TCA 循环）中间体、糖醛酸代谢、尿素循环和胆固醇代谢等在肿瘤发生发展中的作用与机制（Li et al.，2019；Liu et al.，2020；Wang et al.，2019c）；在免疫代谢方面，发现了胆固醇及其类似物、黄嘌呤等对 T 细胞的调控作用，并发现它们在不同的免疫细胞中扮演不一样的角色（Fan et al.，2019；Yang et al.，2016）；从进化角度研究了一些特色动物（如裸鼹鼠、北极熊、大熊猫、驯鹿、羚羊等）的代谢特征，如发现地球上奔跑速度最快的动物之一羚羊，其呼吸链和氧化磷酸化基因受到了强选择，有利于其高效率的能量代谢（Chen et al.，2019a）。

纵观这些研究可以发现，与代谢调控相关的代谢网络运行机制、疾病发生机制仍是人们重点探索的领域，希望借此找到新的健康管理方法与疾病治疗手段。随着生物医学领域新技术体系的建立与交叉融合，从整体上全面分析机体的代谢及其调控过程已成为该领域发展的趋势。

第四节　我国发展战略与重点方向

一、问题与挑战

近年来国际上RNA的新结构和新功能不断被发现，我国也在RNA基础和应用领域取得了一些成绩，但是整体研究水平与美国、欧洲等国家/地区仍存在相当大的差距。想要在非编码RNA研究领域处于领先地位，我国需要投入更多的人力和物力，继续解析RNA代谢、调控、生理功能及与生命健康的关系等前沿性关键科学问题，进一步以人类及其他重要模式生物为研究对象，开展在重要生命活动与健康中的RNA功能及机制研究，进一步提升我国科研的整体水平。

蛋白质是生命活动的主要执行者。对蛋白质结构与功能、相互作用和动态变化的深入研究，将有助于揭示生命现象的本质。蛋白质研究取得的众多成果也将对新药研发、传染病预防与诊断治疗、农作物改良、生物能源转化、工业生物催化等各个领域起到重要作用。但是，与国际蛋白质研究的先进水平相比，我国的蛋白质研究源头创新能力有待加强，重点需要促进多学科交叉合作研究、关注蛋白质研究的原创性新技术和新方法、提升解决重大科学问题的合力。

鉴于糖链在肿瘤等疾病诊断、治疗中的重要意义，糖生物学是研究解决危害我国人民多种疾病问题不可或缺的重要分支。然而我国的糖生物学在早期发展缓慢，近年来我国出台了多个重大计划，覆盖了糖生物学的基础研究与糖工程，但是要吸取之前发展缓慢的教训，需要对糖生物学领域持续资助，支持和鼓励研究人员持续开展糖生物学基础研究，特别是要针对目前糖链精准定性定量分析、糖生物学功能研究的技术瓶颈逐步攻克相关难题。另外，还需要重视研究成果的积累，加强数据库与平台建设，促进研究人员的交流与合作。在应用转化方面，要重视加强糖链诊疗潜在标志物的临床实践转化研究。目前糖蛋白质组学研究还存在诸多技术问题有待解决，迫切需要发展高灵敏、高通量、高选择性的糖蛋白质定性定量分析技术。

由于国家投入的增加，我国在发育和代谢研究领域取得了较大进展，具备了与国际领先研究竞争的实力，但与发达国家相比仍存在一定的差距和不足。目前我国追逐热点的研究较多而转化的成果较少，基础研究向转化方面发展的链条不完善，我国在代谢研究的新技术方面与国外先进水平有较大差距。虽然国内代谢组学研究也在飞速发展，但仍然存在结构松散、规模小、技术储备不足、与临床和功能研究脱节等现状。在临床研究方面，我国医疗部门侧重于流行病学调查、基因组学或转录组学分析，应在研究和改进疾病的诊断与治疗技术的基础上，加大致病机制深入研究的投入力度，注重突破性原创成果。我国的投入纵向比较有较大增长，但横向比较与发达国家仍然有较大差距，我国缺乏对优秀项目的滚动资助机制，缺少对优秀人才的稳定支持机制，难以保障研究的系统性和集成性；经费使用的效率有待进一步提高，拼盘现象依然存在。

二、发展总体思路与发展目标

围绕“三个面向”，瞄准“生物分子与生命调控及健康”这个重大科学问题，从以下 4 个方面开展系列研究：①生物分子生成、代谢与调控；②生物分子的生理功能；③生物分子与疾病、健康；④生物分子前沿技术及其产业化。以解析各种生物分子的结合靶标及结构为基础，以生物分子加工、代谢及识别等为线索，以生物分子调控分子机制和细胞功能为核心，以重要模式生物为研究对象，通过生物学、化学、结构物理学及生物信息学等多学科相互交叉，整合基因组学、转录组学、蛋白质组学、代谢组学、表观组学、化学生物学、RNA 互作组学等关键技术，研发生物分子检测相关新技术及新方法，阐明生物分子代谢与调控的基本机制及生物分子生理功能，为在疾病方面的应用打下坚实基础，为填补生物分子功能认知的缺陷及阐明生命起源与进化的本质提供重要线索。尤其是要结合国家精准医学重大需求，以创新预防和精确诊断与防治为目标，从临床中发现问题，通过基础研究揭示生物分子在人类重大疾病（如恶性肿瘤、心脑血管疾病、神经系统疾病等）及重大传染病的发生和发展过程中的作用与机制，构建这些疾病中的生物大分子作用调控网络，将基础研究的成果用于指导临床诊断与治疗实践。围绕我国重大疾病的预防、临床诊断及防治中的难题，采用体内外模型，筛选和鉴定疾病特异病理状态相关的生物分子及其相互作用分子，探索其对疾病表型的调控作用及机制，开发疾病相关的生物分子精准诊断策略，最终阐明疾病发生发展中生物分子的功能及调控网络，加深对疾病发生、发展机制的认识。同时，开展生物分子药物在重大疾病治疗中的临床前研究。当前正是正处于抢占国际学术和技术前沿领域制高点的关键时期，这些研究既“面向世界科技前沿”，又“面向经济主战场”，同时也“面向国家重大需求”。

三、优先发展领域（或重要研究方向）

1. RNA 的代谢、功能与调控

发现多种新型 RNA，并阐明其生成、加工、代谢及调控的机制；揭示各类 RNA 的加工、转运、代谢及调控机制，解析参与 RNA 生成及调控过程的 RNA 结合蛋白、RNA- 蛋白质复合体等；研究 RNA 结合蛋白在 RNA 加工成熟及执行功能过程中的作用及其相关的生理和病理机制，揭示在生理及应激条件下 RNA 的产生机制及其生物学功能；阐明 RNA 结合蛋白的功能及其结合的 RNA 类型，建立起调控网络；在国际上率先鉴定出一批新型 RNA 化学修饰，绘制修饰图谱，鉴定相关修饰因子，并阐明化学修饰对 RNA 代谢的调控过程；阐明各种亚细胞结构调控 RNA 转录、加工、转运及翻译的过程；阐明外泌体 RNA 的组成、分选入外泌体、分泌出胞等的机制。攻关超微量 RNA 富集、检测以及定量分析技术；攻克核酸药物靶向肿瘤、肝脏等给药技术；建立核酸药物的生产和质量标准体系；开发治疗骨关节炎、乙型肝炎、肿瘤和高血脂等重大疾病和罕见病的核酸创新药物。

2. 蛋白质的结构、功能与设计

研究膜蛋白结构的多样性与动态变化；鉴定重大疾病、传染病、罕见病致病靶点蛋白的结构，设计和筛选基于结构的药物；从头设计蛋白质，人工

设计与拓展功能结构域；建立多样化蛋白质结构测定方法；超级计算机辅助设计蛋白质与模拟分子动态；单分子成像与单分子荧光标记辅助检测蛋白质的构象动态变化；研究脂质分子、糖类分子在蛋白质构象变化中的调节作用；开发与优化分子设计软件，将机器学习应用于蛋白质结构研究。

3. 糖的识别、结构与功能

优先开展对糖基化修饰规律的研究，建立糖型分析平台、糖型生物学信息分析平台，规模化研究疾病相关糖型特点；通过糖生物学着重研究特征性糖型的生物学意义和作用特点；开发、应用具有特征性糖链的糖复合物。

4. 脂代谢网络的运行机制与功能

发展高灵敏、高通量、动态实时的脂质示踪和鉴定技术，研究并揭示细胞、器官和机体对糖脂代谢状态的感知与应答模式，不断完善调节代谢稳态的组织、器官间的信息对话机制及其协同调控网络，研究重要代谢物产生、运输与转化的路径和调控机制及代谢调控与稳态维持的新规律，阐明代谢时空变化在环境适应及生命健康中的作用，揭示代谢稳态失衡在疾病发生发展中的核心机制，为研发新型药物和干预策略提供理论依据。

本章参考文献

Bai R, Yan C, Wan R, et al. 2017. Structure of the post-catalytic spliceosome from *Saccharomyces cerevisiae*. Cell, 171: 1589-1598, e1588.

Barkal A A, Brewer R E, Markovic M, et al. 2019. CD24 signalling through macrophage Siglec-10 is a target for cancer immunotherapy. Nature, 572（7769）: 392-396.

Brauns M, Cramer N. 2019. Efficient kinetic resolution of sulfur-stereogenic sulfoximines by exploiting Cp（Ⅹ）Rh（Ⅲ）-catalyzed C-H functionalization. Angew Chem Int Ed Engl, 58（26）: 8902-8906.

Bunzel H A, Anderson J L R, Mulholland A J. 2021. Designing better enzymes: Insights from directed evolution. Curr Opin Struct Biol, 67: 212-218.

Butterfield G L, Lajoie M J, Gustafson H H, et al. 2017. Evolution of a designed protein assembly encapsulating its own RNA genome. Nature, 552（7685）: 415-420.

Chen L, Qiu Q, Jiang Y, et al. 2019a. Large-scale ruminant genome sequencing provides insights into their evolution and distinct traits. Science, 364（6446）: eaav6202.

Chen L L. 2016. The biogenesis and emerging roles of circular RNAs. Nat Rev Mol Cell Biol, 17（4）: 205-211.

Chen Q, Yan M, Cao Z, et al. 2016. Sperm tsRNAs contribute to intergenerational inheritance of an acquired metabolic disorder. Science, 351（6271）: 397-400.

Chen S, Wang R, Zheng D, et al. 2019b. The mRNA export receptor NXF1 coordinates transcriptional dynamics, alternative polyadenylation, and mRNA Export. Mol Cell, 74（1）: 118-131, e117.

Chen S Y, Dong M, Yang G, et al. 2020. Glycans, glycosite, and intact glycopeptide analysis of N-linked glycoproteins using liquid handling systems. Anal Chem, 92（2）: 1680-1686.

Cherezov V, Rosenbaum D M, Hanson M A, et al. 2007. High-resolution crystal structure of an engineered human beta2-adrenergic G protein-coupled receptor. Science, 318（5854）: 1258-1265.

Chien E Y, Liu W, Zhao Q, et al. 2010. Structure of the human dopamine D3 receptor in complex with a D2/D3 selective antagonist. Science, 330（6007）: 1091-1095.

Choi U, Lee C R. 2019. Antimicrobial agents that inhibit the outer membrane assembly machines of Gram-negative bacteria. J Microbiol Biotechnol, 29（1）: 1-10.

Conn P J, Christopoulos A, Lindsley C W. 2009. Allosteric modulators of GPCRs: a novel approach for the treatment of CNS disorders. Nat Rev Drug Discov, 8（1）: 41-54.

Deng W, Zhu X, Skogerbo G, et al. 2006. Organization of the *Caenorhabditis elegans* small non-coding transcriptome: genomic features, biogenesis, and expression. Genome Res, 16（1）: 20-29.

Duan F, Wu H, Jia D, et al. 2018. *O*-GlcNAcylation of RACK1 promotes hepatocellular carcinogenesis. J Hepatol, 68（6）: 1191-1202.

Epand R M, Walker C, Epand R F, et al. 2016. Molecular mechanisms of membrane targeting antibiotics. Biochim Biophys Acta, 1858（5）: 980-987.

Fan J, Kuai B, Wu G, et al. 2017. Exosome cofactor hMTR4 competes with export adaptor ALYREF to ensure balanced nuclear RNA pools for degradation and export. EMBO J, 36（19）: 2870-2886.

Fan K Q, Li Y Y, Wang H L, et al. 2019. Stress-induced metabolic disorder in peripheral CD4（+）T cells leads to anxiety-like behavior. Cell, 179（4）: 864-879, e819.

Fan L, Tan L, Chen Z, et al. 2020a. Haloperidol bound D2 dopamine receptor structure inspired the discovery of subtype selective ligands. Nat Commun, 11（1）: 1074.

Fan X, Tang D, Liao Y, et al. 2020b. Single-cell RNA-seq analysis of mouse preimplantation embryos by third-generation sequencing. PLoS Biol, 18（12）: e3001017.

Gao W, Mahajan S P, Sulam J, et al. 2020. Deep learning in protein structural modeling and design. Patterns（N Y）, 1（9）: 100142.

Garcia-Nafria J, Lee Y, Bai X, et al. 2018. Cryo-EM structure of the adenosine A2A receptor coupled to an engineered heterotrimeric G protein. Elife, 7: e35946.

Gonzalez-Maeso J. 2011. GPCR oligomers in pharmacology and signaling. Mol Brain, 4（1）: 20.

Gou L T, Kang J Y, Dai P, et al. 2017. Ubiquitination-deficient mutations in human Piwi cause male infertility by impairing histone-to-protamine exchange during spermiogenesis. Cell, 169（6）: 1090-1104, e1013.

Gould K. 2016. Antibiotics: from prehistory to the present day. J Antimicrob Chemother, 71（3）: 572-575.

Hart E M, Mitchell A M, Konovalova A, et al. 2019. A small-molecule inhibitor of BamA impervious to efflux and the outer membrane permeability barrier. Proc Natl Acad Sci USA, 116（43）: 21748-21757.

He H, Cai L, Skogerbo G, et al. 2006. Profiling *Caenorhabditis elegans* non-coding RNA expression with a combined microarray. Nucleic Acids Res, 34（10）: 2976-2983.

He H, Wang J, Liu T, et al. 2007. Mapping the *C. elegans* noncoding transcriptome with a whole-genome tiling microarray. Genome Res, 17（10）: 1471-1477.

He L, Tzarum N, Lin X, et al. 2020. Proof of concept for rational design of hepatitis C virus E2 core nanoparticle vaccines. Sci Adv, 6（16）: eaaz6225.

Hilander T, Zhou X L, Konovalova S, et al. 2018. Editing activity for eliminating mischarged tRNAs is essential in mammalian mitochondria. Nucleic Acids Res, 46（2）: 849-860.

Hilger D, Kumar K K, Hu H, et al. 2020. Structural insights into differences in G protein activation by family A and family B GPCRs. Science, 369（6503）: eaba3373.

Hilger D, Masureel M, Kobilka B K. 2018. Structure and dynamics of GPCR signaling complexes. Nat Struct Mol Biol, 25（1）: 4-12.

Hollingsworth S A, Dror R O. 2018. Molecular dynamics simulation for all. Neuron, 99（6）: 1129-1143.

Hutchings M I, Truman A W, Wilkinson B. 2019. Antibiotics: past, present and future. Curr Opin Microbiol, 51: 72-80.

Im D, Inoue A, Fujiwara T, et al. 2020. Structure of the dopamine D2 receptor in complex with the antipsychotic drug spiperone. Nat Commun, 11（1）: 6442.

Imai K, Takaoka A. 2006. Comparing antibody and small-molecule therapies for cancer. Nat Rev Cancer, 6（9）: 714-727.

Imai Y, Meyer K J, Iinishi A, et al. 2019. A new antibiotic selectively kills Gram-negative pathogens. Nature, 576（7787）: 459-464.

Ji Q Q, Fang Z P, Ye Q, et al. 2017. Self-protective responses to norvaline-induced stress in a leucyl-tRNA synthetase editing-deficient yeast strain. Nucleic Acids Res, 45（12）: 7367-7381.

Jin W, Wang C, Yang M, et al. 2019. Glycoqueuing: isomer-specific quantification for sialylation-focused glycomics. Anal Chem, 91（16）: 10492-10500.

Kato H E, Zhang Y, Hu H, et al. 2019. Conformational transitions of a neurotensin receptor 1-Gi1 complex. Nature, 572（7767）: 80-85.

Koehl A, Hu H, Maeda S, et al. 2018. Structure of the micro-opioid receptor-Gi protein complex. Nature, 558: 547-552.

Korendovych I V, DeGrado W F. 2020. *De novo* protein design, a retrospective. Q Rev Biophys, 53: e3.

Kries H, Blomberg R, Hilvert D. 2013. *De novo* enzymes by computational design. Curr Opin Chem Biol, 17（2）: 221-228.

Kuhlbrandt W. 2014. Cryo-EM enters a new era. Elife, 3: e03678.

Kuhlman B, Bradley P. 2019. Advances in protein structure prediction and design. Nat Rev Mol Cell Biol, 20（11）: 681-697.

Lageveen-Kammeijer G S M, de Haan N, Mohaupt P, et al. 2019. Highly sensitive CE-ESI-MS analysis of N-glycans from complex biological samples. Nat Commun, 10（1）: 2137.

Lajoie M J, Boyken S E, Salter A I, et al. 2020. Designed protein logic to target cells with precise combinations of surface antigens. Science, 369（6511）: 1637-1643.

Latorraca N R, Masureel M, Hollingsworth S A, et al. 2020. How GPCR phosphorylation patterns orchestrate arrestin-mediated signaling. Cell, 183（7）: 1813-1825 e1818.

Latorraca N R, Venkatakrishnan A J, Dror R O. 2017. GPCR dynamics: structures in motion. Chem Rev, 117（1）: 139-155.

Lee H H, Wang Y N, Xia W, et al. 2019. Removal of N-linked glycosylation enhances PD-L1 detection and predicts anti-PD-1/PD-L1 therapeutic efficacy. Cancer Cell, 36（2）: 168-178, e164.

Li L, Mao Y, Zhao L, et al. 2019. p53 regulation of ammonia metabolism through urea cycle controls polyamine biosynthesis. Nature, 567（7747）: 253-256.

Li P, Shi R, Zhang Q C. 2020. icSHAPE-pipe: A comprehensive toolkit for icSHAPE data analysis and evaluation. Methods, 178: 96-103.

Li X, Liu C X, Xue W, et al. 2017. Coordinated circRNA biogenesis and function with NF90/NF110 in viral infection. Mol Cell, 67（2）: 214-227, e217.

Li X, Xiong X, Wang K, et al. 2016a. Transcriptome-wide mapping reveals reversible and dynamic N（1）-methyladenosine methylome. Nat Chem Biol, 12（5）: 311-316.

Li X, Xiong X, and Yi C. 2016b. Epitranscriptome sequencing technologies: decoding RNA modifications. Nat Methods, 14（1）: 23-31.

Li X, Zhu P, Ma S, et al. 2015. Chemical pulldown reveals dynamic pseudouridylation of the mammalian transcriptome. Nat Chem Biol, 11（8）: 592-597.

Lin X, Li M, Wang N, et al. 2020. Structural basis of ligand recognition and self-activation of orphan GPR52. Nature, 579（7797）: 152-157.

Liu C, Bai B, Skogerbo G, et al. 2005. NONCODE: an integrated knowledge database of non-coding RNAs. Nucleic Acids Res, 33（Database issue）: D112-D115.

Liu M Q, Zeng W F, Fang P, et al. 2017a. pGlyco 2.0 enables precision N-glycoproteomics with comprehensive quality control and one-step mass spectrometry for intact glycopeptide identification. Nat Commun, 8（1）: 438.

Liu R J, Long T, Li J, et al. 2017b. Structural basis for substrate binding and catalytic mechanism of a human RNA:m5C methyltransferase NSun6. Nucleic Acids Res, 45（11）: 6684-6697.

Liu R J, Long T, Zhou M, et al. 2015. tRNA recognition by a bacterial tRNA Xm32 modification enzyme from the SPOUT methyltransferase superfamily. Nucleic Acids Res, 43（15）: 7489-7503.

Liu W, Chun E, Thompson A A, et al. 2012. Structural basis for allosteric regulation of GPCRs by sodium ions. Science, 337（6091）: 232-236.

Liu X, Bao X, Hu M, et al. 2020. Inhibition of PCSK9 potentiates immune checkpoint therapy for cancer. Nature, 588（7839）: 693-698.

Liu X, Masoudi A, Kahsai A W, et al. 2019. Mechanism of β_2AR regulation by an intracellular positive allosteric modulator. Science, 364（6447）: 1283-1287.

Lu R M, Hwang Y C, Liu I J, et al. 2020. Development of therapeutic antibodies for the treatment of diseases. J Biomed Sci, 27（10）: 1.

Luo S, Lu J Y, Liu L, et al. 2016. Divergent lncRNAs regulate gene expression and lineage differentiation in pluripotent cells. Cell Stem Cell, 18（5）: 637-652.

Luther A, Urfer M, Zahn M, et al. 2019. Chimeric peptidomimetic antibiotics against Gram-negative bacteria. Nature, 576（7787）: 452-458.

Ma L, Qu L. 2013. The function of microRNAs in renal development and pathophysiology. J Genet Genomics, 40（4）: 143-152.

Maeda S, Qu Q, Robertson M J, et al. 2019. Structures of the M1 and M2 muscarinic acetylcholine receptor/G-protein complexes. Science, 364（6440）: 552-557.

McCorvy J D, Butler K V, Kelly B, et al. 2018. Structure-inspired design of β-arrestin-biased ligands for aminergic GPCRs. Nat Chem Biol, 14（2）: 126-134.

Nastri F, D'Alonzo D, Leone L, et al. 2019. Engineering metalloprotein functions in designed and native scaffolds. Trends Biochem Sci, 44（12）: 1022-1040.

Naylor N R, Atun R, Zhu N, et al. 2018. Estimating the burden of antimicrobial resistance: a systematic literature review. Antimicrob Resist Infect Control, 7: 58.

Ng A H, Nguyen T H, Gomez-Schiavon M, et al. 2019. Modular and tunable biological feedback control using a *de novo* protein switch. Nature, 572（7768）: 265-269.

Nie D, Hu Y, Chen Z, et al. 2020. Outer membrane protein A（OmpA）as a potential therapeutic target for *Acinetobacter baumannii* infection. J Biomed Sci, 27（1）: 26.

Patini R, Mangino G, Martellacci L, et al. 2020. The effect of different antibiotic regimens on bacterial resistance: a systematic review. Antibiotics（Basel）, 9（1）: 22.

Peng H, Shi J, Zhang Y, et al. 2012. A novel class of tRNA-derived small RNAs extremely enriched in mature mouse sperm. Cell Res, 22: 1609-1612.

Peng Y, Wang L, Zhang Y, et al. 2019. Stable isotope sequential derivatization for linkage-specific analysis of sialylated N-glycan isomers by MS. Anal Chem, 91（24）: 15993-16001.

Perlman S. 2020. Another decade, another coronavirus. N Engl J Med, 382（8）: 760-762.

Pin J P, Kniazeff J, Binet V, et al. 2004. Activation mechanism of the heterodimeric GABA（B）receptor. Biochem Pharmacol, 68（8）: 1565-1572.

Polizzi N F, DeGrado W F. 2020. A defined structural unit enables de novo design of small-molecule-binding proteins. Science, 369（6508）: 1227-1233.

Rask-Andersen M, Almen M S, Schioth H B. 2011. Trends in the exploitation of novel drug targets. Nat Rev Drug Discov, 10（8）: 579-590.

Rasmussen S G, DeVree B T, Zou Y, et al. 2011. Crystal structure of the β_2 adrenergic receptor-Gs protein complex. Nature, 477（7366）: 549-555.

Rather I A, Kim B C, Bajpai V K, et al. 2017. Self-medication and antibiotic resistance: Crisis, current challenges, and prevention. Saudi J Biol Sci, 24（4）: 808-812.

Ren S, Zhang Z, Xu C, et al. 2016. Distribution of IgG galactosylation as a promising biomarker for cancer screening in multiple cancer types. Cell Res, 26（8）: 963-966.

Sall A, Walle M, Wingren C, et al. 2016. Generation and analyses of human synthetic antibody libraries and their application for protein microarrays. Protein Eng Des Sel, 29（10）: 427-437.

Shu Q, Li M, Shu L, et al. 2020. Large-scale identification of N-linked intact glycopeptides in human serum using HILIC enrichment and spectral library search. Mol Cell Proteomics, 19（4）: 672-689.

Silva D A, Yu S, Ulge UY, et al. 2019. *De novo* design of potent and selective mimics of IL-2 and IL-15. Nature, 565（7738）: 186-191.

Staus D P, Strachan R T, Manglik A, et al. 2016. Allosteric nanobodies reveal the dynamic range and diverse mechanisms of G-protein-coupled receptor activation. Nature, 535（7612）: 448-452.

Sun S, Shah P, Eshghi S T, et al. 2016. Comprehensive analysis of protein glycosylation by solid-phase extraction of N-linked glycans and glycosite-containing peptides. Nat Biotechnol, 34（1）: 84-88.

Suttapitugsakul S, Sun F, Wu R. 2020. Recent advances in glycoproteomic analysis by mass spectrometry. Anal Chem, 92（1）: 267-291.

Terrillon S, Bouvier M. 2004. Roles of G-protein-coupled receptor dimerization. EMBO Rep, 5（1）: 30-34.

Urfer M, Bogdanovic J, Lo Monte F, et al. 2016. A peptidomimetic antibiotic targets outer membrane proteins and disrupts selectively the outer membrane in *Escherichia coli*. J Biol Chem, 291（4）: 1921-1932.

Venkat Kumar G, Jeyanthi V, Ramakrishnan S. 2020. A short review on antibody therapy for COVID-19. New Microbes New Infect, 35: 100682.

Wacker D, Stevens R C, Roth B L. 2017. How ligands Illuminate GPCR molecular pharmacology.

Cell, 170（3）: 414-427.

Wang J, Chen J, Wu G, et al. 2019a. NRDE2 negatively regulates exosome functions by inhibiting MTR4 recruitment and exosome interaction. Genes Dev, 33（9-10）: 536-549.

Wang J, Sun J, Liu LN, et al. 2019b. Siglec-15 as an immune suppressor and potential target for normalization cancer immunotherapy. Nat Med, 25（4）: 656-666.

Wang P, Xue Y, Han Y, et al. 2014. The STAT3-binding long noncoding RNA lnc-DC controls human dendritic cell differentiation. Science, 344（6181）: 310-313.

Wang S, Che T, Levit A, et al. 2018. Structure of the D2 dopamine receptor bound to the atypical antipsychotic drug risperidone. Nature, 555（7695）: 269-273.

Wang S, Wacker D, Levit A, et al. 2017. D4 dopamine receptor high-resolution structures enable the discovery of selective agonists. Science, 358（6361）: 381-386.

Wang X, Li G, Yang Y, et al. 2012. An RNA architectural locus control region involved in *Dscam* mutually exclusive splicing. Nat Commun, 3: 1255.

Wang X, Liu R, Zhu W, et al. 2019c. UDP-glucose accelerates SNAI1 mRNA decay and impairs lung cancer metastasis. Nature, 571（7763）: 127-131.

Wei W, Ba Z, Gao M, et al. 2012. A role for small RNAs in DNA double-strand break repair. Cell, 149（1）: 101-112.

Wei Y, Shi D, Liang Z, et al. 2019. IL-17A secreted from lymphatic endothelial cells promotes tumorigenesis by upregulation of PD-L1 in hepatoma stem cells. J Hepatol, 71（6）: 1206-1215.

Wu H, Wang C, Gregory K J, et al. 2014. Structure of a class C GPCR metabotropic glutamate receptor 1 bound to an allosteric modulator. Science, 344（6179）: 58-64.

Wu H, Yang L, Chen L L. 2017. The diversity of long noncoding RNAs and their generation. Trends Genet, 33（8）: 540-552.

Xiao H, Chen W, Smeekens J M, et al. 2018. An enrichment method based on synergistic and reversible covalent interactions for large-scale analysis of glycoproteins. Nat Commun, 9（1）: 1692.

Xiao P, Yan W, Gou L, et al. 2021. Ligand recognition and allosteric regulation of DRD1-Gs signaling complexes. Cell, 184（4）: 943-956, e918.

Xiao W, Adhikari S, Dahal U, et al. 2016. Nuclear m（6）A Reader YTHDC1 Regulates mRNA Splicing. Mol Cell, 61（4）: 507-519.

Xu K, Yang Y, Feng G H, et al. 2017. Mettl3-mediated m（6）A regulates spermatogonial differentiation and meiosis initiation. Cell Res, 27（9）: 1100-1114.

Xue Y, Ouyang K, Huang J, et al. 2013. Direct conversion of fibroblasts to neurons by reprogramming PTB-regulated microRNA circuits. Cell, 152（1-2）: 82-96.

Yan J, Qiu Y, Ribeiro Dos Santos A M, et al. 2021. Systematic analysis of binding of transcription factors to noncoding variants. Nature, 591（7848）: 147-151.

Yang F, Mao C, Guo L, et al. 2020a. Structural basis of GPBAR activation and bile acid recognition. Nature, 587（7834）: 499-504.

Yang W, Ao M, Song A, et al. 2020b. Mass spectrometric mapping of glycoproteins modified by Tn-antigen using solid-phase capture and enzymatic release. Anal Chem, 92（13）: 9230-9238.

Yang W, Bai Y, Xiong Y, et al. 2016. Potentiating the antitumour response of CD8（+）T cells by modulating cholesterol metabolism. Nature, 531（7596）: 651-655.

Yin Y, Yan P, Lu J, et al. 2015. Opposing roles for the lncRNA haunt and its genomic locus in regulating HOXA gene activation during embryonic stem cell differentiation. Cell Stem Cell, 16（5）: 504-516.

Yu F, Yao H, Zhu P, et al. 2007. *let-7* regulates self renewal and tumorigenicity of breast cancer cells. Cell, 131（6）: 1109-1123.

Yu J, Gimenez L E, Hernandez C C, et al. 2020. Determination of the melanocortin-4 receptor structure identifies Ca（2+）as a cofactor for ligand binding. Science, 368（6489）: 428-433.

Yuan J, Wu W, Xie C, et al. 2014. NPInter v2.0: an updated database of ncRNA interactions. Nucleic Acids Res, 42（Database issue）: D104-D108.

Yue Y, Meng Y, Ma H, et al. 2016. A large family of *Dscam* genes with tandemly arrayed 5' cassettes in Chelicerata. Nat Commun, 7: 11252.

Zaman S B, Hussain M A, Nye R, et al. 2017. A review on antibiotic resistance: alarm bells are ringing. Cureus, 9: e1403.

Zhan X, Yan C, Zhang X, et al. 2018. Structure of a human catalytic step I spliceosome. Science, 359（6375）: 537-545.

Zhang C S, Hawley S A, Zong Y, et al. 2017a. Fructose-1,6-bisphosphate and aldolase mediate glucose sensing by AMPK. Nature, 548（7665）: 112-116.

Zhang X, Yan C, Hang J, et al. 2017b. An atomic structure of the human spliceosome. Cell, 169（5）: 918-929, e914.

Zhang X O, Wang H B, Zhang Y, et al. 2014. Complementary sequence-mediated exon circularization. Cell, 159（1）: 134-147.

Zhang Y, Sun B, Feng D, et al. 2017c. Cryo-EM structure of the activated GLP-1 receptor in complex with a G protein. Nature, 546（7657）: 248-253.

Zhang Y, Xue W, Li X, et al. 2016. The biogenesis of nascent circular RNAs. Cell Rep, 15（3）: 611-624.

Zhang Y Y, Fu Z Y, Wei J, et al. 2018. A LIMA1 variant promotes low plasma LDL cholesterol and decreases intestinal cholesterol absorption. Science, 360（6393）: 1087-1092.

Zhao L, Zhang F, Ding X, et al. 2018. Gut bacteria selectively promoted by dietary fibers alleviate type 2 diabetes. Science, 359（6380）: 1151-1156.

Zhao S, Gou L T, Zhang M, et al. 2013. piRNA-triggered MIWI ubiquitination and removal by APC/C in late spermatogenesis. Dev Cell, 24（1）: 13-25.

Zhao X, Yang Y, Sun B F, et al. 2014. FTO-dependent demethylation of N6-methyladenosine regulates mRNA splicing and is required for adipogenesis. Cell Res, 24（12）: 1403-1419.

Zhao Y, Li H, Fang S, et al. 2016. NONCODE 2016: an informative and valuable data source of long non-coding RNAs. Nucleic Acids Res, 44（D1）: D203-D208.

Zhou B, Li X, Luo D, et al. 2019. GRID-seq for comprehensive analysis of global RNA-chromatin interactions. Nat Protoc, 14（7）: 2036-2068.

Zhu K, Liu M, Fu Z, et al. 2017. Plant microRNAs in larval food regulate honeybee caste development. PLoS Genet, 13（8）: e1006946.

Zhuang Y, Xu P, Mao C, et al. 2021. Structural insights into the human D1 and D2 dopamine receptor signaling complexes. Cell, 184（4）: 931-942, e918.

第五章

生物分子模块的组装与功能性机器

第一节 概 述

细胞中纷繁芜杂的生物学功能并不是由单个生物大分子独立完成的，而是由成千上万种生物大分子通过相互作用、动态组装形成的生物大分子机器来完成的。生物大分子机器是能够相对独立完成特定生物学功能的多亚基复合体，一般是由蛋白质、核酸、脂类、糖类或者小分子组成，它们相对分子质量大、组成复杂、结构精巧，参与细胞的整个生命过程。生物大分子机器具有自组装、自适应、多样性、模块化、可调控等人造机器难以比拟的特点，这也是它们在体内高效运转、发挥多种功能的基础。

（1）生物大分子机器的自组装基本是非外界能量依赖的，主要靠分子间特异性的识别而自发驱动的组装过程。比如，生物体内存在的大量分子机器通常按照内在规律自发组装到一起，即使是在体外重组，也通常可以获得与内源纯化一致的分子构象。这个过程涉及多种不同作用力，如范德瓦耳斯力、静电作用、金属－配体作用、受体－配体识别、共价反应等。

（2）自适应是指生物大分子机器具有一定的分子柔性，可以根据细胞代谢状态不同、所处环境不同，而改变其构象和活性，以发挥所需的不同功能。

（3）多样性是指生物大分子机器通常会有多种活性，可以高效完成多项需要耦合到一起的生物学过程。比如，酵母中的染色质修饰复合物 SAGA 就有 4 种不同的活性，包括：①组蛋白乙酰化活性，负责启动子区域的 H3 乙酰化；②去泛素化活性，负责清除启动子区域组蛋白 H2B 第 123 位赖氨酸的泛素化；③ TATA 结合蛋白（TATA-binding protein，TBP）结合活性，负责与 TBP 的互作；④结合转录激活因子的活性。这些活性共同保证了 SAGA 的转录共激活功能（Soffers and Workman，2020）。

（4）模块化是指可以把单个生物大分子机器看成是发挥一种具体功能的模块，而不同的模块（即生物大分子机器）可以以多种方式组合起来、协同作用发挥一种复杂功能。一个典型的例子是存在于细菌和真菌中的非核糖体多肽合成酶（nonribosomal peptide synthetase，NRPS）（Reimer et al.，2018）。NRPS 具有典型的模块化组织结构，每个模块负责一轮的氨基酸单体的延伸，每个模块至少含有 3 个必需功能域：负责氨基酸激活的腺苷酰化（adenylation，A）结构域、负责共价结合激活的氨基酸链的肽酰载体蛋白（peptidyl carrier protein，PCP，）结构域、负责催化肽键的缩合（condensation，C）结构域。此外，NRPS 还含有负责释放多肽产物的硫酯酶（thioesterase，TE）结构域，以及众多修饰功能模块，包括差向异构化（epimerization，E）结构域、*N*- 甲基化（*N*-methylation，NMT）结构域、环化（cyclization，Cy）结构域、还原（reduction，R）结构域和氧化（oxidation，Ox）结构域，以增加产物结构多样性。模块化的 NRPS 合成产物时，从起始模块开始，模块内各结构域依次发挥功能，形成顺次接龙式装配线，直至 TE 释放产物。组成 NRPS 的功能模块可以以不同的顺序和数量组合起来，从而决定合成肽段的基本序列和长短。

（5）可调控是指生物大分子机器可以受到多种调节因子的调控而发生结构和功能的改变。生物大分子的构象可变性是其固有性质，包括但不限于底物结合、调节分子（小分子或者生物大分子）结合或者某种化学修饰等，这些通常会改变生物大分子的构象，而增强或者抑制其活性。这种可调控性也为我们人工操控生物大分子机器的活性提供了可能。生物大分子机器不同的结构决定了其在不同的生命过程中发挥不同的生物学功能，包括能量代谢、物质运输、信号感知与应答、遗传信息的传递与维护、蛋白质的合成与降解，以及微生物宿主互作与免疫应答等过程。

（1）能量代谢。细胞通过能量代谢过程获得化学能，并为其他生命活动提供能量，是细胞内最基础的生命过程之一。在细胞线粒体内膜上分布着大量的与产能相关的生物大分子机器，包括 NADH 脱氢酶复合物、泛醌氧化还原酶、细胞色素复合物、细胞色素 c 氧化酶复合物以及 ATP 合酶等。绿色植物、藻类和光合细菌等通过光合作用将光能转变成化学能，这一过程由叶绿体的类囊体薄膜上的光系统Ⅰ、光系统Ⅱ、细胞色素复合物以及 ATP 合酶等共同完成。

（2）物质运输。细胞内膜上存在受体和通道，可实现细胞器之间的物质交换。例如，核膜上的核孔复合物控制着蛋白质和蛋白质复合物进出细胞核以及 mRNA 等分子出核的过程。细胞中的马达蛋白质，如驱动蛋白和动力蛋白，它们具有 ATP 酶的活性，能够将 ATP 的高能磷酸键的化学能转换为机械能，在细胞内起到运载货物的作用，参与有丝分裂、减数分裂和细胞迁移等过程。同时，细胞与细胞、细胞与外界环境之间不断发生物质交换，维持生命活动。细胞膜上存在不同的通道蛋白和蛋白质复合物，控制不同的大分子或小分子进出细胞。例如，细胞膜上的钠离子通道和钾离子通道分别严格控制钠离子内流和钾离子外流，从而控制着动作电位的产生和神经冲动的传导；钙离子通道通过电压或配体进行门控，在膜去极化时打开通道使钙离子内流。

（3）信号感知与应答。细胞膜和细胞内膜上存在着能够感知和接受外界信号的生物大分子机器，当细胞内和细胞外的各种信号分子浓度或活性发生变化时，相应的生物大分子机器接收到信息，并将信息传递到效应分子。例如，细胞内氨基酸等营养物质水平的变化能够激活相应的感受器，进而激活 mTOR 信号通路，最终改变细胞内的能量代谢。再如，不同受体通过感知细胞外信号分子（如激素、神经递质、细胞因子、第二信使等）的浓度变化，激活细胞内相应的效应蛋白，进而改变细胞内的生命过程。其中，G 蛋白耦联受体是细胞表面受体中最大的超级膜蛋白家族，介导众多细胞信号通路，与人类的多种疾病密切相关。

（4）遗传信息的传递与维护。细胞内遗传信息保存在基因组上。DNA 的复制和损伤修复，以及 RNA 转录、加工和出核等过程都涉及大量不同的生物大分子机器。同时，染色体和RNA上不同修饰的建立、识别和移除，染色体构象的改变等过程也需要不同的生物大分子机器参与，从而调控基因的表达。例如，原核和真核细胞中普遍存在的染色体结构维持蛋白（structural maintenance of chromosome protein，SMC）复合物包括黏连蛋白（cohesin）、凝缩蛋白（condensin）以及SMC5/6，在细胞分裂过程中对染色体的修复与分配至关重要（Yatskevich et al.，2019）。其中，黏连蛋白包裹DNA环并通过拓扑结合DNA来介导姐妹染色单体的粘连（Shi et al.，2020），参与染色质拓扑关联结构域的形成，调控染色质区室结构；凝缩蛋白促进有丝分裂期染色质凝聚成X型染色体；SMC5/6复合物参与DNA的损伤修复。再如，mRNA的加工和出核涉及一系列高度动态变化的生物大分子机器，包括RNA剪接体、TREX复合体和Tho复合物等，随着转录、剪接、修饰等的发生，相关的蛋白质-核酸复合物和蛋白质因子等精确地发生顺序结合、解离，最终完成这一复杂过程（Chi et al.，2013；Puhringer et al.，2020；Wan et al.，2020）。

（5）蛋白质的合成与降解。蛋白质的生物合成是生物体内的重要过程，核糖体在其中发挥重要作用。蛋白质的翻译后折叠、修饰以及降解过程也涉及大量生物大分子机器，如分子伴侣、蛋白酶体、泛素化修饰酶等。蛋白质的正确折叠与其功能密切相关。分子伴侣素通过结合部分折叠或未折叠的蛋白质分子，可避免这些蛋白质分子被其他酶水解，从而提供蛋白质折叠成天然构象的环境。蛋白酶体降解错误折叠的蛋白质，同时调节细胞内重要蛋白质（如p53等）的数量，从而影响相关的信号通路，该过程与癌症、炎症、神经退行性疾病等的发生相关。同时，蛋白质的翻译后修饰是化学修饰，包括磷酸化、糖基化、泛素化、亚硝基化、甲基化、乙酰化、脂质化等，是动态、可逆的，修饰基团的添加和移除由一系列相关的酶介导，这些修饰与蛋白质的活性、功能和稳定性都密切相关。蛋白质的化学修饰以指数方式增加了蛋白质组的多样性和复杂性。

（6）微生物宿主互作与免疫应答。包括病毒、致病菌等在内的病原微生物一直威胁着人类健康。病原微生物与宿主的相互作用决定了微生物侵染致病的分子基础。以病毒为例，在入侵细胞的过程中，病毒颗粒需要通过与宿主细胞膜上的受体及辅助蛋白的特异性相互作用，形成一系列的阶段性生物大分子机器，进而完成从识别、附着到基因组释放的动态过程（Fung and

Liu，2019）。病毒 DNA 或 RNA 进入宿主后，需要利用宿主细胞提供的原料、能量和场所来完成自身的转录、复制以及病毒蛋白的翻译、折叠和组装，此过程同样有赖于病毒的转录复制复合物、蛋白水解酶等和宿主相关蛋白形成的生物大分子机器的精确调控。另外，宿主细胞识别病毒的入侵并触发天然免疫防御系统和适应性免疫系统来抵抗病毒感染，同样有赖于诸多生物大分子机器，如感受病原体相关分子模式并被激活的炎症小体、抗病毒 MAVS 超大分子信号转导复合体、诱导干扰素产生的 STING 小体、识别病毒抗原的 T 细胞受体（TCR）复合物等。

由此可见，生物大分子机器贯穿细胞生命的整个过程，在物质交换、能量代谢、信息交流和基因表达等方方面面都发挥重要功能，从细胞、亚细胞、分子和原子水平对生物大分子机器的研究，可以阐明生命过程的分子机理，是当今生物学研究的热点和前沿。随着科学技术的发展，目前对于生物大分子机器的研究，国内外已经具备了相应的基础。X 射线衍射技术、冷冻电镜技术和核磁共振技术的综合使用，结合超分辨显微成像技术、质谱技术、光遗传学技术、化学遗传学技术等，让越来越多的生物大分子机器的组装、修饰、转运、动态变化、调控以及协同工作等过程呈现在人们眼前，使得越来越多的生命现象得到合理的解释。生物大分子机器与整个生命过程息息相关，生物大分子机器的异常将导致生命过程无法正常进行。因此，对于生物大分子机器的研究，不仅有助于从分子或原子水平上理解生命过程，还能揭示相关疾病的发生、发展过程，为疾病的诊断和治疗、相关药物的开发和设计等奠定基础。此外，基于这些信息对生物大分子机器的模拟、设计或者基于调控机制的设计，也能极大地推动了生物学研究的发展，为人类疾病的治疗和干预做出重大贡献。

第二节　关键科学问题

如上所述，生物大分子机器具有自组装、自适应、多样性、模块化、可调控等特性，这些特性是它们在体内参与生命活动、高效运转的基础。对生物大分子机器的研究主要包括组装、修饰、调控、转运、定位、协同作用、实时观测、分子设计等方向，本章将对这些方向分别展开详述。

一、生物大分子机器组装的基本规律

生物大分子机器的组装是指基本结构单元基于非共价键的相互作用自发地组织或聚集为一个稳定、具有一定规则和集合外观，并且表现出特异性功能的结构。从分子水平阐明生物大分子机器中各亚基之间特异性识别和相互作用、精确组装过程的调控，以及发挥功能时的动态变化及其调控，是生物大分子机器组装过程研究中的关键科学问题。这些关键科学问题的解决，不仅依赖于相关技术手段的发展和革新，也依赖于不同技术手段的联用。

首先，各亚基之间特异性识别和正确组装是生物大分子机器发挥功能的基础。细胞中的蛋白质、核酸、脂类和糖类等分子，是如何通过分子间的非共价键相互特异性识别，自发地、正确地组装成复杂的生物大分子机器的呢？对这一过程的分子机制的阐述，主要依赖于近原子高分辨率结构信息的测定。X 射线衍射技术、冷冻电镜技术和核磁共振技术是目前常用的三种结构生物学技术，近年来借助这三项技术，研究人员已经揭示了大量重要的生物大分子机器的组装机制，但是仍有许多重要的生物大分子机器缺乏关键的高分辨率信息，如核孔复合物。核孔复合物是已知的细胞内最大、最复杂的分子机器，在不同生物体内尺寸并不完全相同，对它的结构研究已经进行了 20 多年，但仍未获得完整的核孔复合物的高分辨率结构信息。这是因为结构生物学在高分辨率信息的测定上仍具有一定的局限性：X 射线衍射技术的瓶颈在于难以获得具有高衍射能力的生物大分子机器晶体；核磁共振技术可以观察到分子在溶液中结构变化的高分辨率信息，但是检测分子通常需要小于 50 kDa；固态核磁共振技术也可以测定生物大分子的结构，但因其样品的分子快速运动、化学位移各向异性等作用使谱线增宽等技术限制，与液体核磁共振技术相比难以获得高分辨率结构信息；冷冻电镜技术作为近年来快速发展的技术，在测定生物大分子机器的高分辨率结构信息上有着巨大的优势，但超大复合物样品的制备、冷冻断层重构技术的高分辨率结构获取等尚不成熟。如何获得超大分子机器的高分辨率结构信息，是研究者要面对的重大挑战。结构生物学的发展十分依赖于技术的革新，因此发展和完善相关的结构生物学技术手段是解决高分辨率结构信息测定的核心。同时，多种结构生物学技术联用，结构生物学技术与生物化学、生物物理等技术联用，也是提供生物大分子机器高分辨率结构信息的重要方式。除了传统的检测蛋白质相互作用的技术（如免疫共沉淀、恒温等热滴定、表面等离子共振、酵母双杂交、荧光共振能量转移等），新兴技术的发展也极大地加速了对生物大分子机器组装机制的研究，例如交联质谱技术可辅助研究肽段中两个特定氨基酸残基之间的最大距离，在没有高分辨率结构信息的情况下分析并获得氨基酸之间的相互作用信息。

其次，许多生物大分子机器在发挥功能的过程中并不是单一不变的，而是在不断变化的。特定的亚基或小分子在特定的阶段结合或解离，组装成不同的生物大分子，在不同阶段发挥不同功能，其组装过程受到精确的调控。目前的结构生物学研究技术大多获得的是静止的三维结构信息，在动态研究上存在显著的局限性：处于动态过程中的分子往往很难结晶，无法通过X射线衍射技术测定结构；核磁共振技术只能观察到较小分子在溶液中动态信息；冷冻电镜技术的样品快速冷冻过程具有较高的时间分辨率，在生物大分子机器的动态过程研究中有较大的优势，但对动态变化较快的部分仍难以获得高分辨率结构信息。如何获得动态的三维结构信息以及如何联用不同的技术手段是阐明生物大分子机器组装调控机制的关键，是研究者面临的重大挑战，也是当前领域内的重要前沿。通过改进和完善结构生物学技术，提高时间分辨尺度，可以获得不同状态下的高分辨率三维结构信息。同时，发展和联用生物物理学、计算生物学等领域已有的或新兴的技术手段，对生物大分子机器组装的动态过程进行监测，研究不同状态转换的动态过程，利用动态模拟等计算生物学手段对动态过程进行模拟，再通过实验手段进行分子动力学信息的分析和验证，可以更深入地了解动态变化过程。例如，通过氢氘交换质谱技术可以研究溶液中蛋白质表面的氨基酸序列、蛋白质结构动态变化、蛋白质间或蛋白质与小分子药物之间的结合动态，以及鉴别蛋白质表面活性位点等；通过单分子超分辨率显微成像技术和单分子荧光成像技术可以发现新结构，研究生物分子的动态相互作用、定量分子计数、分子动态信息等。

最后，生物大分子机器在发挥功能时，其动态变化受到细胞内其他信号或环境的精确调控。要观察生物大分子机器在细胞内的动态变化过程，以及其对周围环境和其他生物大分子机器的信号响应和交互作用，需要在生理状态下原位观察生物大分子机器的组装过程。冷冻断层重构技术虽然可以在细胞器水平上观察生物大分子机器的组装和功能，成为连接结构生物学和细胞生物学的重要工具，但是目前仍无法获得高分辨率信息，相关技术的发展需求十分迫切，是生命科学领域的重要前沿。此外，在实验条件下模拟生理或病理状态下的细胞环境，借助现有的生物物理学、生物化学、结构生物学和细胞生物学实验技术手段以及计算生物学模拟方法，观察和探索生物大分子机器对外力、信号分子或其他生物大分子机器的感受和响应，获得这些过程中的动态信息，也有助于理解生物大分子机器在细胞内的动态变化过程。

二、生物大分子机器的化学修饰和动态调控

目前对生物大分子机器的化学修饰和动态调控研究虽然较为深入，但还是在诸多方面存在认识不足，需要进一步明确生物大分子机器的化学修饰的生物学功能，进一步阐明化学修饰在细胞内的动态调控机制，进一步认识化学修饰的动态变化与疾病发生发展的关系。是否还存在新的化学修饰？如何更有效地鉴定新的修饰酶和识别修饰的分子机器？化学修饰建立、维持和解读的动态调控机制是什么？是否可以对疾病相关的化学修饰变化进行干预从而进行相关疾病治疗？这些关键科学问题的解决需要在以下几方面有新的技术发展和科学突破。

一是发展和提升探测化学修饰的新技术。目前主要依赖于质谱检测发现新的化学修饰类型，依赖于抗体来检测化学修饰的动态变化。未来，需发展先进质谱技术，提高检测精度和灵敏度，以更好地发现在不同细胞代谢状态、不同生理病理过程中的新修饰并检测修饰的动态变化；发展更好的针对特定化学修饰的单分子测序和成像技术，以精确地表征细胞内的化学修饰及其变化与功能的关系。目前对修饰的功能研究十分依赖于抗体，抗体的质量不高严重制约了相应的研究。将来需要开发一些新的化学工具，以特异性识别某

种修饰，而不依赖于抗体。

二是发展鉴定生物大分子修饰酶和识别蛋白的新策略。以前主要通过遗传筛选和生化分离来鉴定修饰相关蛋白，针对目前还未鉴定的未知酶（如精氨酸去甲基化酶、DNA 主动去甲基化酶等），现有的传统研究手段难以为继。随着高通量筛选和基因编辑等技术手段的发展和成熟，高灵敏度报告系统的建立和完善，有可能发现新的修饰酶和识别蛋白。另外，质谱技术的发展、新型生物正交反应的应用，也将有助于发掘新的修饰酶和识别蛋白。

三是解析生物大分子动态修饰的调控机制。虽然对生物大分子修饰的调控机制研究较多，但目前对修饰本身是如何调控生物大分子构象的还知之甚少。原因之一在于缺少简易有效的修饰生物大分子的化学标记技术，现有的化学半合成技术门槛较高。化学生物学相关技术和工具的发展，将使研究人员能更好地制备带有修饰的生物大分子以开展结构解析和功能剖析，揭示化学修饰如何从分子层面上影响生物大分子的结构和功能。同时，也要借助新发展的精确特异的化学标记检测手段和工具，开展生物大分子动态修饰的原位检测，揭示动态修饰调控的规律，并辅以细胞生物学、生物化学、生物信息学等方法，综合阐释生物大分子修饰酶和识别蛋白在动态调控中的作用及其分子机制。另外一个值得开展的研究方向是探究生物大分子修饰对相变的调控功能。目前已经发现蛋白质的修饰可以调控相变，但修饰影响相变过程的具体分子机制，特别是修饰如何动态调控蛋白质参与相变形成或从相变颗粒中解离，都还不清楚，这些都是值得研究的新方向。

四是探索生物大分子修饰的有效干预。生物大分子修饰酶和识别蛋白的突变与许多疾病的发生发展息息相关，因此这些蛋白质及大分子复合物日趋成为研究人员感兴趣的新靶标。针对特定的结构靶点，进行活性化合物高通量筛选、药物分子理性设计等，可获得高选择性的小分子作为先导化合物；通过对 Degron 快速降解系统在降解突变修饰酶或识别蛋白过程中的功能和有效干预研究，可为相关疾病的治疗提供可能性。另外，应用 CRISPR 技术靶向修饰系统进行特异位点的精确修饰干预，可以阐释生物大分子动态修饰的干预与功能定向改变的密切关系。上述研究，可为药物研发提供潜在干预小分子和新靶标。

三、生物大分子机器在细胞内的定位和转运

不同的生物大分子机器在不同的细胞场所发挥各自的功能，研究生物大分子机器的定位和转运规律，对进一步深入理解相关生物大分子机器的功能有着重要的意义（Arimura et al.，2004）。生物大分子机器的定位可以被定义为其发挥生物学功能时所在的空间和时间场所（Glynn et al.，2008）。空间场所包括细胞膜、细胞质、核膜等，如 G 蛋白耦联受体发挥功能的场所为细胞膜，蛋白酶体通常在细胞质内发挥功能，而核孔复合物发挥功能的场所为核膜。在时间层面上，可以被定义为在生命体或者细胞的某个特定时间节点，生物大分子机器会集中出现在其发挥功能的空间位置（Praefcke and McMahon，2004）。例如，在需要细胞合成大量蛋白质并保证其合成蛋白质折叠成天然构象时，一型分子伴侣素 Gro-EL/Gro-ES、二型分子伴侣素 TRiC 以及 HSP 家族蛋白会富集并定位在内质网和高尔基体附近，帮助合成的蛋白质前体折叠（Jin et al.，2019，2018）。因此，生物大分子机器的定位包括空间和时间的定位，是一个极其复杂但有序的过程（Benninger et al.，2007）。在细胞内，生物大分子机器如何感知环境中的信号，找到正确的时间和空间场所发挥功能，是研究者希望解决的重大科学问题。有些生物大分子机器的空间定位相对固定，如 ATP 合酶定位在线粒体膜上；有些生物大分子机器的空间定位是动态变化的，如 DNA 双链交联损伤的修复机器 FA 复合物只在 DNA 损伤区域发挥功能。这些生物大分子机器的空间定位机制，需要结合生物化学、分子生物学、结构生物学、细胞生物学、计算生物学等多种手段，研究生物大分子机器自身的信号肽、修饰、组成等因素与环境因素之间的关系。生物大分子机器的时间定位，则需要在时间尺度上观测和研究生物大分子机器与环境因素、胞内信号等之间的关系。生物大分子机器对信号的响应时间不同，如何在毫秒甚至更短的时间以及数小时或数天等更长的时间上，以超高分辨率水平，实时、连续地观测单个生物大分子机器在活细胞内的变化，是研究者面临的重大挑战。

生物大分子机器的转运与定位密切相关，生物大分子机器的转运可以被定义为通过一系列特定的信号肽或者信号分子，以分子马达提供运输的动力，以细胞骨架作为轨道或者通过自由扩散和随机碰撞，通过一系列运输途

径，将已经组装完成的生物大分子机器或正在组装的生物大分子机器的前体从细胞的一个位置运送到细胞的另一个位置，从而发挥相应的功能（Monti，2020）。真核细胞内生物大分子主要存在两种转运模式，一是直接的共翻译的跨膜或者入膜的转运，如内质网上的转运；二是组装好的生物大分子机器的翻译后转运。第一种转运模式更多的是依赖生物大分子自身的特性；第二种转运模式则涉及生物大分子机器与其他蛋白质或者生物大分子机器之间的交互作用。第二种转运模式先通过筛选蛋白指导生物大分子机器转运到细胞器表面，然后特定的通道复合体（如过氧化物酶体、质体、线粒体和核膜上的通道蛋白）与生物大分子结合，进而对生物大分子进行整合和运输。此外，生物大分子机器也可以通过膜泡运输，在不同的位置直接穿梭，如从内质网到高尔基体的直接运输（Howard，2014）。在绝大多数情况下，高尔基体是膜泡运输的筛选和转运的中心。通常，生物大分子机器是通过特定的靶向信号来转运的，靶向信号需要与细胞器表面的特定受体结合。靶向信号通常是拥有保守序列或者拥有特殊的化学性质的氨基酸短肽，在多数情况下会在生物大分子机器发挥功能前被剪掉移除（Kiatpongsan et al.，2006）。生物大分子机器通过特定的靶向信号找到相应的受体继而发生转运，以及转运结束后靶向信号被移除的过程，是研究人员关注的关键科学问题，其中包括促使生物大分子机器转运的因素、生物大分子机器的靶向转运信号如何被精确识别、通道蛋白如何实现生物大分子机器的运输、生物大分子机器运输的能量来源、膜泡的形成和运输过程如何发生、转运信号如何被移除等。

对活细胞或亚细胞器内的生物大分子机器进行实时观测，是阐述物大分子机器定位和转运过程等关键科学问题的直接方法，然而现有的单分子超高分辨率成像技术所能达到的分辨率，对观察单个生物大分子机器的结构和分子间相互作用还远远不够，如何提高成像的分辨率是研究者面临的重大挑战。同时，在活细胞三维大尺度的成像，将产生太字节（terabyte，TB）量级的数据处理需求，高效的图像处理算法和软件的开发与改进也是目前面临的挑战。对成像技术的发展或新技术的开发，对标记探针和标记方法的开发与优化，不同观测技术的结合与优势互补等，是提高分辨率的努力方向。

四、生物大分子机器在体内的高度协同工作规律

在生物体内，多种生物大分子机器通常需要高度协同工作完成某一个特定生物学过程。这些生物大分子机器形成一个高度复杂的超分子复合体来有序完成特定生物学过程的各个步骤。解析超分子复合体的结构，揭示生物大分子机器之间协同作用的机制，是领域内关注的热点问题。另外，目前流行的相分离已被视为生物大分子机器高效协同作用的一种通用机制，但该机制是否在活细胞内真实存在以及相分离如何受到精确调控，也是亟待解决的关键科学问题。

在超分子复合体的纯化和结构解析方面存在着很大的挑战。由于这些超分子复合体组成复杂，异源表达系统往往无能为力，因此现有的超分子复合体的纯化还是以内源复合体为主。而内源复合体存在产量低、异质性高等特点，阻碍了后续的结构解析。因此，还需要发展新的系统来纯化超分子复合体，并发展一些新策略以增强其稳定性。同时，对超分子复合体的结构解析仍存在较大的挑战。单个生物大分子机器，即使是具有很多亚基的复杂大分子机器（如核糖体等），通常具有比较固定的组成和构象，目前已经有相当成熟的研究方法揭示其结构。对于不同分子机器模块组装而成的超大分子机器集合体，虽然冷冻电镜技术可用于其结构解析，但其结构的高度动态性给后续的三维结构计算带来了很大的挑战。因此，如何发展新的方法捕捉动态的结构，是今后结构生物学家面临的一个主要挑战。

解析动态的结构，一个可行的途径是发展整合型的结构生物学，集合多种实验方法和手段共同揭示其动态变化规律（Rout and Sali，2019）。冷冻电镜、晶体学、核磁共振等技术可以提高单个蛋白质机器或者核心组分的高分辨率结构解析精度。X 射线小角散射、相干成像等技术虽然只能提供低分辨率的形貌信息，但对于整合各个高分辨率的模块结构来说是不可或缺的。质谱相关的实验技术，包括交联质谱、氢氘交换质谱、非变性质谱等，在揭示模块间的联系方式和动态变化等方面发挥着重要作用。目前的挑战是如何规范不同实验方法产生的数据，使之标准化，从而在不同的实验模块之间实现互通，加速结构解析。

相分离理论为解释生物大分子机器的组装和高效协同作用提供了一个较为自洽的理论框架（Banani et al.，2017）。虽然对于相分离是否在体内真实存在、是否发挥如此大的作用，还存在一些争议，但总体上相分离理论得到了科学界的广泛接受和认可。然而，目前相分离的研究主要依赖于体外重构相分离体系（Alberti et al.，2019）。相分离的成像方法还有一些局限，如细胞内成像实验往往也依赖人工添加的蛋白标签和过表达蛋白，观察到的细胞内的聚集结构由于显微成像分辨率的限制阻碍了对其动态变化和组装的理解。因此，需要开发更多更准确的技术方法对天然状态下的相分离特性进行表征。目前相分离还存在着很多尚待解决的科学问题，如相分离在诸多生理过程中发挥作用的具体分子机制、细胞内相分离的精确调控机制、相分离在各种疾病发生发展中的作用，以及如何利用调控相分离来进行疾病治疗等。对这些问题的解答将帮助我们进一步理解生物大分子机器的组装和高效协同机制。

五、生物大分子机器的原位和动态观测

许多生物大分子机器和大型细胞超微结构（如纤毛等）的相关结构形式可能仅存在于天然环境中，也可能存在瞬时结合伴侣及特定环境中的构象和其他背景信息。生物大分子机器发挥功能有着高度的时空特异性。对生物大分子机器在体的原位实时和动态观测技术是揭示生物大分子机器在体结构、定位、相互作用网络和动态性，理解其如何参与生命活动、发挥功能的重要手段。

冷冻电子断层扫描（cryo-ET）可以对未扰动的细胞进行成像，保留了每个分子的构象和相互作用的完整网络，成为可对细胞“天然分子景观”进行完整结构描述的核心手段。但与冷冻电镜单颗粒重构技术相比，cryo-ET 技术还处于早期阶段。cryo-ET 技术最大的优势在于可对生物大分子机器进行原位结构测定和原生系统中的三维分类，具有在近原子分辨率水平提供生物大分子机器在细胞中的构象及分布的潜力。然而，细胞质内的极端拥挤环境极大地妨碍了该技术区分单个蛋白质和蛋白质复合物的能力，因此该技术尚有许

多需要探索和优化之处。新近的技术进步，如直接电子探测器、相位板、低温聚焦离子束减薄及软件的开发，极大地提高了 cryo-ET 图像的质量和三维重构的分辨率，扩大了 cryo-ET 的应用范围（Mahamid et al.，2015；Zhang，2019）。在 cryo-ET 基础上还可以对子断层扫描平均（sub-tomogram averaging）以提高分辨率，目前已达到亚纳米分辨率水平（个别体系达到近原子分辨率水平）（Wan and Briggs，2016；Zhang，2019）。

此外，将 cryo-ET 与荧光显微镜及其衍生技术联合使用的多尺度联合成像技术成为对生物大分子机器原位实时和动态观测的主要技术手段，为生物医学研究中对生物大分子机器结构 - 功能关系的探索开辟了新的途径（Ando et al.，2018）。在细胞和亚细胞层面直接可视化的能力决定了人们对生物过程的了解程度。在各种显微镜技术中，荧光显微镜可以通过分子特异性标记来观察特定的细胞成分，是应用最广泛的技术之一。然而，传统的荧光显微镜由于光的衍射而被限制在相对较低的空间分辨率上，无法用于详细观察许多亚细胞结构。单分子定位显微技术是目前荧光显微镜技术中分辨率最高、应用也颇为普遍的技术（Li and Vaughan，2018）。常用的技术还包括使用光激活荧光蛋白的光激活定位显微术（photo-activated localization microscopy，PALM），使用有机染料的随机光学重构显微术（stochastic optical reconstruction microscopy，STORM）等（Jacquemet et al.，2020）。目前，多尺度联合成像技术中最常见的是结合光学显微镜 / 荧光显微镜和能高分辨率解析结构的电子显微镜的光电联合成像术（correlative light and electron microscopy，CLEM）（Ando et al.，2018）。极有应用潜力的超高分辨率荧光显微镜与电子显微镜的结合，有望精确定位在电镜图像中无法标明的分子，并利用电镜进行高分辨功能成像。

cryo-ET 可用于开展生物大分子机器在细胞 / 组织原位的高分辨率、动态结构分析。这项技术需要克服的核心科学技术问题包括：①简洁快速地获得完整且接近细胞内环境的生物样品；②提高 cryo-ET 原始图像的衬度、更好地保存成像中的高分辨率信息；③在极其拥挤的细胞内环境中准确定位目标大分子复合体；④实现对子断层图像的高效精确的中和平均，提高分辨率；⑤实现对诸多的 cryo-ET 重构图像进行分类并区分生物大分子机器的构象。此外，生物大分子机器在溶液中执行其功能的过程中往往会呈现多种构象状态，但构象变化通常是短暂的，如何应用冷冻电镜技术捕获溶液中生物大分子机器的多种构象状态？这些都是需要解决的科学问题。

六、生物大分子机器的模拟、操控和设计

人们对生物大分子机器的结构和功能的理解已经取得了巨大的进步。利用计算机分子动力学方法可以模拟生物大分子机器的结构形变和功能，有助于在生物物理和生物化学方面总结相应的规律，从而使人们在分子层面对生命活动具有更深的理解。分子动力学方法自从20世纪70年代被发展以来，在算法、软件和硬件上都取得了巨大的进步，现已可以对具有上亿原子的生物大分子机器实现近微秒的仿真模拟。在准确度上，生物大分子机器的模拟结果不仅已经可以与实验结果做精确地比照，并且可以对复杂大分子动力学系统的结果做出正确的预测。现在分子动力学模拟已经被认为是在微观尺度上观察生物大分子机器运行的计算式的“显微镜”（Dror et al., 2012）。但遗憾的是，目前对于超大型蛋白质分子机器无法长时间有效地进行动力学模拟和研究，急需发展和优化软件系统（如建立准确的力场模型来模拟复杂体系，构建自适应的多尺度模型，改进采样算法等）与硬件系统，专注在蛋白质分子机器大空间尺度和长时间尺度（微秒乃至毫秒）的动力学模拟。

在生命活动中，生物大分子机器在结构和功能上受到精确的调控。研究生物大分子机器的调控机制对于理解生物生理过程和疾病发生发展过程具有至关重要的作用。同时，研究分子的调控机制会极大地帮助研究人员设计生物活性更高和具有新颖功能的生物大分子机器来研究生物学功能。基于生物的分子调控机制而设计的具有新颖功能的生物大分子机器，将会使遗传学、细胞生物学、神经生物学和细胞疗法等生物医学领域得到空前的发展，并为治疗人类疾病做出巨大的贡献。但如何准确地调控和设计生物大分子机器是一个具有挑战性的难题。如何系统性地从头设计生物大分子机器实现所需功能的准确调控，目前甚至还没有一个具体的概念上的框架。

第三节　重要研究方向的发展现状与趋势分析

近几十年来，生命科学领域飞速发展，随着研究的深入和技术的发展，许多生物大分子机器参与细胞生命过程、发挥特定功能的机制被逐步揭示，基于生物大分子机器的设计也在多个领域得到广泛应用，本节将深入探讨几个重要研究方向的发展现状及未来研究趋势。

一、生物大分子机器组装的基本规律

目前，对于生物大分子机器组装的分子机制研究主要通过体外的结构生物学、生物化学与分子生物学等技术手段，结合体内细胞生物学和遗传学等技术开展。受限于技术手段，当前结构生物学研究主要局限于生物大分子机器某一固定状态的研究，对于组装动态过程的研究还需要进一步加深和拓展。值得肯定的是，目前已有很多生物大分子机器的结构得到解析，高分辨率的信息清晰地展示了生物大分子机器的组装和构象，部分揭示了组装的分子机制和规律，例如，与能量代谢相关的 ATP 合酶、mTOR 复合物，将光能转变成化学能的光系统Ⅰ和光系统Ⅱ，与物质运输和信息传递相关的离子通道、转运蛋白、G 蛋白耦联受体、分子马达，与基因表达调控相关的 RNA 聚合酶、转录起始复合物，与 mRNA 加工有关的剪接体、核糖体复合物、染色质重塑复合物、核孔复合物、组蛋白修饰复合物，与 DNA 损伤修复相关的蛋白质 -DNA 复合物，与蛋白质修饰降解相关的蛋白酶体复合物、泛素化修饰复合物等。已有多位科学家凭借对生物大分子复合物结构的研究工作而获得诺贝尔奖，这些研究工作成果揭示了生物大分子机器组装和发挥功能的机制，为其相应的功能研究奠定了基础，在本质上揭示其所执行的生物功能的基本规律。因此，对于生物大分子机器的研究工作，将推进和加深人类从分子水平和原子水平上对生命过程本质的认知与理解，引领现代生命科学的发展。

同时，对生物大分子机器的研究，还将极大地推动和促进相关医药领域、疾病研究领域的发展，对具有重要功能的生物大分子机器的深入研究有着极大的应用前景和价值。

未来对于生物大分子机器组装的研究，除了对生物大分子机器组装的结构生物学进行研究，还向生物大分子机器组装的动态变化以及生物大分子机器在细胞内的原位组装过程两个方向拓展。现有研究大多通过人为手段将生物大分子机器稳定在某一特定状态，通过结构生物学和其他手段对其结构进行表征，从而揭示该生物状态下生物大分子机器的组装机制。通过加入不同分子或改变条件等方式，可使生物大分子机器处于不同的生理状态，从而获得不同状态下的分子信息，揭示各构象之间的动态变化。但是这种方式获得的仍是一系列静态信息，如何在动态变化过程中获得高分辨率原子信息以及不同时间点的原子动态变化信息是未来研究的重要方向之一。此外，生物大分子机器在细胞内的组装过程受到其他因素的调节，生物大分子机器之间也存在相互联系，生物大分子机器内有些蛋白质与蛋白质之间的相互作用短暂、快速，还可能伴随着蛋白质折叠和构象上的变化。例如，在非编码 RNA 转录后的调控过程中，非编码 RNA 分子可能同时存在多种不同构象，参与多个生物大分子机器介导的生命过程，同时 RNA 与 RNA、DNA 和蛋白质之间都有短暂的相互作用，RNA 分子的二级结构变化也将调节其相应的功能。如何在精确时间维度上和活细胞水平上获得这些生物大分子机器组装的动态过程中的原子水平高分辨率信息，也将是未来研究的一个重要方向。

目前，有多项新兴技术的发展有望能快速推动这些方向的研究。单分子成像、超高分辨率荧光成像以及荧光探针等技术可用于研究生物大分子机器在细胞内组装的动态过程，但这些技术在活细胞上的应用不够广泛，且在时间维度和分辨率上仍存在不足，对某些快速运动的观察受限，无法达到原子分辨率水平。cryo-ET 能够原位观察细胞内蛋白质与蛋白质之间的相互作用，这一技术在保持生物体活性和结构特征细节等方面具有独特优势，能够更好地解析一些分子量大、结构复杂、组装难度大的生物大分子机器的结构，在获得生物大分子机器结构信息的同时，还能获得它们在细胞环境中的空间分布信息及其与其他分子相互关系的信息。目前这项技术主要在培养的细胞或单细胞生物中进行，不过已有适用于经典模式生物如线虫和果蝇的新方案被开发，未来有望应用于多细胞生物的各项研究。通过对包含不同构象样品的数据进行三维重构和分类有可能得到具有时间分辨率的信息，从而阐明生物大分子机器组装及发挥功能的机制。目前这一技术仍有几个重要问题有待解决：①数据收集时样品台倾转角度有限，部分角度上的信息缺失；②样品倾转导致不同的位置相对于焦点具有不同的高度，为数据处理带来了困难；③这项技术目前分辨率为 4～10 nm，难以获得精细的动态组装信息。

二、生物大分子机器的修饰和动态调控

生物大分子修饰是指蛋白质、核酸、糖脂等生物大分子的双向可逆的化学修饰。生物大分子的动态化学修饰会影响生物大分子的结构和功能，在各种生命活动中发挥着重要作用，例如表观遗传学的发端就来源于对组蛋白和 DNA 的化学修饰研究。组蛋白的共价修饰可通过影响组蛋白与 DNA 双链的亲和性，而改变染色质的松散或凝集状态，或者招募下游效应分子，来调节基因的表达，因此组蛋白的修饰研究一直是过去二十多年生命科学的热门领域之一。近些年来对 RNA 的修饰研究（如 m6A）也成为科学界追逐的一个焦点。

本部分主要讨论蛋白质的化学修饰。蛋白质修饰的研究热点主要集中在以下三个方面。

1. 新修饰的鉴定

这方面的研究内容包括了新修饰位点的揭示和新修饰形式的鉴定。常见化学修饰包括磷酸化、甲基化、乙酰化、泛素化等。目前已经绘制了细胞内很多蛋白质的常见修饰谱，但可以预见，随着蛋白质组学的进步和发展，特定蛋白质在特定位点的修饰仍会被不断地鉴定出来，这些修饰在生理和病理过程中的作用也将被揭示出来。目前鉴定出来的蛋白质化学修饰已经达 200 多种，该数字还在不断增长中。比如近些年发现了一系列受细胞代谢物控制的新型组蛋白赖氨酸修饰（Zhao et al.，2018），包括甲酰化（Kfo）、丙酰化（Kpr）、丁酰化（Kbu）、丙烯酰化或巴豆酰化（Kcr）、2- 羟基异丁酰化（Khib）、β- 羟基丁酰化（Kbhb）、琥珀酰化（Ksucc）、丙酰化（Kmal）、戊二

酸化（Kglu）、苯甲酰化（Kbz）、长链脂肪酰化以及乳酸化（Klac）等。这些新型修饰从不同方面参与生物体内的各种生命活动，如转录、染色质运动和发育等。在未来的研究中需要发展和提升探测修饰的新技术，进一步揭示新的化学修饰，揭示在不同细胞状态（如不同发育阶段、不同代谢环境、不同生理病理过程等）下修饰类型和丰度的动态变化，并与功能表征密切联系起来。

2. 修饰相关酶和识别蛋白的鉴定与功能阐释

蛋白质的化学修饰需要有“书写器”（writer）加上特定修饰，“擦除器”（eraser）去除特定修饰，以及“阅读器”（reader）识别修饰并产生相应的功能输出。对这些蛋白质修饰相关酶和识别蛋白的鉴定一直是生物学研究的热点。对这些蛋白质的研究，能够帮助我们了解修饰是怎样被建立、识别最后被擦除的这样一个动态过程，也能帮助我们理解修饰在多种生命活动中的功能和调控机制。目前，大多数蛋白质修饰酶和识别蛋白已经被鉴定了出来，但有些修饰相应的修饰酶仍然未找到，如目前还没有一个大家普遍接受的精氨酸去甲基化酶。目前，对修饰酶和识别蛋白的结构与功能的研究虽然已相当深入，并系统阐明了它们在正常生理活动中的重要功能以及在疾病发生发展中的作用，但它们的具体作用机制还有待进一步阐释。例如，缺失某种修饰酶产生的功能变化到底是化学修饰本身变化导致的，还是缺少修饰酶这个蛋白导致的，就是一个容易陷入争论的科学问题。鉴定新的修饰酶，解析它们的调控和功能机制，仍将是生物学研究的热点。

3. 动态修饰的调控机制与有效干预

蛋白质的化学修饰是在不改变蛋白质序列的基础上增强其功能的多样性和可塑性，因此修饰的一个特性就是动态。动态修饰表现为双向可逆和时空特异。双向可逆即修饰不是一成不变的，而是可以被添加和去除；时空特异即修饰的状态随着细胞类型、环境、发育阶段、细胞周期等的变化而变化。蛋白质修饰的动态性在个体发育和正常生理活动中发挥决定性的调控作用。因此，过去对蛋白质动态研究最关注的问题是发现和阐明化学修饰的动态特性，揭示其生物学效应。在今后的研究中，生物大分子机器动态修饰的原位检测、动态修饰调控机制的阐释，将成为研究的主流。另外，探索生物大分子机器修饰的有效干预，也将有助于蛋白质修饰相关疾病的药物研发。

三、生物大分子机器在细胞内的定位和转运

生物大分子机器在细胞内的定位和转运是其发挥相应功能的基础，控制生物大分子机器在细胞不同区间的运输是细胞调节生命活动过程的一个直接而又极为重要的手段（Kartsogiannis and Ng，2004；Walters，2004）。众多生物大分子机器在复杂的细胞大环境内实现其各自功能都依赖于特定的环境。在真核生物细胞内，各种生物大分子机器组装和发挥功能的位点往往并不相同。生物大分子机器是通过何种途径精准定位并各自转运至相应的位点以发挥生物学作用的，其中涉及一系列复杂的细胞生物学过程（Cabianca and Gabellini，2010；Hause and Fester，2005）。研究生物大分子机器的定位和转运规律，可以揭示相关生物大分子机器结构和功能的动态变化规律，阐明其在细胞内外环境中的变化、其在维持细胞稳态和决定细胞命运中的作用与分子机制，对进一步了解细胞的生命活动、阐明生物大分子机器定位和转运过程中的异常在人类重大疾病发生发展中的关键作用具有重大意义（Virshup and Kaldis，2010）。近年来，随着相关技术的发展和研究的推进，人们对部分生物大分子机器（如G蛋白耦联受体和RNA聚合酶Ⅱ等）的定位和转运过程有了一定的认识（Peeters et al.，2011）。

G蛋白耦联受体和G蛋白在内质网合成后，经细胞内膜泡运输系统运输并定位到细胞膜上（Xiao et al.，2011）：合成好的GPCR和G蛋白以膜蛋白的形式与COP Ⅱ包被膜泡融合并运输到高尔基体，初步折叠成天然构象，随后由网格蛋白/接头蛋白膜泡包裹运输到细胞膜上。在某些特殊的细胞器中，GPCR的运输由特殊的蛋白复合物介导（Henry，2004）。例如，纤毛膜表面的GPCR GPR161和生长制素受体3（somatostatin receptor 3，SSTR3）是通过巴尔得–别德尔综合征蛋白［Bardet-Biedl syndrome（BBS）protein，BBSome］复合体与纤毛内转运（intraflagella transport，IFT）复合物的结合，利用驱动蛋白Ⅱ作为马达蛋白质，沿着微管被运输到纤毛膜表面的（Duan et al.，2021；Klink et al.，2020；Singh et al.，2019；Takahara et al.，2019）。

RNA 聚合酶Ⅱ（Pol Ⅱ）主要负责合成 mRNA 和大量非编码 RNA（Wild and Cramer，2012），其转运机制也得以系统揭示。研究表明，Pol Ⅱ在 HSP90、RPAP1 等蛋白的辅助下在细胞质内完成组装（Boulon et al.，2010；Zeng et al.，2018），随后通过转运因子 Iwr1 提供核定位信号，通过核孔复合物进入细胞核内发挥转录功能。随后，Pol Ⅱ与 DNA 结合并在各种转录因子的作用下开始转录，最后 Pol Ⅱ与 Iwr1 及各种组装辅助蛋白分离，通过核输出信号（nuclear export signal，NES）在转运蛋白出核因子 1（exportin 1，CRM1）的协助下离开细胞核进入细胞质中，参与下一轮 Pol Ⅱ的组装和转运（Czeko et al.，2011）。有研究表明，在酵母细胞中也存在其他转运途径与 Iwr1 依赖的途径协同作用，实现 Pol Ⅱ从细胞质到细胞核的稳定转运（Gomez-Navarro and Estruch，2015）。目前，其他 RNA 聚合酶组装和转运的研究仍有许多问题亟待阐明：RNA 聚合酶Ⅰ和 RNA 聚合酶Ⅲ是如何完成生物合成以及转运的？在 RNA 聚合酶生物合成及转运的过程中是否存在其他的中间形态？在此过程中是否还有其他辅助因子参与？这一过程是如何被调节的？RNA 聚合酶Ⅱ是如何与 RNA 聚合酶Ⅰ、RNA 聚合酶Ⅲ的组装与转运过程相互协调的（Wild and Cramer，2012）？

生物大分子机器结构和功能的复杂性以及提纯组装的困难等，极大地限制了人们对生物大分子机器定位与转运动态过程和机制的了解。至今，大量生物大分子机器的定位与转运过程仍然不清晰，何种因素导致了这些生物大分子机器的定向转移？生物大分子机器的转运具体是通过何种生物学机制来完成的？定位与转运过程涉及何种细胞成分？转运因子和生物大分子机器之间如何完成识别和选择？转运过程是通过何种能量来源驱使的？是否存在多种转运途径协同作用以完成高效的生物大分子机器的定向转运？这些问题仍然急需解答。除此以外，部分生物大分子机器也可以在不同的空间位置发挥相同的功能，如 DNA 修复蛋白等可以在含基因组的位点（细胞核、线粒体、叶绿体）以及细胞质中均发挥作用（Bauer et al.，2015），生物大分子机器的多处分布现象也越来越多地被发现。这些生物大分子机器的多处分布现象在遗传和进化过程中具有高度保守性，同时有证据表明这些生物大分子机器在不同细胞器中的分布并不均一。是什么因素决定了生物大分子机器的多处分布？又是什么因素在调控生物大分子机器的多处定位和转运？这些问题目前也都不清楚。

生物大分子机器作为重要的生命活动参与者，其一旦在细胞内遭遇错误定位或转运就可能引发一系列疾病，如G蛋白耦联受体家族视紫红质的错误定位会导致视网膜色素变性（Mendes et al.，2005）；抗利尿激素受体V2的错误定位会引发肾性尿崩症（Robben et al.，2006）；信号通路相关的生物大分子机器的错误定位会导致肿瘤的发生和转移，部分亲核生物大分子机器的错误定位会导致抑癌基因的表达失活进而导致癌症发生（Fabbro and Henderson，2003）；酶类生物大分子机器的错误定位可导致代谢相关疾病。部分生物大分子机器在错误的空间场所具有毒性作用，与神经退行性疾病的发生有重要联系。近年来，越来越多的证据表明，生物大分子机器的早期定位过程失调会导致阿尔茨海默病、亨廷顿病、帕金森病、肌萎缩侧索硬化等神经退行性疾病的发生。另外，亲核蛋白转运相关的转运复合物的定位障碍，使得核转运功能失调，也是导致阿尔茨海默病发生的重要因素（Hung and Link，2011）。

随着人们对众多疾病发病机制的研究和认识的深入，以及各种成像和定位技术的进步，靶向生物大分子机器的定位和转运过程也可能成为治疗人类疾病的一种有效方法。然而，靶向生物大分子机器定位过程的临床药物的发展和临床前研究目前还非常局限。高质量抗体的研发以及特异性定位的荧光染色技术等的发展对病理学相关的生物大分子机器的精确定位工作有重要推动作用（Hung and Link，2011）。生物大分子机器的区域定位或许能成为更多人类疾病预测及诊断的重要生物标志。鉴于肿瘤的发生越来越多地被认为是信号通路相关因素的可逆性错误定位的结果，靶向生物大分子机器定位的策略或许会对新型抗肿瘤治疗产生更加直接的影响（Butler and Overall，2009）。

四、生物大分子机器在体内的高度协同工作规律

生物大分子机器在某种程度上可以看成是一块块乐高积木，它们必须遵循一定的规律组合起来，而组合起来的模块又可以和其他的模块拼接起来，组成一个更大的几何体，具有独特的结构和功能。这使得生物大分子机器兼具灵活性和可塑性。生物大分子机器在体内通常并不“单打独斗”，多种生物大分子机器之间通常需要高度协同工作完成某一个特定生物学过程。比如转

录过程的起始，就需要 RNA 聚合酶Ⅱ、中介体复合物（mediator），以及通用转录因子复合物 TFⅡ-A、TFⅡ-B、TFⅡ-D、TFⅡ-E、TFⅡ-F 和 TFⅡ-H 等分子机器的共同参与，同时也需要染色质重塑复合物将转录起始位点的核小体去除以产生可供转录的裸露 DNA。这些生物大分子机器模块是如何被调控组装到一起完成转录起始的？它们之间又是如何高度协同作用而不是各自为战的？这些问题目前还没有明确的答案。破解生物大分子机器之间协同作用的机制，是生命科学研究的前沿。

一方面，生物大分子机器之间通过物理上的直接联系，形成一个功能性的超级复合物。这方面的代表性工作是对转录过程中形成的功能性转录机器的解析。例如，转录前起始复合物（pre-initiation complex，PIC）和中介体复合物组成的超级复合物，包含 46 条多肽，分子量高达 2M Da，帕特里克•克拉默（Patrick Cramer）研究组解析了其 5.8 Å 分辨率的冷冻电镜结构。这类功能性的超级复合物中不同生物大分子机器之间的联系通常是非常动态的，处于不断的结合－解离循环之中，高度动态性的结构给研究这类超级复合物带来了很大的挑战。虽然运用现在发展相当成熟的冷冻电镜解析方法，能够对不同状态的生物大分子机器进行区分，但对这类超级复合物的结构解析通常只能揭示其相对固定的核心结构，而对与功能调控相关的动态区域往往无能为力。剪接体就是一个典型的超大功能性复合体。目前科学家已经解析了剪接通路中各个主要状态的剪接体的高分辨率三维结构，为理解 RNA 剪接的分子机制提供了清晰且全面的结构信息（Wan et al.，2020）。但值得注意的是，所有解析的结构中分辨率最高的是明确且稳定的催化 U2/U6RNP 核心，其由 U5 snRNP 组分与包含 RNA 和蛋白质组分构成；其外周区域的分辨率较低，还有一些动态区域甚至完全没有电子密度。在最后的结构模型中，通常有 1/3 的蛋白质序列是缺失的。剪接体的高度动态性是实现其功能的关键，大量组分需要在剪接过程中调整与核心组分的结合强度和方式，而这些外周组分的低分辨率解析乃至缺失，在很大程度上影响了我们对剪接体完整工作机制的理解。因此，发展新的方法捕捉动态区域的结构，发展整合型结构生物学来揭示生物大分子机器的动态变化规律，将是今后结构生物学的主流发展方向。

另一方面，不同生物大分子机器可通过空间上的邻近关系来相互协同作用。细胞内存在大量的无膜区室，如核内的核仁将负责核糖体生成的生物大分子机器富集在一起有效地合成核糖体前体。尽管无膜细胞器广泛存在且具有重要功能，但目前人们对无膜细胞器的形成机制、识别机制和动态重塑机制了解甚少。目前“液–液相分离”理论的发展成熟为解释这些机制带来了曙光。生物大分子机器的液–液相分离可以说是当下最具突破性的研究领域，给人们提供了一个全新的视角来看待生物大分子机器的组织和调控形式。生物大分子机器相分离被发现广泛存在于细胞内，并参与驱动多种生物学过程。例如，组蛋白修饰通过促进相分离来调节异染色质形成和染色质区室化，共激活因子在超级增强子处的相分离可富集转录机器从而调控转录，相分离促进神经突触形成和信号传递，相分离在固有免疫信号传递过程中发挥功能等。近年来，蛋白质突变而导致相分离紊乱也用于解释多种人类重大疾病（如神经退行性疾病、肿瘤及自身免疫病等）的发病机制。但目前相分离理论还存在着很多尚待解决的技术瓶颈和科学问题，如相分离的检测手段发展、相分离的物理规律探寻、相分离的精确调控机制、相分离在各种疾病发生发展中的作用机制等。

五、生物大分子机器的原位和动态观测

在复杂多变的细胞环境内，生物大分子机器在正确的时间和正确的空间与底物或者相互作用因子结合，这对它们发挥生物功能来说至关重要。由于细胞样品过厚、样品倾转移动、低电子剂量导致的信噪比/衬度极低以及缺失锥（missing cone）等技术问题，目前应用 cryo-ET 并结合子断层扫描平均法对生物大分子机器进行结构解析通常还难以突破亚纳米分辨率水平。制约因素包括：生物大分子机器或细胞超微结构难以提纯及冷冻制样易碎，细胞减薄不易实现，目标大分子难以准确定位；低电子剂量导致图像数据衬度极低、高分辨率信息缺失，导致子断层扫描平均法的精确度和速度不高，需要大量的计算资源；准确的欠焦量确定和衬度传递函数（contrast transfer function，CTF）校正具有挑战性；cryo-ET 大量高质量的倾斜系列数据收集极其耗时等。近年来发展起来的螺旋相位板（vortex phase plate，VPP）相位板技术、直接电子探测技术、球差校正技术、应用低温聚焦离子束（cryo-FIB）对细胞/组织减薄技术，以及子断层扫描平均法计算策略上的优化，为突破上述限制

因素、实现大分子复合体的原位亚纳米分辨率三维重构和动态结构分析奠定了基础。

研究的进步使得 cryo-ET 能在亚纳米分辨率进行大分子机器原位结构测定的三个主要方向是：样品制备、数据收集和图像处理。新的样品制备方法能打破细菌和大型真核细胞冷冻电镜原位成像时的样品厚度限制（0.5 μm）。其中以低温聚焦离子束进行微加工，产生 150～250 nm 厚度的细胞薄层，能使我们观察真核细胞内的任意位置。在数据采集方面，直接电子探测器、相位板和能量过滤器的发展和应用大大增强了极低电子剂量下 cryo-ET 图像的信噪比，并且使用剂量对称方案进行系列倾斜数据收集可确保有限电子剂量的最优化利用。此外，可开发新的算法用于三维 CTF 矫正、缺失锥问题的弥补，通过人工智能高效区分并提取目标生物大分子机器或细胞器用于结构渲染和分析，提高 cryo-ET 倾转数据集的对中和重构精度及效率等，并将其整合到 cryo-ET 数据处理软件中。上述 cryo-ET 的技术进步将极大地推动我们对细胞“天然分子景观”进行高分辨率、动态性的完整结构描述，促进我们对生物大分子机器生物功能的深入理解。

低温光电联用对于使用低温聚焦离子束靶向感兴趣区域至关重要，但目前其精度限制在成像平面内几百纳米，在 *Z* 轴方向上则更差。进一步增强的超分辨率低温光电联用有潜力达到单分子水平的关联，将使 cryo-ET 中生物大分子机器的定位和识别成为可能。2014 年诺贝尔化学奖得主斯蒂芬•黑尔（Stefan Hell）结合受激发射损耗显微技术（stimulated emission depletion，STED）和光激活定位显微技术（photoactivated localization microscopy，PALM）技术原理进一步开发的最小光子流显微技术分辨率可以达到纳米级定位精度。但到目前为止，成功的超分辨率低温光电联用只能通过使用低温保护剂和 / 或外涂高聚物的网格处理，这对冷冻电镜来说都不是理想的样品条件。因此，迫切需要一种普适的解决方案，打破目前对样品制备的限制，使超分辨率荧光显微镜与冷冻电镜成像完全兼容。此外，我国科学家也在开发新型光转化荧光蛋白探针，并将其应用于新型活细胞超分辨率成像方法的开发。要在极其拥挤的细胞环境中对目标生物大分子机器进行准确定位，开发其他种类的探针技术也非常必要，如标签－特定抗体、金属结合蛋白、纳米金颗粒的可克隆电镜标记技术等，用于在冷冻电镜条件下实现细胞中的单分子识别和精确定位。

应用冷冻电镜技术可以捕获溶液中大分子的多种构象状态。构象状态的变化可以由多种生物学反应触发：将配体添加到酶中，将多组分大分子机器的成分混合在一起，或添加 ATP 或 GTP 形式的能量等。构象变化通常是短暂的，但可以在反应开始后的特定时间通过快速玻璃化（vitrification）样品，并使用电子显微镜成像捕获。将这种处理样品的方法应用于研究发生在毫秒级的构象变化时通常被称为"时间分辨的 cryo-EM"（time-resolved cryo-EM）。近期发展的 Spotiton 冷冻制样技术和设备，只需 50 pl 左右样品，可捕获 100 ms 时间尺度的生物反应，为发展时间分辨的 cryo-EM 冷冻制样技术奠定了基础。

推进 cryo-ET 技术的发展，探索稳定高效的样品制备技术和光电联合技术，高分辨率、高衬度的成像技术和高效的数据集采集技术，精确的目标生物大分子机器定位技术，同时发展准确高效的 cryo-ET 数据集三维重构、子断层扫描平均法、目标生物大分子机器的自动识别和相互作用网络分析及构象聚类的计算方法和软件，最终建立可对细胞"天然分子景观"进行原位、高分辨率、实时动态结构描述的核心手段，将推进我国在冷冻原位结构研究领域的发展，极大地促进我们对正常生理条件和病理条件下生物系统机制的理解。

六、生物大分子机器的模拟、操控和设计

分子动力学模拟在对生物大分子机器的物理和化学过程的理解以及科学技术的发展过程中发挥重要的作用。这种模拟方法通过数值求解相互作用粒子系统的牛顿运动方程来分析分子的物理运动。1977 年，对小蛋白质牛胰蛋白酶抑制剂（体系大约 400 个原子，58 个氨基酸）在真空中实现了不到 10 ps 的模拟，是世界上第一个蛋白质的分子动力学模拟（McCammon et al., 1977）。之后，分子动力学模拟在算法、软件和计算机硬件上都得到了长足的发展，且应用于广泛的科学技术领域。分子动力学模拟通过求解系统的牛顿运动方程来探索大分子的结构及能量表面，其结果的质量对大分子起始结构

模型和力场具有高度敏感性。40 多年的技术发展已使全原子动力学模拟的空间和时间尺度都比以前增长了几个数量级。现在常用的处理生物大分子模拟的动力学软件有 Amber、CHARMM、GROMACS、NAMD 和 Desmond。

构象变化对于生物系统中生物大分子机器的活性通常很重要，因为生物大分子机器在多种催化反应和跨膜运输过程中都必须经历大的构象变化。然而，大多数分子动力学模拟的轨迹并不是各态历经的。由于复杂的生物分子的自由能表面崎岖不平以及有限计算资源导致的对于生物尺度较短的模拟时长，人们可能需要增强的采样方法，而不是直接的无偏向性的动力学模拟。在过去的几十年中，研究人员已经为此目的开发了多种方法，其中使用比较多的有模拟退火、副本交换分子动力学和元动力学（Bernardi et al.，2015）。模拟退火方法类比冶金中的退火过程，最初在足够高的温度下进行采样，以使模拟可以轻松克服高自由能壁垒，然后随着模拟的进行温度降低，从而使模拟平稳地达到局部能量最小值。副本交换分子动力学采用并行独立模拟的副本在几个不同温度之间进行蒙特卡洛随机游走，根据所选模拟之间的温度和能量差，交换由原子位置定义的系统状态。元动力学则人为增加模拟系统在已搜索构象的自由能，当系统能量高于分隔两个最小值的势垒点时，系统将以新的最小值进入较低能量状态，再次搜索许多可能的构象。

生物大分子机器的主要构象变化通常发生在微秒至秒的时间范围内。现在动力学模拟的技术已经可以实现具有数万个原子的微秒级时长的模拟，并且毫秒级时标仿真也即将出现。长时长的模拟过程不仅得益于算法的提高，而且硬件的改进也为其提供了很大的帮助。大型超级计算机和动态负载平衡的并行化方案极大地提高了分子动力学模拟的效率。专门为分子动力学模拟而设计的第二代专用超级计算机 Anton 2，可以模拟 2 万多个原子的基准系统（二氢叶酸还原酶，大约 160 个氨基酸），达到每天 85 μs 的速率，比其他商用硬件平台或通用超级计算机快 180 倍（Shaw et al.，2014）。

现有的模拟技术已经可以应用于超规模的生物大分子机器和机器集团，包括核糖体和病毒，甚至细胞器。核糖体的分子动力学模拟系统有大约 250 万个原子，它的多种动力学过程和作为靶点与对药物的结合已经被成功模拟。整个病毒颗粒也已成为分子动力学模拟的系统，包括卫星烟草花叶病毒（约 100 万个原子，50 ns）、卫星烟草坏死病毒（约 120 万个原子，约 1 μs）、南方豆花叶病毒（约450万个原子，100 ns）、乙肝病毒（约600万个原子，1 μs）、脊髓灰质炎病毒（280万～400万个原子，50 ns）、人类免疫缺陷病毒（约6400万个原子，约200 ns）和流感病毒（约1.6亿个原子，约121 ns）（Perilla et al.，2015）。至今模拟的最大细胞器是能进行光合作用的细菌载色体（约1.36亿个原子，0.5 μs），且可清楚地看到模拟系统中的能量传输过程（Singharoy et al.，2019）。因此，分子动力学模拟能够将原子细节与超分子配合物的功能联系起来，这是较小规模的模拟或仅现有的实验方法无法实现的任务。

生物大分子机器的动力学模拟虽然在软件和硬件上已有巨大的进步，然而在分子力场上，特别是静电作用的分子力场上，还存在相当的不足。表示静电作用中的可极化力场，特别是生物中过渡金属的力场，几乎还没有得到开发（Jing et al.，2019）。可极化力场的成功开发可极大地帮助人们预测和设计具有高生物活性的金属蛋白酶。

能够在不同层面上对生物大分子机器进行调控和设计的能力极大地推动了生物医学的进步。基因调控在生命的形成、发育和衰老过程中至关重要。自20世纪60年代发现乳糖操纵子的基因调控机制以来，许多可调控的基因表达系统被设计了出来。这些技术和重组DNA技术结合，革命性地改变了生命科学的研究方法，并且推动了合成生物学的发展。以前基因调控的改造集中于低等生物中；近些年，通过对CRISPR-Cas系统调控机制的研究，以及对Cas9、Cas12a和Cas13蛋白的靶向性与可调控性等功能上的设计和改造，在高等动物和植物中已经可以轻易地实现基因工程。CRISPR-Cas的应用革命性地增强了人类研究生物学的能力，也使得很多遗传病的治疗成为可能（Pickar-Oliver and Gersbach，2019）。

荧光蛋白的应用极大地推动了生物学的发现进程。20世纪90年代荧光蛋白结构和机制的解析，使设计具有调控功能的荧光蛋白的技术日新月异。基于所有蛋白质结构组分结合起来才会发光的原理而设计的互补荧光蛋白可以研究复杂的蛋白质－蛋白质相互作用。同时，具有小分子或蛋白质结合功能的结构域已被设计与荧光蛋白组成融合蛋白用来感受各种生物刺激。设计的荧光蛋白也可以感受钙离子浓度、温度、pH、氧化还原电势和电压等环境因素（Zhang et al.，2002），并且可以通过改造的光学性质实现超分辨率荧光成像（Shcherbakova et al.，2014）。荧光蛋白结构和功能的设计也推动了生物发光蛋白相似功能的改造（Rowe et al.，2009）。

光遗传学和化学遗传学极大地推动了神经生物学的发展。光遗传学通过调控一种源自无脊椎动物的 GPCR 光敏蛋白质对外在光源的开关反应传输正离子或负离子，改变神经细胞的膜电压从而激活或抑制神经活动。经过 21 世纪初不断地探索和发现，光遗传学现在已可以达到毫秒级别的时间精确度和单个细胞级别的空间精确度。并且，根据光敏蛋白质的结构已经设计出可以调节开关时间和更高光敏感度的光敏蛋白质。这些蛋白质已经被用于研究大脑中各种神经环路的功能和疾病发生机制（Yizhar et al.，2011）。同时，光遗传学已展示出治疗嗜睡症、抑郁症、焦虑、疼痛、帕金森病和视网膜色素变性失明等神经类疾病的可能性（Chow et al.，2013）。化学遗传学通过设计改造 GPCR，使其只接受生物体内不存在的小分子的外界调控，以此来激活 G 蛋白耦联受体下游通路，从而达到可逆地靶向控制神经细胞活动的目的。与光遗传学相比，化学遗传学可以在数小时尺度上长时间控制神经活动。化学遗传学已在信号转导、药物开发和功能基因组学等研究中得到了广泛应用（Sternson and Roth，2014）。

细胞疗法将革命性地改变医学。用嵌合抗原受体（chimeric antigen receptor，CAR）改造的 T 细胞（CAR-T）在治疗晚期难治性 B 细胞恶性肿瘤方面已取得了显著成功。CAR 设计的基本概念是将细胞外配体识别域连接到细胞内信号转导模块，该模块通常包含 CD3 ζ 和共刺激信号 CD28 两基团以在抗原结合后诱导 T 细胞活化。此种设计理念基于对常规 T 细胞活化过程调控的长期研究和理解。常规 T 细胞活化需要 CD3 ζ 和 CD28 作为两种不同来源的信号共同发挥作用。在 CAR 中 CD3 ζ 和 CD28 作为顺式传递信号，导致 T 细胞活化。在新一代 CAR 设计中，研究人员不断地测试新的共刺激信号以完善 CAR-T 疗法（June et al.，2018）。基于功能性结构模块的设计概念也深刻地影响了其他种类细胞疗法中的蛋白质设计（Choe et al.，2020）。

综上所述，仿真模拟生物大分子机器的结构和功能、理解它们的调控机制，以及基于这些调控机制的生物大分子机器设计，极大地推动了生物学的发展，也为人类治疗疾病做出了重大的贡献。

第四节 我国发展战略与重点方向

一、发展总体思路与发展目标

在我国科技和经济飞速发展的时期，国民健康不仅是民生问题，也是重大的政治、经济和社会问题。党的十九大报告将“实施健康中国战略”作为国家发展基本方略中的重要内容，健康中国建设不仅直接关乎民生福祉，而且关乎国家全局与长远发展、社会稳定和经济可持续发展，具有重大的战略意义。人类的出生、发育、衰老等过程的基础是生物大分子机器在各不同时间段和特定区域精准高效地发挥相应功能。疾病的发生和发展过程与生物大分子机器的异常息息相关。

蛋白质、DNA、RNA、糖类、脂类等生物大分子是细胞生命活动的物质基础，只有这些大分子在特定的细胞环境下精确、快速地组装成生物大分子机器，并与众多生命有机分子共同形成超分子系统，进而组装成亚细胞结构、细胞器，才能最终形成具有物质交换和能量转换等新陈代谢能力的细胞。DNA 是超分子自组装的典型，其组装模式相对简单（遵循简单的碱基互补配对原理），经过核小体、30 nm 纤维、染色丝和染色质等多个层次的有序组装构成，基于现有的研究成果，科学家已经可以利用如 DNA 折纸术（DNA origami）自组装成一系列精巧的 DNA 纳米结构和机器。相较 DNA 而言，蛋白质组装过程中有着更为复杂的相互作用，还需要大量的实验数据、理论模型来解释蛋白质折叠和组装的规律，也需要生物学、化学、数学、物理学、材料学、计算机科学等学科的交叉融合。

对生物大分子机器研究的总体思路是围绕生物大分子机器的自组装、自适应、多样性、模块化、可调控等特性开展深入研究，探寻内在规律，建立和发展生物大分子机器特性研究相关的检测与分析的新技术和新方法，改进和推动已有的研究技术和平台，强调新理论、新技术、多学科的交叉融合，基于模型和实验数据进行相应的分子设计，寻找疾病的药物靶点，开发和设计用于疾病诊断、干预和治疗的方法与药物，从根源上寻求疾病的治疗和干预策略。

发展目标是阐明生物大分子机器研究领域的关键科学问题，在规律阐释、技术开发、分子设计等重要研究方向上取得突破性成果，推动相关研究领域的发展，促进相关生命健康研究进入新的阶段，提升我国在此研究领域的国际影响力，在部分重要研究方向上达到国际领先水平，更加深入地了解生命过程的本质，最大限度地保护人民的身体健康。

二、优先发展领域或重要研究方向

目前，我国对生物大分子机器的研究已经具备了良好的基础和技术平台，未来的发展将需要更多地强调不同学科和技术的交叉融合以及不同技术手段的综合使用，结合结构生物学技术、超高分辨率成像技术、质谱相关技术、光遗传学技术、化学遗传学技术等，对生物大分子机器在细胞内参与生命活动的整个动态过程进行追踪。此外，未来还需要关注从原子和分子水平、细胞水平、组织器官水平等不同尺度对生物大分子机器参与生命活动的研究。通过单分子超高分辨荧光成像技术在大尺度观察生物大分子机器的整体信息和动态变化，通过 X 射线衍射技术、冷冻电镜技术和核磁共振技术等获得局部的精细结构信息，通过质谱相关技术获得生物大分子机器表面特定氨基酸信息，通过单分子成像技术追踪生物大分子机器中单个分子动态变化信息，通过 cryo-ET 技术和荧光成像技术研究生物大分子机器在细胞水平的分子动态变化，通过光学显微镜对细胞和组织进行成像，通过计算机辅助 X 射线断层扫描来研究组织器官的形态，最终将所得信息进行整合以获得生物大分子机器在发挥生物作用过程中的组装、调控、运输等动态变化过程。

为更好、更快地推动我国在该研究领域的发展，提升我国在相关研究领域的国际影响力，建议优先发展以下三个研究方向。

1. 生物大分子机器的从头设计

随着对生物大分子机器的深入研究，我们越来越清晰地了解到这些生物大分子机器所发挥功能的多样性以及相应的工作机制，并发现它们具有自组装、自适应、多样性、模块化、可调控等特性。然而，如何利用这些特性和内在规律在参数空间上从头设计生物大分子机器是一个还没有触及的难题。现有的从头设计的蛋白质在结构和机制上比生物大分子机器要简单得多，同时现在生物大分子机器的设计通常只是对已有蛋白质结构的局部改造。能够在参数（几何参数或功能参数）空间上从头设计生物大分子机器不仅可以检验现有机制对生物大分子机器解释的正确性，而且可以在很大程度上拓展生物大分子机器功能上的多样性和提高可调控参数的操作性。基于生物物理学和生物化学机制上的第一性原理，在连续的几何参数空间中，设计在功能上具有连续可调控参数的生物大分子机器将是一个具有挑战性的全新研究。生物大分子机器自组装、模块化和可调控的特性可以通过蛋白质从头设计来实现。自组装结构的从头设计也完全可以通过蛋白质–蛋白质界面从头设计来完成（Zhang et al.，2018），并且从头设计的自组装结构因为与天然蛋白质相互作用界面不同，可以与生物原有的自组装的生物大分子机器互不干预，从而实现结构上的正交化。生物大分子机器具有多种多样的DNA序列或蛋白质结构域的结构和功能模块，而现在已有的从头设计的功能性模块为双组分正交的螺旋束和具有变构作用的蛋白开关（Kuhlman and Bradley，2019）。根据功能需要，自然界具有的结构和功能模块计从头设计的结构和功能模块的多样性组合可以极大地拓宽生物大分子机器的结构和功能空间。自然界存在的模块可以通过突变来调节自身功能的强度，从头设计的模块可以通过再设计实现功能的可调控性，因此生物大分子机器作为不同模块的组合，通过简单的构建模块的功能调节，可实现复杂的可编程的调节能力，以及所需要的理想生物信号的输入、输出功能。生物大分子机器的自适应性在从头设计层面上还没有涉及，现今蛋白质从头设计的目标多为设计刚性的非常稳定的结构集团，然而生物大分子机器的自适应性多来自于结构的柔性。由于目前对蛋白质的构象柔性与自适应性的关系还没有很好的了解，因此从头设计柔性的可调控功能的蛋白质是生物大分子机器设计中非常具有挑战性的一个课题。合成生物学在能源、制药和医疗方面显示出不可替代的重要性，而从头设计的生物大分子机器的应用深深植根于合成生物学。综上所述，基于第一性原理从头设计具有自组装、自适应、多样化、模块化、可调控等特性的生物大分子机器，不仅会极大地推动基础生物学的发展，还可以为国民经济发展和人民健康起到不可估量的作用。

2. 病毒与防御系统的研究

病毒与宿主细胞识别和相互作用机制的研究是分子生物学和生物医学领域的前沿热点，也是推动药物研发转化、保护人民健康的最重要的抓手。病毒与宿主细胞相互作用，完成病毒颗粒的附着、内吞、融合、基因组释放，然后在宿主细胞内完成复制、转录以及合成病毒蛋白，最后病毒组装为子代病毒释放并感染新的宿主细胞。这一系列过程涉及诸多生物大分子机器。目前，人们对这些生物大分子机器已经有了相当深入的结构和功能研究，但对于病毒入侵、组装和释放的具体动态过程仍缺乏分子水平的认识，病毒蛋白和宿主蛋白的相互作用以及它们在病毒生命过程中的动态调控机制还未得以完全阐释。除了对病毒本身的认识外，还存在很多尚待解决的问题，如宿主细胞是如何识别病毒入侵进而迅速触发抗病毒的天然免疫和适应性免疫应答的，而病毒又是如何逃避宿主免疫应答的。天然免疫模式识别受体分子如何识别相关病毒核酸分子而不识别自我核酸，并激活下游信号转导通路等问题，一直是免疫学领域研究的焦点和瓶颈。另外，病毒会主动对宿主细胞的某些蛋白质因子进行修饰而逃逸宿主细胞天然免疫反应，这一过程中的重要病毒蛋白的具体功能是什么，它们具有怎样的特殊结构以帮助它们实现这些功能，它们与哪些宿主分子共同作用来完成免疫逃逸，这些问题都没有得到很好的阐述。

3. 生物大分子机器原位结构研究技术和方法的建立

对生物大分子机器自组装、自适应、模块化和可调控特性的研究在很大程度上依赖于技术的革新和发展。全面推进 cryo-ET 技术在我国的发展，探索发展稳定高效的样品制备技术、光电联合技术、高分辨率高衬度的成像技术、高效的数据集采集技术和精确的目标分子定位技术，同时，发展准确高效的 cryo-ET 数据集三维重构技术、子断层扫描平均法、目标大分子机器的自动识别和相互作用网络分析技术，以及构象聚类的计算方法和软件，最终在我国建立可对细胞“天然分子景观”进行原位、高分辨率、实时动态结构描述的核心技术，将极大地促进我们对生理条件和病理条件下生物系统机制的理解。

本章参考文献

Alberti S, Gladfelter A, Mittag T. 2019. Considerations and challenges in studying liquid-liquid phase separation and biomolecular condensates. Cell, 176（3）: 419-434.

Ando T, Bhamidimarri S P, Brending N, et al. 2018. The 2018 correlative microscopy techniques roadmap. Journal of Physics D: Applied Physics, 51（44）: 443001.

Arimura S, Yamamoto J, Aida G P, et al. 2004. Frequent fusion and fission of plant mitochondria with unequal nucleoid distribution. Proceedings of the National Academy of Sciences of the United States of America, 101（20）: 7805-7808.

Banani S F, Lee H O, Hyman A A, et al. 2017. Biomolecular condensates: organizers of cellular biochemistry. Nature Reviews Molecular Cell Biology, 18（5）: 285-298.

Bauer N C, Doetsch P W, Corbett A H. 2015. Mechanisms regulating protein localization. Traffic, 16（10）: 1039-1061.

Benninger Y, Thurnherr T, Pereira J A, et al. 2007. Essential and distinct roles for cdc42 and rac1 in the regulation of Schwann cell biology during peripheral nervous system development. The Journal of cell biology, 177（6）: 1051-1061.

Bernardi R C, Melo M C R, Schulten K. 2015. Enhanced sampling techniques in molecular dynamics simulations of biological systems. Biochimica et Biophysica Acta, 1850（5）: 872-877.

Boulon S, Pradet-Balade B, Verheggen C, et al. 2010. HSP90 and its R2TP/Prefoldin-like cochaperone are involved in the cytoplasmic assembly of RNA polymerase Ⅱ. Molecular Cell, 39（6）: 912-924.

Butler G S, Overall C M. 2009. Proteomic identification of multitasking proteins in unexpected locations complicates drug targeting. Nature Reviews Drug Discovery, 8（12）: 935-948.

Cabianca D S, Gabellini D. 2010. The cell biology of disease: FSHD: copy number variations on the theme of muscular dystrophy. The Journal of Cell Biology, 191（6）: 1049-1060.

Chi B, Wang Q, Wu G, et al. 2013. Aly and THO are required for assembly of the human TREX complex and association of TREX components with the spliced mRNA. Nucleic Acids Research, 41（2）: 1294-1306.

Choe J H, Williams J Z, Lim W A. 2020. Engineering T cells to treat cancer: the convergence of immuno-oncology and synthetic biology. Annual Review of Cancer Biology, 4（1）: 121-139.

Chow B Y, Boyden E S. 2013. Optogenetics and translational medicine. Science Translational Medicine, 5（177）: 177ps.

Czeko E, Seizl M, Augsberger C, et al. 2011. Iwr1 directs RNA polymerase Ⅱ nuclear import. Molecular Cell, 42（2）: 261-266.

Dror R O, Dirks R M, Grossman J P, et al. 2012. Biomolecular simulation: a computational microscope for molecular biology. Annual Review of Biophysics, 41: 429-452.

Duan S, Li H, Zhang Y, et al. 2021. Rabl2 GTP hydrolysis licenses BBSome-mediated export to fine-tune ciliary signaling. The EMBO Journal, 40（2）: e105499.

Fabbro M, Henderson B R. 2003. Regulation of tumor suppressors by nuclear-cytoplasmic shuttling. Experimental Cell Research, 282（2）: 59-69.

Fung T S, Liu D X. 2019. Human coronavirus: host-pathogen interaction. Annual Review Microbiology, 73: 529-557.

Glynn J M, Froehlich J E, Osteryoung K W. 2008. Arabidopsis ARC6 coordinates the division machineries of the inner and outer chloroplast membranes through interaction with PDV2 in the intermembrane space. The Plant Cell, 20（9）: 2460-2470.

Gomez-Navarro N, Estruch F. 2015. Different pathways for the nuclear import of yeast RNA polymerase Ⅱ. Biochimica et biophysica acta, 1849（11）: 1354-1362.

Hause B, Fester T. 2005. Molecular and cell biology of arbuscular mycorrhizal symbiosis. Planta, 221（2）: 184-196.

Henry B. 2004. GPCR allosterism and accessory proteins: new insights into drug discovery, 17 July 2004, Glasgow, Scotland. IDrugs: the Investigational Drugs Journal, 7（9）: 819-821.

Howard J. 2014. Quantitative cell biology: the essential role of theory. Molecular Biology of the Cell, 25（22）: 3438-3440.

Hung M C, Link W. 2011. Protein localization in disease and therapy. Journal of Cell Science, 124（Pt 20）: 3381-3392.

Jacquemet G, Carisey A F, Hamidi H, et al. 2020. The cell biologist's guide to super-resolution microscopy. Journal of Cell Science, 133（11）: jcs240713.

Jensen O N. 2004. Modification-specific proteomics: characterization of post-translational modifications by mass spectrometry. Current Opinion in Chemical Biology, 8（1）: 33-41.

Jin M, Liu C, Han W, et al. 2019. TRiC/CCT Chaperonin: Structure and Function. Sub-cellular biochemistry, 93: 625-654.

Jin M, Liu X, Jia W, et al. 2018. *ZmCOL3*, a CCT gene represses flowering in maize by interfering with the circadian clock and activating expression of *ZmCCT*. Journal of integrative plant biology, 60（6）: 465-480.

Jing Z, Liu C, Cheng S Y, et al. 2019. Polarizable force fields for biomolecular simulations: recent advances and applications. Annual Review of Biophysics, 48: 371-394.

June C H, O'Connor R S, Kawalekar O U, et al. 2018. CAR T cell immunotherapy for human cancer. Science, 359（6382）: 1361-1365.

Kartsogiannis V, Ng K W. 2004. Cell lines and primary cell cultures in the study of bone cell biology. Molecular and Cellular Endocrinology, 228（1-2）: 79-102.

Kiatpongsan S, Tannirandorn Y, Virutamasen P. 2006. Introduction to stem cell medicine. Journal of the Medical Association of Thailand = Chotmaihet thangphaet, 89（1）: 111-117.

Klink B U, Gatsogiannis C, Hofnagel O, et al. 2020. Structure of the human BBSome core complex. eLife, 9: e53910.

Kuhlman B, Bradley P. 2019. Advances in protein structure prediction and design. Nature Reviews Molecular Cell Biology, 20（11）: 681-697.

Li H, Vaughan J C. 2018. Switchable fluorophores for single-molecule localization microscopy. Chemical Reviews, 118（18）: 9412-9454.

Mahamid J, Schampers R, Persoon H, et al. 2015. A focused ion beam milling and lift-out approach for site-specific preparation of frozen-hydrated lamellas from multicellular organisms. Journal of Structural Biolology, 192（2）: 262-269.

McCammon J A, Gelin B R, Karplus M. 1977. Dynamics of folded proteins. Nature, 267（5612）: 585-590.

Mendes H F, van der Spuy J, Chapple J P, et al. 2005. Mechanisms of cell death in rhodopsin retinitis pigmentosa: implications for therapy. Trends in Molecular Medicine, 11（4）: 177-185.

Monti M. 2020. Essential Current Concepts in Stem Cell Biology. European Journal of Histochemistry, 64（4）: 3201.

Peeters M C, van Westen G J, Guo D, et al. 2011. GPCR structure and activation: an essential role for the first extracellular loop in activating the adenosine A2B receptor. FASEB journal: Official Publication of the Federation of American Societies for Experimental Biology, 25（2）: 632-643.

Perilla J R, Goh B C, Cassidy C K, et al. 2015. Molecular dynamics simulations of large macromolecular complexes. Current Opinion in Structural Biology, 31: 64-74.

Pickar-Oliver A, Gersbach C A. 2019. The next generation of CRISPR-Cas technologies and applications. Nature Reviews Molecular Cell Biology, 20（8）: 490-507.

Praefcke G J, McMahon H T. 2004. The dynamin superfamily: universal membrane tubulation and fission molecules? Nature Reviews Molecular Cell Biology, 5（2）: 133-147.

Puhringer T, Hohmann U, Fin L, et al. 2020. Structure of the human core transcription-export complex reveals a hub for multivalent interactions. eLife, 9: e61503.

Reimer J M, Haque A S, Tarry M J, et al. 2018. Piecing together nonribosomal peptide synthesis. Current Opinion in Structural Biology, 49: 104-113.

Robben J H, Knoers N V, Deen P M. 2006. Cell biological aspects of the vasopressin type-2 receptor and aquaporin 2 water channel in nephrogenic diabetes insipidus. American Journal of Physiology. Renal Physiology, 291（2）: F257-F270.

Rout M P, Sali A. 2019. Principles for integrative structural biology studies. Cell, 177（6）: 1384-1403.

Rowe L, Dikici E, Daunert S. 2009. Engineering bioluminescent proteins: expanding their analytical potential. Analytical Chemistry, 81（21）: 8662-8668.

Shaw D E, Grossman J P, Bank J A, et al. 2014. Anton 2: Raising the bar for performance and programmability in a special-purpose molecular dynamics supercomputer. International Conference for High Performance Computing, Networking, Storage and Analysis: 41-53.

Shcherbakova D M, Sengupta P, Lippincott-Schwartz J, et al. 2014. Photocontrollable fluorescent proteins for superresolution imaging. Annual Review of Biophysics, 43: 303-329.

Shi Z, Gao H, Bai X C, et al. 2020. Cryo-EM structure of the human cohesin-NIPBL-DNA complex. Science, 368（6498）: 1454-1459.

Singh M, Garrison J E, Wang K, et al. 2019. Absence of BBSome function leads to astrocyte reactivity in the brain. Molecular brain, 12（1）: 48.

Singharoy A, Maffeo C, Delgado-Magnero K H, et al. 2019. Atoms to phenotypes: molecular design principles of cellular energy metabolism. Cell, 179（5）: 1098-1111, e1023.

Soffers J H M, Workman J L. 2020. The SAGA chromatin-modifying complex: the sum of its parts is greater than the whole. Genes Development, 34（19-20）: 1287-1303.

Sternson S M, Roth B L. 2014. Chemogenetic tools to interrogate brain functions. Annual Review of Neuroscience, 37: 387-407.

Takahara M, Kunii M, Nakamura K, et al. 2019. C11ORF74 interacts with the IFT-A complex and participates in ciliary BBSome localization. Journal of biochemistry, 165（3）: 257-267.

Virshup D M, Kaldis P. 2010. Cell biology. Enforcing the Greatwall in mitosis. Science, 330（6011）: 1638-1639.

Walters J R. 2004. Cell and molecular biology of the small intestine: new insights into differentiation, growth and repair. Current Opinion in Gastroenterology, 20（2）: 70-76.

Wan R, Bai R, Zhan X, et al. 2020. How is precursor messenger RNA spliced by the spliceosome? Annual Review Biochemistry, 89: 333-358.

Wan W, Briggs J A. 2016. Cryo-electron tomography and subtomogram averaging. Methods in Enzymology, 579: 329-367.

Wild T, Cramer P. 2012. Biogenesis of multisubunit RNA polymerases. Trends in Biochemical Sciences, 37（3）: 99-105.

Xiao X, Wang P, Chou K C. 2011. GPCR-2L: predicting G protein-coupled receptors and their types by hybridizing two different modes of pseudo amino acid compositions. Molecular BioSystems, 7（3）: 911-919.

Yatskevich S, Rhodes J, Nasmyth K. 2019. Organization of chromosomal DNA by SMC complexes. Annual Review of Genetics, 53: 445-482.

Yizhar O, Fenno L E, Davidson T J, et al. 2011. Optogenetics in neural systems. Neuron, 71（1）: 9-34.

Zeng F, Hua Y, Liu X, et al. 2018. Gpn2 and Rba50 directly participate in the assembly of the Rpb3 subcomplex in the biogenesis of RNA polymerase Ⅱ. Molecular and Cellular Biology, 38（13）: e00091-18.

Zhang J, Campbell R E, Ting A Y, et al. 2002. Creating new fluorescent probes for cell biology. Nature Reviews Molecular Cell Biology, 3（12）: 906-918.

Zhang P J. 2019. Advances in cryo-electron tomography and subtomogram averaging and classification. Current Opinion in Structural Biology, 58: 249-258.

Zhang S Q, Huang H, Yang J, et al. 2018. Designed peptides that assemble into cross-alpha amyloid-like structures. Nature Chemical Biology, 14（9）: 870-875.

Zhao S, Zhang X, Li H. 2018. Beyond histone acetylation-writing and erasing histone acylations. Current opinion in structural biology, 53: 169-177.

第六章

亚细胞结构的形成与相互作用

第一节　概　　述

真核细胞的一切生命活动都是通过亚细胞结构（包括各种膜包裹的细胞器如内质网、溶酶体、线粒体等），在特定的空间有序高效地介导进行的。每种细胞器均有其特化的功能，但同时又相互作用，通过相互协调来完成一系列重要的生理功能。细胞器的精细分工、相互协作和密切接触，形成细胞器互作网络，实现快速的物质交换和信息交流，执行不同条件下细胞生命活动的多种生物学过程。细胞器互作网络的紊乱与多种疾病的发生发展密切相关。然而，目前我们对细胞器互作的方式、机制和功能尚知之甚少。

内膜系统是真核细胞中功能密切关联的一系列膜结构，包括内质网、高尔基体、溶酶体、内吞体、自噬体和质膜等，主要介导细胞内的物质合成、分选、转运及降解等。以前的研究认为，内膜系统的物质交换和功能维系主要依赖于经典的囊泡介导的膜转运。近年来新的研究表明，内膜系统的细胞器之间也通过膜接触的形式实现相互作用，这种方式的协同互作在维持内膜系统结构和功能的稳态中起到重要作用。未来，内膜系统研究将从单个的细胞器研究转向多个细胞器的集成研究，将更多地聚焦内膜系统协同互作的精细调控机制和生理功能及意义；内膜系统膜结构的完整性维持、修复与降解也将成为新的研究热点和前沿。同时，内膜系统紊乱的病理相关性研究，尤其在发育缺陷和神经退行性疾病的发生发展等方面，将受到持续关注。

除了膜性细胞器，细胞内还存在大量的无膜细胞器——生物大分子凝聚体，在不同的生理过程中发挥重要的功能（Banani et al.，2017；Boeynaems et al.，2018；Kaganovich，2017；Shin and Brangwynne，2017；Zhang et al.，2020a）。近年来，科学家发现无膜细胞器主要通过液－液相分离形成，并进一步通过“相变”形成具有不同物理性质的结构如液晶态、凝胶态及固态等结构而发挥功能。生物大分子“相分离”和“相变”已经成为基础生命科学研究中最为前沿和重要的研究领域之一。此外，生物大分子凝聚体的动态调控还与多种人类重大疾病（如神经退行性疾病、肿瘤、糖尿病等）的发生发展密切相关，是极其重要的疾病治疗的潜在靶点。因此，对生物大分子“相分离”“相变”领域的研究，是关系到前沿基础研究发展的重大战略需求。

细胞核是细胞遗传与代谢的调控中心，繁杂的细胞核成分高度组织化，形成了众多形状各异和功能不同的细胞核无膜细胞器核体（nuclear body，NB），如核仁、旁斑等（Mao et al.，2011；Stanek and Fox，2017）。这些细胞核无膜细胞器是细胞遗传与代谢功能发生的主要场所，保障了细胞生理活动的正常进行，其结构或功能的失调与几十种人类疾病相关（Morimoto and Boerkoel，2013）。针对细胞核无膜细胞器的研究主要围绕其精细的结构、从头组装的过程和多样性的生物学功能展开，涉及显微成像技术，高精度且高特异的分离、纯化及相应的生物学分析技术，系统且多样的生物学模型研究等。该领域相关研究进展有助于补充细胞核内无膜细胞器的相关理论基础，并推动长非编码 RNA 代谢与功能、相分离调控以及高分辨率纤维成像的研发等前沿基础性研究。

蛋白质稳态维持与动态变化对细胞生命活动至关重要，在特定空间下细胞内单个蛋白质或整个蛋白质组的稳态水平取决于其合成与降解过程之间的动态平衡。蛋白酶体（proteasome）是细胞内最重要的蛋白质降解场所，人体中存在 26S、PA28 和 PA200 三种不同类型的蛋白酶体，且每个类型均是由几十个亚基组成的酶复合物，以严格受控的方式调控降解大多数细胞蛋白质，进而参与多种重要的生命过程，包括细胞周期、细胞凋亡、细胞应激、免疫反应等。因此，蛋白酶体是多种疾病的重要治疗靶点，而其本身的功能异常又与多种疾病密切相关，如肿瘤、神经退行性疾病等。

细胞的活动需要线粒体提供能量。线粒体呼吸链活动中不可避免产生的活性氧自由基，是细胞自由基的主要来源之一（Zorov et al.，2014）；同时，线粒体是细胞凋亡的调控中心，通过其释放的细胞色素 c 激活胱天蛋白酶（caspase）促进细胞凋亡的发生（Liu et al.，1996）。线粒体也是细胞内炎症反应与抗病毒的重要中转站。线粒体不断进行分裂和融合，产生新的线粒体，也通过自噬清除多余线粒体，以维持线粒体数量的动态平衡（Gui et al.，2019；Youle，2019）。同时，线粒体还不断与细胞质和细胞核进行物质和信息的交换，以维持线粒体稳态和调控细胞活动（Shao et al.，2020）。可见，线粒体在细胞代谢、氧化应激、信号转导和细胞凋亡调控等诸多生理过程中具有重要作用，损伤线粒体的积累和稳态异常与衰老及相关疾病的发生密切相关。

细胞骨架是细胞内的纤维状网架结构，主要包括微管、微丝和中间丝三种组分。微管由微管蛋白亚基组成，微丝由肌动蛋白组成，不同种类的中间丝则是由不同的中间丝蛋白组成。细胞骨架具有动态性，参与细胞的形态维持、细胞迁移、细胞分裂、胞内物质运输、信号转导、组织和个体发育甚至病原体感染等生物学过程。细胞骨架异常与多种疾病的发生发展密切相关。例如，微管功能异常与肿瘤发生和转移、神经系统发育缺陷以及退行等疾病有关；微丝功能异常与听力损害、免疫失调和发育畸形等疾病有关；中间丝异常与皮肤、心肌、神经退行和感染等疾病相关。此外，微管细胞骨架也是抗肿瘤药的重要靶点。因此，细胞骨架研究不仅可以加深对其生理功能的理解，而且可促进对疾病发生机理的认识，并为疾病治疗提供理论基础和新策略。

细胞膜的基本作用是维护细胞内微环境的相对稳定，并参与同外界环境进行物质交换、能量和信息传递。细胞膜的破裂导致细胞坏死，表现为细胞胀大、细胞内容物外溢。细胞膜的破裂诱导因子可以归为两类，一类是细胞外毒素；另一类是细胞内源蛋白。细胞膜的修复是与细胞膜裂解相抗衡的一个重要的生物学过程。为了维持细胞内稳态、防止细胞死亡，真核细胞进化出强大的膜修复能力，主要包括修复斑、膜张力调节、胞吐－内吞循环和内吞体分选转运复合体（endosomal sorting complex required for transport，ESCRT）介导的出芽4种方式（Andrews and Corrotte，2018）。细胞质膜裂解及修复不仅涉及质膜的动力学变化、膜脂分子的重塑和蛋白质与膜脂的相互作用等基础生化理论问题，膜裂解及修复的动态平衡被打破后引起的细胞坏死又涉及微生物感染、免疫应答、肿瘤发生发展、组织损伤修复和神经退行性病变等重大疾病。膜裂解及修复的深入研究不仅能够更好地理解蛋白质－膜相互作用的基础理论，还对相关疾病的预防和治疗具有重要的临床意义。

亚细胞结构的研究依赖于显微镜技术的发展，光学显微镜和电子显微镜是获得生物样品内部超微结构及其在特定生物过程中动态变化等重要生物学信息不可或缺的技术方法。近年来发展的超分辨光学显微镜和高通量三维电镜等技术极大促进了生命科学研究中的亚细胞结构和功能的解析，成为当今生命科学进步的重要驱动力。

第二节　关键科学问题

虽然我国细胞器互作研究取得了一定成绩，但是我们也要清醒地认识到，目前我国的研究距世界顶尖水平还存在一定的差距，亟待解决的问题还很多。

一、内膜系统细胞器的结构功能与相互作用

该研究方向的关键科学问题包括：内膜系统细胞器如何通过调控其形态来维持功能；内膜系统细胞器之间如何交互调控发挥协同功能；内膜细胞器如何感应和维持其膜结构的完整性，如何在破损后实现修补或必要的降解，膜完整性的生理相关性又如何体现；一些单基因的突变是如何通过内膜系统

紊乱来导致人类疾病的。

二、无膜细胞器形成和动态平衡的调控机制

生物大分子凝聚体相分离和相变的生物学功能、形成的理化机理及其动态调控机制；生物大分子凝聚体与疾病发生发展的关系，发现疾病相关的全新治疗靶点，并开发相应的化学、基因干预等诊断治疗手段，同时开发一系列用于生物大分子“相分离”“相变”研究的新方法和新技术。

三、核内无膜细胞器的结构、组装和功能调控

细胞核无膜细胞器的功能与其正确的结构和组装密切耦联，因此，系统探索不同细胞核无膜细胞器的精细结构以及相应的组装过程是首要解决的科学问题。此外，随着研究的深入，人们发现不同的细胞核无膜细胞器之间存在结构、数量、定位分布等差异，同时各细胞核无膜细胞器具有细胞、组织、个体和物种的特异性。因此，构建全面的高精度的无膜细胞器分布是当前研究者面临的另一重大挑战。

四、不同类型蛋白酶体及其亚基间的相互作用与靶向干预

蛋白酶体是最重要的蛋白质降解场所，但不同类型蛋白酶体及其亚基间的相互作用机制及生理病理意义仍不清楚。蛋白酶体途径和细胞内另一个重要的降解途径自噬可交叉调控，其机制也有待阐明。泛素化是蛋白酶体最主要类型 26S 蛋白酶体底物识别的标签，但 PA200 蛋白酶体却不能识别泛素化底物，是否存在其他蛋白质修饰可能影响蛋白酶体底物识别和不同类型蛋白酶体及其亚基间的相互作用。靶向干预不同类型蛋白酶体或亚基可能在治疗肿瘤或其他疾病中发挥重要作用，新的特定类型蛋白酶体或亚基抑制剂的开发是未来的重要研究方向。

五、线粒体稳态调控机制及其在衰老和相关疾病中的作用

线粒体生物发生需要线粒体和细胞核的精密调控，深入了解线粒体和细胞核互作的分子机制，对于解析生命奥秘至关重要。受损伤线粒体和多余的线粒体通过线粒体自噬机制及时清除，以维持线粒体数目和质量稳态。线粒体蛋白稳态异常和功能异常蛋白的积累导致细胞衰老与凋亡，并与多种衰老相关疾病的发生密切相关。深入了解线粒体蛋白和细胞器水平的稳态调控，分析线粒体与细胞核、细胞质互作的分子细节，阐明其在疾病发生中的作用。为此，需要解决一些关键技术问题：①线粒体蛋白复合体装配的实时动态观察；②线粒体基因组编辑与线粒体 DNA 疾病治疗；③活体水平检测线粒体动态和功能状态。

六、细胞骨架的动态调控及其与生理功能和疾病的相关性

微管在发挥广泛的生物学功能中呈现高度的动态性，因此需要深入研究微管的动态调控机制，解析微管的动态变化与不同生理和病理过程的关系，开发针对微管的高效、低副作用抗肿瘤药。微丝结构与细胞功能和疾病之间的联系是本领域的关键科学问题，细胞质微丝对细胞器的调控、细胞核微丝对转录的调节以及微丝在心血管疾病和肿瘤转移中的影响已成为前沿领域。中间丝在物理化学刺激、病原体感染等应激条件下具有高度敏感性，需要全面研究中间丝变化与不同病理过程的关系，阐明中间丝的动态变化规律及与微管和微丝互作的调控机制，重点解析病原体与宿主中间丝的交互应答和利用策略，开发针对中间丝的抗感染药。由于细胞骨架的微小性和动态性的特点，实时观测分析和操控骨架动态性是该领域急需解决的技术问题，超分辨活细胞成像结合细胞骨架特异性标记探针的发展是探索本领域的密钥。

七、细胞形态形成和维持的调控机制

形态多样性是细胞作为构建复杂生命的单元的重要基础。细胞形态既与

细胞类型有关，又有较强的可塑性。前者是表观遗传调控的结果，后者则是细胞内部结构自主调控及其与环境相互作用的表现。虽然我们对调控细胞分化、形态变化的分子机理已有较为深刻的了解，但对使不同类型细胞形成其标志性形态的控制程序及其作用机理却缺乏认识。揭示参与各种类型细胞形态发生、维持和调节的分子及其作用机理，完善对相关调控网络、形态－功能关系和环境影响的认识，逐步建立理论模型（数字细胞）模拟、预测细胞的形貌变化，实现细胞形态的工程化理性设计和改造，是在今后十五年或更长时间里重要的科学问题。

八、细胞膜结构动态维持、裂解及修复

该领域仍然存在许多亟待解决的问题：各种方式膜破裂的动态过程是如何被调控的？膜脂分子，如胆固醇、磷脂酰肌醇、鞘磷脂、脂筏等，是如何参与膜破裂的？几种以膜裂解为标志的细胞坏死，如细胞坏死性凋亡（necroptosis）、细胞焦亡（pyroptosis）等，其破膜机制有何异同？与之相对应的膜修复过程又有何异同？不同形式的膜破裂造成的生理病理后果有何异同？细胞膜的支撑离不开细胞骨架，细胞骨架处于不断地解聚和聚合的过程中。而细胞骨架的动态组装又是如何参与膜的破裂和修复的？这一系列问题的解决有助于我们对细胞膜结构的完整性的保持及维护建立全面而深刻的认识。

九、亚细胞结构在体高时空分辨解析技术

多细胞生物内部的光学非均匀性大大增加，引起激发光和信号光波前畸变，导致超分辨成像技术所依赖的物理效应在多细胞生物中不能完美实现。因此，开发对生物样品的实时像差矫正技术，实现复杂样本的大视场超分辨成像是当前光学成像领域的难点和热点。如何将超分辨成像技术特别是适于活细胞成像的结构光超分辨成像技术与光片成像技术融合，开发超分辨光片显微镜是显微成像领域的前沿技术问题。这一技术突破可使两者优缺点互补，

将超分辨成像由单细胞层次提升到多细胞模式生物层次，为当前生命科学研究的诸多前沿问题带来新的观测技术方法。因此，实现在体高时空分辨解析亚细胞结构，既是今后生命科学研究的重要需求，同时也是显微成像技术领域的发展趋势。

第三节　重要研究方向的发展现状与趋势分析

围绕细胞器研究的国际前沿领域，结合我国的优势，我们提出了亚细胞结构未来发展的 9 个方向。

一、内膜系统细胞器的结构功能与相互作用

1. 研究背景

内膜系统由功能高度关联的内质网、高尔基体、溶酶体、内吞体等细胞器组成，在细胞的分泌等生命活动中扮演关键角色。重要的是，近年来的国内外前沿研究开启了内膜系统领域的新篇章。

内质网是由管状和片状等形态组合成的连续膜系统（Shibata et al., 2006），是蛋白质和脂质合成的重要场所。内质网中储存了高浓度的钙离子，其低氧化还原电势为蛋白质的糖基化、氧化折叠等修饰和加工提供了必要的条件。由于形态复杂、功能多样，内质网成为生物膜结构与功能关联研究的重要对象。另外，内质网的稳态调控也一直是细胞生物学研究的热点。内质网应激及其诱发的未折叠蛋白响应方面持续有新的研究发现（Walter and Ron, 2011）。内质网与多种其他细胞器也存在着广泛的互作（Zhang and Hu，2016）。

高尔基体是由分为顺式、中间、反式等区域的扁平囊堆叠形成的结构，主要实现如糖基化、磷酸化、蛋白质水解等分泌蛋白的加工功能（Farquhar and Palade，1998）。反式高尔基体是蛋白质分选的重要场所，其分选能力在

极性细胞中尤为明显（Guo et al.，2014b）。研究表明，高尔基体不仅是物质转运的重要枢纽，也在细胞信号传递过程中发挥重要作用。

溶酶体是由单层膜包被的细胞器，其内部含有大约 60 种水解酶，其在酸性条件下协同作用，降解几乎所有类型的大分子物质。在传统意义上，溶酶体仅被认为是细胞的“垃圾处理站”，负责将新陈代谢所产生的废物降解，并将降解产物转运到细胞质中循环使用。近年来的研究显示，研究人员对溶酶体的作用认知逐渐从“垃圾处理站”转为“信号中心”。这一已被广泛认同，也使得溶酶体重新成为细胞生物学研究的热点（Lawrence and Zoncu，2019）。

内吞体介导的跨膜转运是细胞与环境之间物质交换和信号传递的重要途径（Scott et al.，2014），可分为网格蛋白介导和非网格蛋白介导两种，而后者又包括陷窝蛋白介导的内吞、吞噬和巨胞饮等途径（Mayor and Pagano，2007）。内吞体经过管化、分裂、融合等一系列膜动态变化，逐渐向细胞中心移动，酸化并成熟。内吞的组分或被溶酶体降解，或被回收利用。胞吞过程在营养摄取、抗原呈递、信号转导、突触调节、极性建立等多种生理过程中具有重要作用（Thottacherry et al.，2019）。

此前研究认为，内膜系统这些细胞器依赖膜转运进行物质交换（Lee et al.，2004），内质网通过包被蛋白Ⅱ（coat protein II，COP Ⅱ）衣被小泡向高尔基体转运物质，高尔基体则通过包被蛋白Ⅰ（COP Ⅰ）衣被小泡实现与内质网和膜囊之间的交换，在反式高尔基体和质膜，网格蛋白介导的运输起主导作用。然而，内膜细胞器之间也可以通过膜接触进行物质交换和信号同步。这种多元化的协同为内膜系统的相关研究打开了新的局面。

2. 国际研究现状与趋势

目前，国际上对内质网的研究主要聚焦于形态与功能的关联、内质网稳态调控、内质网参与的细胞器互作等几个方向。哈佛大学医学院的 Rapoport 实验室开创了管状和片状内质网塑形的新研究领域（Powers et al.，2017）；加利福尼亚大学的 Walter 实验室主导了内质网应激新通路的鉴定（Costa-Mattioli and Walter，2020）；加利福尼亚大学 Dillin 实验室对细胞非自主未折叠蛋白响应机制进行了探索（Frakes et al.，2020）；剑桥大学 Ron 实验室发现了内质网应激的早期应答途径（Preissler and Ron，2019）；科罗拉多大学的 Voeltz

实验室观测了内质网与线粒体和内吞体的相互作用等（Wu et al.，2018）；耶鲁大学的 De Camilli 实验室等对内质网介导的细胞器互作功能进行了初步探索（Saheki and De Camilli，2017）。针对高尔基体的最新研究则通常与膜转运等通路相结合。西班牙基因组研究中心的 Malhotra 实验室等聚焦高尔基体介导的蛋白质分选（Malhotra et al.，2015），密歇根大学的 Wang 实验室鉴定了高尔基体囊膜拴连的分子机制和调控方式（Ahat and Wang，2019）。然而，内质网和高尔基体的塑形特征与其特异性功能之间的内在联系还未有系统性的报道，内质网和高尔基体协同调控蛋白质稳态的新研究方向也未有尝试。

长期以来，溶酶体只是作为细胞的“垃圾处理站”被熟知。近期，Hall 实验室发现雷帕霉素靶蛋白 1（target of rapamycin 1，TOR1）可定位于酵母液泡膜上（等同于溶酶体）（Sturgill et al.，2008）。Sabatini 实验室报道，氨基酸可诱导 mTOR 定位至溶酶体，并在溶酶体膜被激活，证明 mTOR 向溶酶体膜的募集是调控细胞生长及增殖的关键步骤（Sancak et al.，2008）。自此，溶酶体成为细胞感知环境变化，特别是营养条件变化，并协调应答反应的“信号中心”（Lawrence and Zoncu，2019）。

早期的溶酶体发生研究聚焦在溶酶体相关蛋白的运送过程。Ballabio 研究组发现了转录因子 EB（transfer factor EB，TFEB）及其在溶酶体发生中的关键作用，揭示了溶酶体及自噬的转录调控机制（Sardiello et al.，2009）。此后，TFEB 相关研究成为溶酶体及自噬领域的热点。另外，Lenardo 实验室报道了溶酶体通过自噬溶酶体的再生过程，解释了营养匮乏条件下细胞启动自噬，利用溶酶体降解提供营养，并最终协调溶酶体再生以补充消耗的溶酶体的完整过程（Yu et al.，2010）。溶酶体是具有高度动态变化的细胞器，可沿微管运动，与多种囊泡融合、分裂，溶酶体在细胞中分布的变化与其功能密切相关，包括分泌、抗原呈递、胁迫应答、细胞膜修复等。溶酶体分布的调控在近期被广泛研究，主要通过小 G 蛋白和微管马达蛋白质的协同调控实现（Ballabio and Bonifacino，2020）。

溶酶体膜的完整性对其功能及细胞存活至关重要。早期研究主要聚焦依赖溶酶体的细胞死亡，而目前相关研究主要利用靶向溶酶体的化合物造成膜损伤，发现自噬途径可清除损伤溶酶体，而 ESCRT 复合物可修复轻微损伤的溶酶体（Daussy and Wodrich，2020）。病原菌和蛋白聚集体也可以造成内吞

体 / 溶酶体损伤，但机制并不清楚（Daussy and Wodrich，2020）。另外，在生理及病理条件下，溶酶体损伤是否发生以及如何发生，机体和细胞如何应对均不明晰。

与溶酶体密切关联的内吞体也具有独特的发生机制和特定的生理功能。哈佛大学 Kirchhausen 研究组、耶鲁大学 De Camilli 研究组和加利福尼亚大学 Drubin 研究组鉴定了多种参与网格蛋白组装的元件（Doyon et al.，2011；Ferguson et al.，2007；Wang et al.，2017a）；多伦多大学 Grinstein 研究组和密歇根大学 Swanson 研究组解析了吞噬与巨胞饮过程的分子机制（Freeman et al.，2020；Yoshida et al.，2015）；斯隆 - 凯特琳癌症研究所 Thompson 研究组和纽约大学 Bar-Sagi 研究组发现了巨胞饮在肿瘤细胞生长中的重要作用（Commisso et al.，2013；Palm et al.，2015）。

近年来，细胞自噬领域在国内外成为研究热点。大隅良典（Yoshinori Ohsumi）因在细胞自噬方面的发现于 2016 年获得诺贝尔生理学或医学奖。细胞自噬过程同样涉及了多个细胞器的共同参与：多细胞器间协同进行的自噬体膜构建、转运与融合过程，线粒体和内质网的直接接触位点对于细胞自噬的发生有重要作用，自噬体与溶酶体融合形成自噬性溶酶体，自噬性溶酶体再生等。

内膜系统之间的膜转运与膜接触的联动方式在国际上已形成成熟的研究体系和研究团队。不过，膜转运的特异性调控，尤其是不同衣被小泡的货物分选细节此前未得到足够重视。此外，膜接触的形成与调控的分子机制虽广受关注，但这种细胞器互作方式的相应生理功能一直是该领域的痛点和难点。最后，内膜细胞器的膜完整性研究方兴未艾，国际上的研究力量还未形成规模，这为我国研究人员的领先突破提供了良好的契机。

3. 我国研究现状与水平

目前，我国关于内质网 / 高尔基体的研究在结构与功能的关联以及蛋白质稳态调控等方面形成了一定特色。中国科学院生物物理研究所的胡俊杰研究组首次建立了管状内质网的组分分析体系（Wang et al.，2017b），并在膜塑形和膜融合等领域得到国际同行的高度认可（Hu and Rapoport，2016）。武汉大学的宋保亮和刘勇课题组分别在内质网相关的脂代谢与应激反应方面取得新的突破（Huang et al.，2019；Zhang et al.，2018d）。中国科学院生物物理研究所的王志珍和王磊研究组在内质网氧化还原态调控蛋白质稳态方面积累了重要的研究经验，并发现内质网－高尔基体协同调节内质网应激的新途径（Yu et al.，2020；Zhang et al.，2018b），王立堃研究组则在内质网应激信号的调控和生理功能研究方面处于领域的前沿（Ghosh et al.，2014）。

作为疏水的小分子，胆固醇的胞内运输是细胞生物学的基本问题，一直以来备受关注但始终没有得到解决。宋保亮课题组发现，过氧化物酶体在胆固醇的胞内运输中发挥了重要作用，过氧化物酶体通过与溶酶体形成膜接触，介导胆固醇从溶酶体膜转运至过氧化物酶体膜。该接触是由溶酶体上的 Syt Ⅶ蛋白结合过氧化物酶体上的脂质分子 PI（4,5）P2 来实现的（Chu et al.，2015）。同时，他们的研究还发现过氧化物酶体膜上 PI（4,5）P2 也可与内质网上的 E-Syts 蛋白结合，从而促进胆固醇向内质网的转运（Xiao et al.，2019）。这表明，细胞器互作是主要的脂质运输机制，决定了脂质在细胞内的精确分布。

内质网向高尔基体转运物质的重要载体是 COP Ⅱ 衣被囊泡。此前领域里研究人员认为，这一通路的核心组分已被鉴定完全，对 COP Ⅱ 的研究也被渐渐忽略。然而，COP Ⅱ 的货物分选机制要比此前想象的复杂很多，涉及一系列未被重视的研究的货物受体。香港科技大学的郭玉松研究组建立了通透细胞的原位 COP Ⅱ 出芽实验体系（Niu et al.，2019），由此突破了此前的研究瓶颈，打开了一系列新的研究方向。北京大学的陈晓伟研究组也在 COP Ⅱ 分泌调控的生理功能方面取得了新的突破（Liu et al.，2019）。此外，内质网在形成 COP Ⅱ 衣被囊泡过程中膜完整性的维持也逐渐受到胡俊杰研究组等的关注和探索（Niu et al.，2019）。

我国在溶酶体稳态平衡方面的研究尚不广泛，但有特色和优势（Yang et al.，2017）。在溶酶体发生（杨崇林，云南大学）（Li et al.，2016b）、再生（俞立，清华大学）（Du et al.，2016；Rong et al.，2012）、自噬调控（张宏，中国科学院生物物理研究所）（Guo et al.，2014a；Tian et al.，2010）、自噬体－溶酶体融合机制（张宏、钟清，上海交通大学）（Diao et al.，2015；Wang et al.，2016）、能量代谢的溶酶体调控机制（林圣彩，厦门大学）（Zhang et al.，2014，2016b）等方面均处于领域的前沿。除巩固目前的研究优势外，今后应努力拓展溶酶体稳态平衡的研究范围，包括机体水平和体外重组体系的建立与应用。

我国溶酶体膜完整性的相关研究比较少，杨福愉 / 魏涛涛研究组（中国科学院生物物理研究所）关注溶酶体通透性改变所引起的细胞死亡；王晓晨研究组（中国科学院生物物理研究所）建立了以线虫为模式的溶酶体完整性在体研究系统，鉴定了溶酶体完整性的维持基因、调控机制及溶酶体损伤的细胞响应机制（Li et al.，2016a）。目前国外溶酶体完整性的相关研究多集中在清除及修复方面，以细胞系研究为主。今后可能的突破方向是维持机制，细胞及机体的响应机制的解析，以及溶酶体完整性维护的生理意义及病理相关性。

溶酶体在机体发育、衰老及胁迫应答中的作用，目前国内外对其的相关研究都不多，是溶酶体相关领域具有发展潜力的研究方向。溶酶体相关疾病的研究从早期的溶酶体贮积症、代谢疾病，到近期的神经退行性疾病和肿瘤等，国内外对其的相关研究还都比较分散。溶酶体贮积症属于罕见病，国外对其的研发投入相对少。如果建立医院和基础研究的有效联系平台，我们可以争取在此领域取得突破。

我国科学家对胞吞胞吐领域的研究也做出了重要贡献。清华大学陈晔光研究组在内吞体介导信号转导（Zuo et al.，2013），中国科学院生物物理研究所王晓晨研究组在吞噬的分子机制及其在发育过程中的生理功能（Cheng et al.，2015；Wang et al.，2003），中国科学院生物物理研究所蔡华清研究组在巨胞饮的分子机理和细胞功能（Jiao et al.，2020），云南大学杨崇林研究组在调控早－晚期内吞体转换的分子作用机制（Liu et al.，2016a；Liu et al.，2017a）方面的研究均取得了一系列原创性成果。中国科学院生物物理研究所徐涛研究组在介导胰岛素分泌的致密核心囊泡研究方面取得了很好的研究优势（Kang et al.，2006；Zhou et al.，2007）。

二、无膜细胞器形成和动态平衡的调控机制

1. 研究背景

真核细胞各种膜包裹的细胞器（如内质网、溶酶体、线粒体），通过生物

膜与外部环境分开，将行使某一特定功能的生物大分子集合在某一特定区域，催化特定生化反应在特定的时间、特定的空间有效进行。细胞内还存在大量的无膜细胞器——生物大分子凝聚体（如核仁、着丝粒、核孔复合物、应激颗粒及生殖颗粒等），在不同的生理过程中发挥重要的功能（Banani et al., 2017；Boeynaems et al., 2018；Kaganovich, 2017；Shin and Brangwynne, 2017；Zhang et al., 2020a）。生物大分子凝聚体主要是RNA与蛋白质结合所形成的大的核糖核蛋白颗粒（RNP颗粒），富含RNA及RNA识别结构域和固有无序区域；也有一些凝聚体由固有无序蛋白或多肽组成，或者通过稳定折叠的蛋白质相互结合介导形成。这些分子通过液-液相分离凝集在一起，在细胞中形成液滴状的颗粒，具有液体扩散融合的特性。如同油在水中那样形成一个个液滴，这些液滴可进一步通过相变成凝胶态（hydrogel-like）、液晶态（liquid crystal）及固态（solid）等结构。不同生物大分子凝聚体具有的不同物质性质在细胞生命活动中是可调控的，与其功能密切相关。

2. 国际研究现状与趋势

生物大分子“相分离”与“相变”属于全新的交叉研究领域，至今仅有十几年的发展历史。2009年，Hyman研究组报道线虫早期胚胎中的P颗粒呈液滴状，首次明确提出细胞中存在生物大分子的相分离。2012年，Rosen研究组在体外成功重构出与胞内生物大分子凝聚体类似的生物大分子液滴，并揭示了多价相互作用驱使相分离发生。随后相继发现，多种无膜细胞器如核仁、各种细胞体和RNA颗粒，以及膜下信号转导复合体的形成都是由特定蛋白多价互作介导的相分离所驱动的。近十年，生物学家发现细胞中的生物大分子凝聚体具有重要的生物学功能，如介导细胞特定区域的信号转导，赋予生物体应对胁迫环境的能力及参与细胞内质量控制等，而且如同物理学家和高分子科学家研究多年的软物质那样体现在感应微小的外环境而产生剧烈的理化特性变化，发生相分离与相变，即可以液态、凝胶态、液晶态、固态等形式存在于细胞中，这些新发现的相分离与相变实际上是生物体功能调控的一种极为普遍的方式。蛋白质凝聚体的相分离、相变受多种因子调控。转录后修饰，包括蛋白质磷酸化、甲基化、乙酰化、泛素化、小类泛素修饰物蛋白（SUMO）化修饰等，都会直接调控蛋白质凝聚体的物理状态。比如在阿尔茨海默病中 tau 蛋白质的磷酸化是一个显著的病理特征，而磷酸化又会促进 tau 蛋白质形成聚集体并发生相变转化。分子伴侣在调节蛋白质凝聚体相变中发挥着极其重要的监控作用，分子伴侣功能缺失导致多种蛋白质凝聚体相变异常，进而导致相关疾病的发生。

细胞中无膜的生物大分子凝聚体的液 - 液相分离和相变在老年性疾病的发生、发展中具有重要的作用。例如，细胞衰老或细胞对病原体（病毒或细菌）感染、缺氧、化学药品、辐射等各种应激条件响应，形成应激颗粒，这是一种生理条件下的相分离现象，而长期的应激刺激后，或关键组分发生病理性突变，应激颗粒的组分就会发生病理性的纤维化相变（固化）。例如，在阿尔茨海默病中的 β 淀粉样蛋白（Aβ）、帕金森病的 α- 突触核蛋白（α-synuclein）、亨廷顿舞蹈症的 PolyQ 无序结构域、肌萎缩侧索硬化即渐冻症的 Fus、2 型糖尿病的胰岛淀粉素（amylin）等疾病相关蛋白质或多肽都会纤维化。而这些纤维化与疾病发生、发展密切相关。另外，机体也会形成抗病毒应激颗粒作为固有免疫的作用平台。比如细胞针对病原体的感染做出响应，在细胞质中形成含有 cGAS（cyclic G/AMP synthase）和病原体 DNA 的凝聚体，以激活 cGAS 下游的固有免疫反应。而在一些类型的肿瘤发生过程中，有研究表明原癌基因或肿瘤抑制因子的致病性突变破坏正常的蛋白质凝聚体相变和功能，促进肿瘤发生。在超过 98% 急性早幼粒细胞白血病中，15 号和 17 号染色体的异位重组产生融合的原癌基因 *PML-RARα*，从而破坏了由 PML 蛋白组装的 PML 核体（PML-NB）的正常装配和功能，导致白血病发生。还有研究表明，生物大分子凝聚体的相变调控直接影响其被细胞内的质量控制系统（如自噬）所清除的效率。因此从累积和清除这两个方面来看，对老龄化导致蛋白质凝聚体相变异常的研究，是了解和治疗这类疾病的最关键步骤。对蛋白质凝聚体的相变调控研究，也将为多种疾病的发生和治疗提供全新的理论靶点。因此研究蛋白质凝聚体的相变调控，对于理解疾病发生机理和开发治疗策略具有极为重要的科学意义。

研究者以“相变”的视角重新审视传统生命科学研究中众多长期悬而未决的科学问题，取得了大量重要的原创性成果，并开创了一系列全新的研究方向。生物大分子“相分离”和“相变”已经成为基础生命科学研究中最为前沿和重要的研究领域之一。因此，对生物大分子“相分离”和“相变”领域的研究，是关系到前沿基础研究发展的重大战略需求。

3. 我国研究现状与水平

我国科学家在生物大分子“相分离”和“相变”领域，如神经系统发育和信号转导、细胞自噬、表观遗传调控、神经退行性疾病等方向，已经取得了一系列重要的原创性成果，建立了一批具有深厚知识累积和良好工作基础的高水平研究团队。利用体外重构等方法，系统证明了兴奋性突触前、后突触致密层以及抑制性突触后致密层都是蛋白质相分离凝聚体（Wu et al.，2019b；Zeng et al.，2018，2019，2016；Zhu et al.，2020）；从相分离的角度重新阐释了细胞自噬降解蛋白质凝聚体的机理（Sun et al.，2018；Zhang et al.，2018a）；研究了 RNA 加工、RNA m6A 修饰、表观遗传调控等核内无膜细胞器的相分离调控机制（Gao et al.，2019；Huo et al.，2020；Wang et al.，2019b，2020b；Yao et al.，2019）；发现了神经干细胞不对称分裂过程中蛋白质相分离调控细胞极性的建立（Shan et al.，2018）；探索了蛋白质相分离异常与神经发育障碍类疾病及退行性疾病的相关性（Duan et al.，2019；Gui et al.，2019；Liu et al.，2020b，2020c；Wang et al.，2020a）；以突触前膜突触致密层、自噬小体为例，探讨了有膜与无膜细胞器的互作；发展了基于原子力显微镜、双色荧光互相关谱等多种新型相分离表征手段（Zhang et al.，2020a）。

三、核内无膜细胞器的结构、组装和功能调控

1. 研究背景及现状

哺乳动物细胞的细胞核高度组织化，形成了众多形状各异和功能不同的细胞核无膜细胞器，包括核仁（nucleolus）、旁斑（paraspeckle）、核斑（nuclear speckle）、细胞核应激小体（nuclear stress body）等（Mao et al.，2011；Stanek and Fox，2017）。其中最有代表性的是细胞核仁和旁斑，它们的结构、组装和功能研究较为清晰。在结构方面，研究人员发现核仁是最大的细胞核无膜细胞器，直径在 3～5 μm。电子显微镜的观察结果指出，高等

真核生物细胞核仁具有独特的形态特征，包含了三个相对独立的亚区域，分别是低电子密度的纤维中心（fibrillar center，FC）区域、高电子密度的致密纤维组分（dense fibrillar component，DFC）区域和具有颗粒状结构的颗粒组分（granular component，GC）区域（Boisvert et al.，2007）。这种三层结构的核仁几乎存在于所有羊膜动物中，而在非羊膜动物中，核仁的FC与DFC区域相互融合，仅具有两层结构的核仁（Thiry et al.，2005）。核仁的结构与功能高度偶联：在核糖体RNA（rRNA）的生物学生成过程中，不同的加工步骤发生在核仁的不同的亚区域中（Nemeth and Grummt，2018）。旁斑的超微结构已被解析。它们在共聚焦显微镜下呈不均匀的点状分布，数量依细胞类型而定（Fox et al.，2005；Fox et al.，2002）。早期利用透射电子显微镜观察旁斑发现：旁斑中电子密度高，富含RNA；位置上靠近染色质间颗粒相关区（interchromatin granule-associated zone，IGAZ）；在空间结构上高度有序，大多为细长的扩展结构，短轴约为360nm，长轴则可以达到1～2 μm（Cardinale et al.，2007；Prasanth et al.，2005；Souquere et al.，2010）。近年来利用超高分辨率显微成像和多色旁斑组分成像进一步揭示了其有序的超微结构（West et al.，2016）。

研究人员发现核仁与旁斑的组装都与RNA的转录相关。核仁结构的组装与rRNA的转录高度耦联。在小鼠胚胎发育过程中，核仁在细胞分裂末期形成，与rRNA转录激活保持同步。低剂量放线菌素D（actinomycin D，ActD）抑制rRNA转录，核仁结构被破坏和解体（Fulka and Aoki，2016），同时核仁的形变影响转录以及DNA修复。异位整合核糖体DNA（rDNA）序列研究发现，整合了上游结合因子（upstream binding factor，UBF）结合位点的非核仁组织区（nucleolar organizing region，NOR）携带的染色体上出现异位次缢痕，被称为伪NOR（pseudo-NOR）（Mais et al.，2005）。伪NOR在蛋白组分上与核仁FC相似，能够招募Pol I转录机器以及pre-rRNA加工因子，但是不含rDNA启动子，也不支持转录（Prieto and McStay，2007）。在研究旁斑的从头组装过程中，研究人员提出旁斑的结构RNA核旁斑装配转录物1（nuclear paraspeckle assembly transcript 1，NEAT1）开始转录后会作为“骨架分子”招募蛋白质并起始旁斑组装。利用NEAT1转录的延时成像技术，冷泉港实验室科学家发现NEAT1的转录活动伴随着旁斑蛋白的招募和后续组装（Mao et al.，2011）。旁斑蛋白多为 RNA 结合蛋白，富含多种蛋白结构域，包括 RNA 识别基序、NOPS 结合域和 CC 结构域，这些蛋白结构域可有效促进蛋白二聚化以及可逆的蛋白多聚化。利用 CRISPR-Cas9 技术对 NEAT1 进行系统性的逐段缺失，日本科学家鉴定出了 NEAT1_2（NEAT1 长异构体）的中间序列（8～16.6 kb）招募旁斑蛋白的二聚体-起始旁斑的组装（Yamazaki et al.，2018）。旁斑蛋白在 NEAT1_2 上不断聚集并进一步促进这些蛋白的多价结合，起始核壳结构中核心层的建立，最终通过液-液相分离的方式形成核壳状的旁斑。

细胞核无膜细胞器种类繁多，它们维持了细胞正常的生理活性，参与细胞增殖调控、压力应激响应、蛋白质质量控制等众多生理过程（Boisvert et al.，2007；Iarovaia et al.，2019）。研究发现细胞核无膜细胞器的结构和功能与多种严重的人类疾病的发生发展密切相关（Morimoto and Boerkoel，2013），如核仁与特雷彻•柯林斯综合征（Calo et al.，2018；Farley-Barnes et al.，2019），旁斑与多种癌症（Adriaens et al.，2016）以及神经退行性疾病如肌萎缩侧索硬化（amyotrophic lateral sclerosis，ALS）和阿尔茨海默病（Alzheimer disease，AD）（Fox，2018；Lee et al.，2011）；PWS 小体与人类遗传疾病普拉德-威利综合征（Prader-Willi syndrome，PWS）（Wu et al.，2016；Yin et al.，2012）。因此，细胞核无膜细胞器在体内受到严格的调控，以保证其发挥正常的生理功能。其中一个重要方式是通过非编码 RNA 调控这些无膜细胞器的结构而影响其发挥功能。核仁在热休克或酸中毒刺激下，分割 rDNA 转录单元的基因间间隔（intergenic spacer，IGS）区域所转录的长非编码 RNA（IGS-RNA）上调，并靶向核仁滞留序列（nucleolar detention sequence，NoDS）肽段的蛋白因子，将这类蛋白滞留在细胞核仁中（Audas et al.，2012；Jacob et al.，2013）。IGS-RNA 的聚集与核仁中形成可逆的核仁亚结构滞留中心（detention center）结构相关联，该核仁亚结构能够调控核仁因子的重新分布：当 IGS-RNA 被敲低后，NoDS 相关蛋白的流动性升高并调控 rRNA 合成（Mekhail et al.，2004）。

结构 RNA 的产生和丰度直接影响细胞核无膜细胞器的数量与形态。旁斑的组装和形态受 NEAT1 转录与加工的双重调控。NEAT1 的消除显著减少旁斑的数量（Chen et al.，2020a；Clemson et al.，2009；Fox et al.，2005），而响应不同细胞应激条件时，NEAT1 表达上调促进旁斑形成（Adriaens et al.，

2016；Hirose et al.，2014；Imamura et al.，2014；Wang et al.，2018）。此外，NEAT1 的加工异常还会改变旁斑的形态，NEAT1_1 转录活性的终止会导致 NEAT1_1 在核内形成较小且分布均匀的点状信号，称为微斑（microspeckle）（Li et al.，2017b）。而蛋白酶体抑制剂处理（Hirose et al.，2014）、线粒体功能缺陷（Wang et al.，2019a）及 NEAT1_1 多聚腺苷化位点的缺失或修饰（Li et al.，2017b）导致 NEAT1_2 比例增加，显著增加旁斑的数量及更多长条形旁斑的形成。

另一个细胞核无膜细胞器的调控方式基于相分离模式。旁斑中的 NEAT1_2 具有相分离特性，促进了旁斑蛋白二聚体的结合和进一步的多价组装（Yamazaki et al.，2018）。核仁的组装是由液－液相分离所驱动的，这个过程在核仁形成过程中不可或缺。核仁的液态特征的观测最早来源于 Hyman 实验室对非洲爪蟾卵母细胞的研究，他们发现邻近的核仁发生液滴状的分离与融合（Brangwynne et al.，2011）。在 2016 年，Brangwynne 实验室利用荧光标记的 NPM1、FBL 和 POLR1E 分布指示核仁 GC、DFC 以及 FC 区域，成功在体外模拟了核仁的融合，并模拟 DFC 融合形成一个大液滴，包裹多个 FC 区域（Feric et al.，2016）。单独的 NPM1 和 FBL 在体外可形成液滴结构，但当这两种成分混合时，FBL 会被 NPM1 包裹形成类似于核仁的结构，而不同液滴之间的表面张力控制了这一过程（Feric et al.，2016）。

2. 我国研究现状与水平

我国科学家在细胞核无膜细胞器方面的研究处于国际领先水平，取得了一系列重大的研究成果。国内目前对于人类细胞核无膜细胞器的研究主要集中于核仁、旁斑和核斑，包括超微结构解析、转录调控、动力学研究、功能研究等。

中国科学院科研人员发现长非编码 RNA SLERT 通过调控核仁超微结构影响前核糖体 RNA（pre-rRNA）的转录活性（Xing et al.，2017）；利用超高分辨率显微成像技术结合生化分析，系统研究了人类活细胞当中核仁的超微结构（Yao et al.，2019）。清华大学科研人员发现 RNA 解旋酶 DDX18 与 PRC2 相互作用，通过拮抗 PRC2 的活性维持 rDNA 低 H3K27 甲基化机制，该过程对干细胞的多能性维持十分重要（Zhang et al.，2020b）；通过开发鉴定核仁相关结构域的高通量测序方法 Nucleolar DNA-seq，发现 L1 重复序列显著富集在核仁中，揭示了重复序列对宿主基因组结构和功能的调控作用（Lu et al.，2020）；基于肝癌多组学队列数据的计算分析，发现肿瘤中高表达的长非编码 RNA LETN 与核仁蛋白 NPM1 互作，并通过控制 NPM1 的寡聚化组装调控核仁的形态结构、rRNA 的合成以及染色质的凝聚，进而影响细胞增殖（Wang et al.，2021）。

以旁斑为研究对象，中国科学院科研人员揭示了含有 IRAlus 序列的 mRNA 从旁斑转运出核的两种调控途径（Hu et al.，2015）；进一步地，发现结构 RNA 自身的转录调控与细胞核无膜细胞器的功能及其与核外细胞器之间的协作关联，通过调控旁斑的数量和形态对 mRNA 的出核质量调控并反馈调节核外细胞器的稳态（Wang et al.，2018）；开发了普适性更好、标记效率更高的活细胞 RNA 标记系统 CRISPR-dCas13（Yang et al.，2019），用于标记旁斑骨架 RNA NEAT1，并发现旁斑通过碰撞－融合 / 分裂的形式进行形态转换；中山大学研究人员开发了 MS2 一步敲入编辑的系统，可有效地对 RNA 进行长时间标记并可联合 MS2-MCP 系统进行靶向 RNA 复合体的分离纯化（Chen et al.，2020a）。

聚焦于核斑，中国科学院科研人员开展了有关 RNA 出核转运的研究。RNA 的出核转运命运决定受到不同蛋白质和复合体参与，并受到 RNA 转录、加工和降解的影响。NRDE2 是 RNA 外泌体复合物（RNA exosome complex）的负调控因子，集中在核斑中。NRDE2 利用一段氮端保守区与 RNA 解旋酶 MTR4 结合，通过广泛抑制 MTR4 的招募、MTR4 与外切复合体的互作，阻止核斑中 RNA 的降解，并确保其出核转运（Wang et al.，2019a）。出核受体 NXF1 也协同调控转录延伸、mRNA 的可变加尾及出核（Chen et al.，2019），这种反向调控可能对于基因的精准表达尤为关键（Chen et al.，2019；Fan et al.，2019）。

四、不同类型蛋白酶体及其亚基间的相互作用与靶向干预

1. 研究背景及现状

蛋白酶体是一种由几十个蛋白亚基组成的庞大蛋白质复合物，具有多种

蛋白水解酶活性，负责细胞内大多数蛋白质的降解，对各种细胞活动都至关重要。蛋白酶体的功能紊乱会打破细胞内蛋白质的动态平衡，与肿瘤、病毒感染、神经退行性疾病等许多疾病的发生发展关系密切（Li et al.，2019d；Zhang et al.，2010）。对蛋白酶体通路的干预将可能成为多种疾病的新的治疗方法。例如，蛋白酶体抑制剂硼替佐米（Bortezomib，亦称 Velcade）已广泛投放市场，用于治疗多发性骨髓瘤和套细胞淋巴瘤（Adams，2004；Goldberg，2005）。蛋白酶体抑制剂在其他肿瘤治疗中的应用正在广泛的研究中（Jiang et al.，2021b，2020）。蛋白酶体激活剂或抑制剂在肿瘤以外的许多疾病的治疗中也有很大的潜力，这些疾病包括病毒感染、寄生虫病、细菌/真菌感染、神经退行性疾病和自身免疫疾病等（Cao et al.，2020）。迄今发现的蛋白酶体可分为三种类型。一是普遍存在于各种组织中的 26S 蛋白酶体，由 20S 核心催化颗粒和 19S 调节颗粒组成，其催化颗粒中的三个催化亚基（β1、β2 和 β5）催化泛素化底物的降解。二是 PA28 蛋白酶体，包括由 PA28α/PA28β 作为调节颗粒并具有三种可诱导的催化亚基 β1i、β2i 和 β5i 的免疫蛋白酶体，以及由 PA28γ 作为调节颗粒的 PA28γ 蛋白酶体。免疫蛋白酶体能够加快多肽形成进而促使抗原递呈（Rock and Goldberg，1999），而 PA28γ 则可以依赖（Zhang and Zhang，2008）和不依赖（Li et al.，2007）于泛素化的两种模式降解多个重要细胞核蛋白。三是由蛋白酶体活化因子 PA200 作为调节颗粒的 PA200 蛋白酶体（包括含有睾丸特异表达的 α4s 亚基的生精蛋白酶体）。PA200 蛋白酶体可介导由乙酰化（而非泛素化）介导的核心组蛋白降解，该结果颠覆了科学界关于体细胞组蛋白不降解的理论（Hauer et al.，2017；Jiang et al.，2021a；Lin et al.，2020；Mandemaker et al.，2018；Qian et al.，2013）。

26S 蛋白酶体是哺乳动物存活所必需的，但其中某些功能接近的亚基之间存在一定的互补性。敲除 19S 调节颗粒中的泛素受体亚基 Rpn10 或去泛素化酶 UCH37，可导致小鼠胚胎致死（Al-Shami et al.，2010；Hamazaki et al.，2007）。敲除 19S 调节颗粒中另一个泛素受体亚基 Rpn13 的小鼠仅少数可出生，且在出生后一天内死亡（Hamazaki et al.，2015）。在小鼠肝脏中单独敲除 Rpn10 或 Rpn13，对肝脏的功能影响很小，但同时敲除 Rpn10 和 Rpn13 会使小鼠肝脏严重受损并伴随大量泛素化底物的积累（Hamazaki et al.，2015）。

免疫蛋白酶体及PA200蛋白酶体并不是哺乳动物存活所必需的，但在疾病发生和治疗中可能起着重要作用。同时敲除免疫蛋白酶体的全部三个催化亚基虽然极大地改变了组织相容性复合体（MHC）Ⅰ型抗原递呈，但是小鼠仍能存活（Kincaid et al.，2011）。*PA200*基因敲除的小鼠也能存活，其精细胞变长阶段的组蛋白降解延迟，雄性小鼠生育率虽然降低，但仍可育（Qian et al.，2013）。单独敲除PA28γ亦不影响存活且对精子发生影响不大，同时敲除PA200和PA28γ的小鼠能存活，但雄性小鼠不育（Huang et al.，2016a）。全身敲除睾丸特异表达的20S催化颗粒亚基α4s，小鼠其他器官或部位正常，但雄性不育。检查蛋白酶体亚基组成发现，敲除α4s降低了成熟睾丸精母细胞中蛋白酶体激活因子PA200及普通催化亚基β1、β2和β5的含量，但增加了在免疫蛋白酶体中同源亚基β1i和β5i的含量。α4s通过调控生精蛋白酶体组装，促进精母细胞减数分裂过程中姐妹染色体互换后DNA修复时的组蛋白降解，从而保障减数分裂和精子发生的正常进行，并为精子发生所必需（Zhang et al.，2020c）。这些现象提示，PA200蛋白酶体与其他类型的蛋白酶体（包括普通26S蛋白酶体和免疫蛋白酶体）在特定的生理或病理条件下可能存在互补性。不同类型蛋白酶体及其亚基间的相互作用机制与生理病理意义仍不清楚。蛋白酶体途径和细胞内另一重要降解途径自噬可交叉调控（Bartel，2015；Jiang et al.，2019），其机制也有待阐明。

蛋白酶体亚基会发生泛素化、磷酸化、肉豆蔻酰化、甲基化、乙酰化等各种翻译后修饰，且同一位点可存在不同的修饰类型，显示这些修饰在蛋白酶体功能调控中存在交叉作用（Kors et al.，2019；Liu et al.，2006；Zong et al.，2014）。另外，蛋白酶体的转录活性也可被调控，从而调整其降解蛋白质的能力。例如，γ干扰素（IFN-γ）能激活细胞的JAK-STAT和AKT-mTOR信号通路，增强β1i、β2i、β5i、PA28α/PA28β等蛋白酶体亚基基因的表达；氧化应激也会通过NRF2激活PA28α/PA28β和20S核心颗粒多个蛋白酶体成分的表达（Kors et al.，2019）。对蛋白酶体亚基翻译后修饰及其转录调控机制的研究将有利于发现靶向蛋白酶体通路的新靶点。

在目前已知的三种蛋白酶体中，26S蛋白酶体的组成、结构、作用机制等已相对清晰，并且已有多种应用于临床治疗多发性骨髓瘤和套细胞淋巴瘤的针对其催化亚基的蛋白酶体抑制剂，但这些抑制剂在其他肿瘤或疾病中的作用仍有待研究。即使针对多发性骨髓瘤和套细胞淋巴瘤，其药物耐受性或抗性也仍是一大难题。另外，现在临床使用的蛋白酶体抑制剂也同时抑制其他两种类型的蛋白酶体。亟须解析 26S 蛋白酶体的调控因子或机制及其与疾病发生的相关性，研发针对其 19S 调节颗粒或调控因子等的更具选择性的靶向药物。抑制蛋白酶体活性是化疗增敏和克服肿瘤耐药性的有效策略，此方面的研究也前景广阔。近期研究显示，蛋白酶体抑制剂与其他药物，如免疫检查点 PD-1 抑制剂、来那度胺（Lenalidomide）、地塞米松（Dexamethasone）等，联用能够显著地提高多发性骨髓瘤的治疗效果（Badros et al.，2017；Boccon-Gibod et al.，2020）。

PA28 蛋白酶体（包括含 PA28 α/PA28 β 调节颗粒的免疫蛋白酶体和含 PA28 γ 调节颗粒的 PA28 γ 蛋白酶体）的底物识别机制仍不清楚，这一问题的解决对传染性疾病的预防或治疗可能非常重要。PA200 蛋白酶体的研究起步较晚，将来研究的空间很大。PA200 蛋白酶体在不同疾病包括肿瘤和免疫系统疾病中的作用和机制，识别乙酰化底物的机制、调控因子或机制，及其与疾病发生的相关性等方面仍有待阐明。

2. 我国研究现状与水平

蛋白酶体发现于 1986 年（Tanaka et al.，1986），我国对于蛋白酶体的研究起步较晚，第一篇国内学者的论文发表在 1999 年（Ding et al.，1999）。随后，我国关于蛋白酶体的研究逐渐活跃，发表论文数已占全球的 10%；近几年仍在升温，2020 年发表了 91 篇国际论文，占全球的 20%。国内学者的很多研究成果已处于或接近世界领先水平：解析了各种类型的蛋白酶体的结构（Chen et al.，2021；Dong et al.，2019；Guan et al.，2020；Huang et al.，2016b；Luan et al.，2016）；发现 Rpn13 是哺乳动物 26S 蛋白酶体的一个新亚基，并且与蛋白酶体中的去泛素化酶 UCH37 结合（Qiu et al.，2006），Rpn13 和 UCH37 已被国外学者证明为更有选择性的抗癌药物靶标（Anchoori et al.，2013；D'Arcy et al.，2011）；发现了在睾丸特异表达的新蛋白酶体亚基 α 4s 与含有 PA200 和 α 4s 的生精蛋白酶体，揭示了 PA200 蛋白酶体可介导由乙酰化（而非泛素化）介导的核心组蛋白降解，修正了科学界关于体细胞组蛋白不降解的理论（Lin et al.，2020；Qian et al.，2013；Yin et al.，2019；Zhang et al.，2020c）；发现了PA200及其酵母同源物Blm10可维持组蛋白修饰及转录组的稳定性，最终延缓衰老（Chen et al.，2020b；Jiang et al.，2021a）；发现了蛋白酶体亚基磷酸化可调控蛋白酶体活性和含量（Li et al.，2015a；Liu et al.，2006；Liu et al.，2020a）；揭示了PTEN缺失增强胆管癌和胆囊癌对蛋白酶体抑制剂硼替佐米的敏感性（Jiang et al.，2021b，2020）；发现了PA28 γ/REG γ 蛋白酶体的多种靶蛋白及其在肿瘤发生发展等的调控作用（Ali et al.，2013；Tong et al.，2020；Xu et al.，2016）；揭示了PA28 γ/REG γ 蛋白酶体促进自噬标志性蛋白LC3-Ⅱ的前体LC3-Ⅰ通过蛋白酶体降解（Jiang et al.，2019）；发现了免疫蛋白酶体催化亚基β 5i通过降解血管紧张素ATRAP促进心房颤动，通过靶向ATG5的降解从而抑制自噬的活性，在血管紧张素诱导的心肌肥厚中作为重要的正向调节因子（Li et al.，2019a；Xie et al.，2019）；发现砒霜通过结合PML引起致癌融合蛋白PML-RARalpha通过蛋白酶体降解（Zhang et al.，2010）等。

虽然我国在蛋白酶体研究的多个方面已处于国际领先地位，但迄今在蛋白酶体靶向药物研发方面仍没有任何实质性的突破，明显落后于欧美发达国家。蛋白酶体领域的研究内容涉及众多科学领域，亟须联合国内各方面优势团队、交叉融合、取长补短、尽快在该领域取得更重要的突破。

五、线粒体稳态调控机制及其在衰老和相关疾病中的作用

1. 研究背景及现状

线粒体是能量代谢和细胞凋亡等途径的调控枢纽。随着技术的进步和研究的深入，对线粒体的认识也越来越丰富，线粒体与细胞免疫、坏死、铁死亡、代谢重编程等信号途径的关系（Bock and Tait，2020；West et al.，2015），线粒体与其他细胞器的相互作用，也受到越来越多的关注（Elbaz-Alon et al.，2020）。鉴于线粒体对细胞的重要作用，而线粒体功能紊乱会造成生长、发育障碍和各种疾病，加速衰老，目前，阐释线粒体质量控制的分子机制，如线粒体发生、融合、分裂，线粒体应激，线粒体自噬，解析线粒体稳态在相关疾病中的分子机制，是线粒体领域研究的热点问题。

虽然关于线粒体的研究已经取得了长足的发展，但是仍需进一步开发研究线粒体的新技术，在微观层面如亚细胞器水平上探明线粒体结构与功能的分子机制，在宏观层面上探究线粒体功能紊乱导致疾病的分子机制，并研发出相应药物用于线粒体疾病的治疗。

（1）线粒体生物发生和线粒体自噬调控新机制。线粒体是细胞能量的主要来源，是细胞代谢的核心枢纽。线粒体数量的多寡与好坏对细胞的功能有着十分重要的影响。由于线粒体是被氧自由基攻击的主要位点，旧的、受损伤的线粒体需要通过线粒体自噬的方式被及时清除，新的、健康的线粒体通过生物发生的方式重新生成以满足细胞的正常需求（Ploumi et al.，2017）。细胞内线粒体保持新旧更替的动态平衡对于细胞维持稳态十分关键，线粒体生物发生和线粒体自噬异常与肿瘤、神经退行性疾病、糖尿病等多种疾病以及衰老密切相关（Li et al.，2019b；Sliter et al.，2018；Wu et al.，2019a）。深入研究线粒体生物发生和线粒体自噬，探索调控新机制，对于理解线粒体的功能和治疗线粒体相关疾病都有重要意义。

（2）线粒体蛋白质稳态和线粒体与细胞核互作。与其他细胞器不同，线粒体是半自主细胞器，线粒体蛋白质由细胞核基因和线粒体基因共同编码。细胞核编码蛋白和线粒体编码蛋白在基因转录、翻译，蛋白转运、降解等水平上的协同是维持线粒体正常运转的基础。线粒体蛋白质稳态的维持不仅与线粒体内部蛋白质稳态维持机器息息相关（Li et al.，2019c），而且细胞核编码了线粒体的大部分蛋白质，因此需要线粒体与细胞核的密切协作（Bragoszewski et al.，2017；Pfanner et al.，2019；Priesnitz and Becker，2018）。线粒体如何通过逆向调控的方式向细胞核发出信号，以满足自身所需，细胞核又如何通过正向调控的方式向线粒体发出指示，以应对挑战，对于理解不同环境下的线粒体功能调控十分关键。

（3）线粒体与细胞代谢和细胞死亡。一方面，在机体的生殖、发育、生长、衰老、疾病等过程中，细胞的功能和环境不断发生变化，细胞的代谢活动也随之改变，甚至发生细胞死亡，以应对不同需求。大量研究表明，线粒体是细胞代谢和细胞凋亡的调控中心。然而，随着研究的深入，越来越多的细胞死亡方式如坏死、铁死亡、焦亡等被发现，线粒体如何在这些死亡方式中起作用，仍然有很多问题待解决（Bock and Tait，2020）。另一方面，在不同环境下，线粒体如何调控细胞代谢以实现细胞代谢重编程，是理解疾病和发育过程中细胞功能变化的前提（Ma et al.，2015）。深入理解线粒体在细胞代谢和细胞死亡中的作用与机制对于治疗肿瘤等疾病有关键性的影响。

（4）线粒体蛋白结构与功能。解析蛋白质结构是理解其功能的基础。结构生物学的发展极大地促进了对线粒体功能的理解。未来仍需进一步解析线粒体蛋白结构，尤其是线粒体中大或者超大蛋白复合体的结构（Letts and Sazanov，2017）。通过结构生物学的手段，揭秘线粒体大蛋白复合体的组装机制，是进一步理解线粒体功能的基础，也将为线粒体相关疾病的治疗提供新的药物靶点。

（5）线粒体与免疫和衰老及相关疾病。线粒体在免疫调节中有重要作用，线粒体产生的自由基、线粒体 DNA 等物质是重要的免疫激活剂。长期以来的研究表明，线粒体功能异常是衰老及相关疾病的重要致病因素。在衰老的过程中，往往伴随着严重的炎症反应。然而，线粒体在免疫和衰老中的具体分子机制仍然有待进一步的研究（Kauppila et al.，2017；Sliter et al.，2018；West et al.，2015）。目前，我国面临着极大的老龄化压力，严重威胁社会经济发展、人民生命健康和家庭幸福。因此，关于线粒体在免疫和衰老中的作用的研究显得尤为重要。

（6）线粒体 DNA 突变如何导致疾病。线粒体有自己的 DNA，可以编码 13 种多肽、2 种核糖体 RNA、22 种转运 RNA，对于线粒体的功能十分关键。线粒体 DNA 突变是导致衰老以及肿瘤等多种疾病的重要原因（Schon et al.，2012）。线粒体 DNA 突变可通过遗传的方式获得，另外，因易被氧自由基攻击，在生理或者病理性条件下，线粒体 DNA 也容易发生突变，导致线粒体异质性（Sharpley et al.，2012）。但是，线粒体 DNA 突变的致病机制仍然不清楚，研究线粒体 DNA 突变相关疾病的具体分子机制将为这些疾病的治疗带来曙光。

（7）靶向线粒体的小分子化合物筛选和线粒体疾病治疗的新策略。线粒体功能异常可能会影响机体的各个组织和器官，包括大脑、心脏、肝脏、眼睛、肌肉、神经等。线粒体疾病影响人群广泛，往往症状严重，但是却缺乏有效的治疗方法，迫切需要新的靶向线粒体的药物和疾病治疗策略（Russell et al.，2020）。近年来，随着线粒体研究领域的不断深入，对靶向线粒体的小分子化合物筛选提出了新的要求，而技术的快速进步也为开发出线粒体疾病治疗的新策略提供了深厚的土壤。

2. 我国研究现状与水平

线粒体生物学一直是生命科学研究的前沿。从细胞代谢、线粒体蛋白结构与功能来看，线粒体与很多疾病的发生都有密切关系。我国在线粒体领域的研究发展迅速，取得了一大批世界前沿水平的成果，但是在引领世界的科研成果方面仍然有所欠缺，目前还没有形成线粒体研究的科学高地。

六、细胞骨架的动态调控及其与生理功能和疾病的相关性

1. 研究背景及现状

细胞骨架是大多数细胞力学支撑、物质转运、信号转导等多项任务的承担者，有着广阔的研究内涵和外延。近几十年来，随着相关学科研究技术的不断发展，细胞骨架领域的研究已经取得了巨大的进步。然而，我们对微管、微丝和中间丝的了解仍然比较碎片化，每个分支领域各自都有许多关键问题仍亟待解决。深入研究细胞骨架，形成对其动态调控和功能的全面认知，不仅具有重要的理论意义，还将为疾病治疗和药物开发提供坚实的理论依据。

1）微管细胞骨架

微管的动态性是微管执行多种细胞功能的基础，受到微管蛋白亚型、翻译后修饰和微管结合蛋白的调控（Diao et al.，2019；Xie et al.，2019）。然而，由于研究挑战性大、收效较慢，领域内的许多基础问题仍未得到系统性的回答。不同微管蛋白亚型的结构特征、动态性差异和时空分布是怎样的？微管蛋白的翻译后修饰和微管结合蛋白是如何调节微管的动态性的？微管蛋白是否存在新的翻译后修饰和新的结合蛋白？等等。这些问题的解决会大大促进人们对微管结构和功能的理解。以往通过显微成像技术和生物化学技术已经向我们展示了一些体外和细胞内微管动态调控的规律，未来这仍将是我们探索的重点内容。

除了对微管基础调节的研究，深入探索微管动态调控在多种细胞生物学

过程和疾病中的作用是另一个重要的方向。微管参与细胞增殖、运动、信号转导和细胞极化等过程（Janke and Magiera，2020），它们在个体发育尤其是神经系统发育、肿瘤发生和转移以及纤毛病等生理病理过程中的作用需要深入探索；微管在免疫细胞、肌肉细胞、分泌细胞和植物细胞等多种细胞中的功能也缺乏研究。因此，充分利用新技术和新方法，在在体环境或模拟在体环境（类器官等）下深入探索微管动态调控与不同生理过程的关系，将大大拓展我们对生命过程的理解。进一步地，阐明微管动态调控与疾病发生发展的关系，有望为药物开发和疾病治疗提供理论支持。

微管靶向药物的出现是微管研究向临床转化的范例，目前其已成为临床使用最广泛的化疗药物（Steinmetz and Prota，2018），但是在应用上依然有许多问题存在。通常认为药物是通过破坏微管动态性而导致肿瘤细胞死亡，然而临床结果显示这些药物在患者中只导致很小比例的肿瘤细胞发生周期停滞。因此，对于微管靶向药物是否通过干扰细胞周期来杀伤肿瘤这一问题仍存在争议，药物引发肿瘤细胞死亡的机制仍待进一步研究。另外，我们对此类药物药效决定机制的理解非常局限，对化疗后患者的肿瘤耐药机制也知之甚少。对微管细胞功能的探索是解决耐药性等问题的前提之一。

虽然生化、分子和细胞生物学的研究方法在微管研究中仍占据重要地位，但由于微管构成复杂、尺度较小并且动态性高的特点，传统方法也面临不少瓶颈问题。近年来，高速高分辨率显微镜和冷冻电镜等生物物理学技术的进步，数学和模拟计算与细胞生物学、发育生物学的学科交叉融合，正在为微管相关研究带来新的机遇。尤其是微管体外重组技术的发展，简化了研究系统，使许多细胞内无法开展的研究成为可能，取得了大量有启发性的结果。但是正是由于其简化性，体外试验的结论仍然需要回归在体条件进行检验。因此，充分整合利用体内和体外研究各自的优势与特长，进一步阐明微管动态变化与功能的关系，是未来的研究趋势。

2）微丝细胞骨架

微丝骨架的排布方式和动态组装决定了其生理功能，并使微丝骨架在支撑细胞形态和调控细胞运动中扮演着独特的力学信号转导角色（Fletcher and Mullins，2010）。微丝的分支、交联、聚合和解聚在时间与空间上受多种微丝结合蛋白的精细调控。微丝骨架通过细胞间连接将单细胞连成整体，提供了组织和器官的结构基础。然而，微丝结合蛋白如何影响微丝的整体特性？不同亚细胞分布的微丝如何响应细胞外化学和力学信号并参与细胞应激？跨细胞的微丝结构是如何协同细胞的群体行为的？这些重要问题仍亟待系统性的探索和解答，需要生命科学、化学、物理学与计算机科学领域跨学科合作才能取得突破性进展。

微丝参与细胞迁移、胞质分裂、线粒体分裂和细胞力学转导等过程（Cramer，2010；Fededa and Gerlich，2012；Hatch et al.，2014），它们在心血管和骨骼肌发育、神经系统发育、肿瘤转移等生理病理过程中发挥重要作用（Lian and Sheen，2015）。目前仍需要对发育过程和病理条件下组织与细胞进行力学测量及模拟，对微丝进行动态观测和系统功能研究，需要利用有效的手段解析力-化学因素的耦合对个体发育和肿瘤转移等过程的影响，精准地解析肿瘤细胞特异性微丝调控靶点。此外，微丝在免疫和代谢过程中的功能也缺乏研究。微丝及其结合蛋白的靶向药物仍面临从基础研究向临床转化的挑战，如何靶向微丝骨架从而破坏肿瘤细胞与阻止其浸润和迁移目前成为重要的研究方向，加深对微丝结合蛋白的细胞功能的研究是解决微丝药物特异性等问题的前提。阐明微丝的亚细胞分布、动态调控和精细结构，将大大拓展我们对生命过程的理解，有助于探索心血管疾病和肿瘤的新型诊疗手段，为药物开发和疾病治疗奠定坚实的理论基础。

微丝在细胞内具有丰度高、排布复杂、动态性强等特性，决定了本领域的研究对交叉学科和先进技术的需求。我国化学生物学的高速发展和新型的空间特异微丝探针的开发，将推动我们认识微丝对细胞质和细胞核功能的协同调控。近年来细胞核内微丝的发现和功能研究，对高时空分辨率成像技术提出了更新的要求。超分辨光学显微镜、冷冻电镜和光镜-电镜联用等技术的突飞猛进，对系统性地理解微丝细胞骨架的结构和功能，发育过程与疾病中的作用，生命体对环境化学和物理信号的协同应答将具有重要的推动作用。

3）中间丝细胞骨架

组成中间丝细胞骨架的蛋白质种类多样，具有组织表达特异性，它们受到翻译后修饰、细胞信号转导和非极性动态重组模式的调控（Etienne-Manneville，2018；Snider and Omary，2014）。由于中间丝蛋白种类众多，许多基础问题尚不完全清楚。不同中间丝的结构和分布特征是怎样的？中间丝蛋白的保守性和翻译后修饰是怎样的？诸多信号转导通路和中间丝如何交互调控？中间丝如何通过动态变化进而调控特异的细胞功能？等等。这些问题的回答将极大地完善对中间丝结构和功能的认识。

除了对中间丝自身动态结构变化和具有骨架支撑功能等基础问题的研究，系统探索中间丝在多种病理过程和疾病中的非结构支撑功能与机制是另一重要方向。第一，由于中间丝蛋白在不同病原体及在同种病原体不同生命周期阶段中功能的多样性（Ramos et al.，2020），探究中间丝蛋白与病原体及宿主其他重要因子相互作用的分子机制，有助于我们理解病原体入侵机体的过程。第二，确定不同中间丝蛋白如何相互作用，与微管和微丝网络如何协同互作，从而影响肿瘤发生发展进程，对于预防和治疗癌症至关重要（Gao et al.，2019）。第三，由于中间丝蛋白突变与神经退行性疾病之间的紧密关联，探究特定的神经中间丝蛋白及其翻译后修饰与神经递质受体之间的相互作用，有助于更好地理解神经系统功能及相关疾病。第四，中间丝在免疫细胞中的功能以及其参与调控的诸多自身免疫和炎症反应过程尚缺乏研究。第五，需要深入探索中间丝在创伤、渗透压、氧化应激和机械压力等应激下的调控机制。与国际上对中间丝逐渐重视起来的趋势相比，国内针对中间丝以及其与人类疾病发生发展之间的关系的研究还处于初期。对中间丝分子功能及致病机理的研究将对多种疾病的临床诊断和治疗具有指导意义。

基于中间丝与人类疾病的相关性，筛选靶向中间丝蛋白的治疗药物与针对其相关信号网络的治疗策略可能使癌症和神经系统疾病患者受益。此外，病原体入侵过程中中间丝相较于宿主细胞内的其他细胞器或信号通路是较为合适的抗感染靶点。原因有两点：其一，中间丝广泛分布于细胞质和细胞核内，并连接细胞核和细胞膜表面，其参与感染周期从入侵、复制到组装、释放的全过程，而其他细胞器通常只参与一个或几个感染步骤；其二，研究发现，一些中间丝编码基因的缺失在小鼠中无明显表型，但在感染应激下会被病原体挟持利用，其重要功能才被识别。因此，以宿主中间丝蛋白为靶向将筛选出具备广谱抗感染潜力的药物。

基于中间丝的研究，业界主要利用分子和细胞生物学、生物化学等传统技术方法，而进一步突破面临瓶颈。近年来，多种先进的显微成像、组织工程、基因编辑和组学技术的发展，为中间丝的相关研究带来新的曙光。

2. 我国研究现状与水平

细胞骨架是细胞的基本结构组成，细胞骨架调节异常与多种疾病发生发展密切相关，因而细胞骨架动态调控的基础、生理和病理作用研究不仅对细胞生物学基础理论有贡献，而且对于理解致病机制进而有针对性地开展预防和治疗具有应用价值。由于细胞骨架研究内涵广泛，目前我国多数研究方向较为分散，缺乏系统、深度和协同研究。要更加强对细胞骨架本身调控机制的基础研究，发展相关研究技术和方法，系统地开展细胞骨架生理和病理作用的研究，针对前沿研究方向统筹规划和资源配置，推动药物开发。

七、细胞形态形成和维持的调控机制

1. 研究背景及现状

细胞作为生命的最小独立单元，其特征之一是形状的多样性。从进化的角度看，单细胞生物中细菌和真菌由于细胞壁的限制多为棒状，摆脱了细胞壁限制的原生动物演化出千姿百态，出现了纤毛这一自主运动器官，有些种类还展现出巨大的形变能力。多细胞生物（尤其是动物）中细胞沦为个体的建筑模块和功能单元。数百种细胞类型需要与其角色相匹配的形貌，以完成对各种组织、器官和个体整体的搭建及功能的行使，从而使细胞形态变化的能力发挥到极致。除了整体形貌，细胞表面的微绒毛（microvillus）、伪足（lamellipodium/filopodium）、纤 / 鞭毛等突起也是细胞形态和功能的重要组成部分。细胞的塑形主要是由细胞骨架在很多结合蛋白的辅助下与细胞膜共同完成的（Akhmanova and Steinmetz，2015；Bodakuntla et al.，2019；Campellone and Welch，2010；Revenu et al.，2004）。

不同于单细胞生命与生俱来的基本形态，多细胞生命中各种细胞的形态需要从无到有（*de novo*）产生。也就是说，各种组织细胞形态的形成程序（或蓝图）须藏于卵细胞和干细胞基因组中，并在受精卵发育成个体的过程中，伴随细胞分化而相应开启。而且，转分化和 iPSC 等研究表明，这些程序确实是与细胞类型捆绑并可通过表观遗传调控进行重置。细胞形态的发生既有自主性的程序（比如在分离过程中已经变成球形的神经元，仍然能在体外培养过程中形成典型的神经元形态），也受到包括其他细胞在内的环境因素的影响，从而使相同和不同类型的细胞可以通过相互影响来协调一致地构成组织、器官及个体。不言而喻，细胞形态的异常必将导致其功能的异常，从而对个体的健康甚至存活产生不利影响。镰状细胞贫血可能是人类最早知晓的这类例子。

认识特定细胞形态的控制程序，不仅需要厘清不同物种和类型的细胞中形态调节因子的表达谱与动态表达调控，还需要认识它们之间相互作用的方式、时空效应和调节机理，构建数理模型等，因而需要多学科的交叉与紧密合作。并且，既往研究已对多种细胞的形态发生和表面特化结构形成的分子机理有诸多认识，但主要还是描述性的定性知识，缺少量化，其精细程度远不能用于建模模拟及预测形态发生的位置和过程。缺乏定量数据一直是生命科学研究的短板。这一方面归因于生命活动的高度动态性和复杂性，另一方面则限于研究工具，尤其是成像技术。最常用的光学显微镜受光学分辨率（约 200 nm）的限制，成像速度和持续时间也受设备限制。冷冻电镜断层技术在分辨率方面有很强的优势，但难以获取实时动态信息（Ma et al.，2019；Nicastro et al.，2006）。这些都大大局限了时空信息的精细程度，成为研究的瓶颈。尽管超分辨荧光显微镜在一定程度上突破了分辨率屏障，但其实远未达到能够清晰观察蛋白质和复合物的动态活动细节的程度。

2. 我国研究现状与水平

我国在细胞形态发生的研究方面已经具备一支不断壮大的有国际竞争力的队伍，在纤毛发生和功能研究中做出了重要的发现（Cao et al.，2012；Huang et al.，2018a；Liang et al.，2014；Liang et al.，2018；Long et al.，2016；Ran et al.，2020；Sun et al.，2019；Xie et al.，2020；Yan et al.，2020；Yu et al.，2019；Zhang et al.，2016a，2018c，2019；Zhao et al.，2013），在超分辨快速成像方面也取得可喜成果，搭建出世界领先的 GI-SIM 和 Hessian-SIM 荧光显微镜（Guo et al.，2018；Huang et al.，2018b）。将来需要更多的学科交叉和相互合作，以及对研究对象广度和深度的继续开拓。

八、细胞膜结构动态维持、裂解及修复

诱导质膜裂解的成孔毒素（pore forming toxin，PFT）是孔形成蛋白（pore forming protein，PFP）成员最多的家族，其通过特异性受体，如糖类、脂质和蛋白质，识别靶细胞并与质膜结合，使得PFT局部浓度增加，从而导致PFT的寡聚化，进而形成跨膜孔道（Dal Peraro and van der Goot，2016）。PFT根据在成孔过程中跨膜孔道的二级结构，被分为α-PFT和β-PFT两大类。α-PFT利用α-螺旋形成膜孔结构，主要包括ClyA家族和actinoporin家族（Cosentino et al.，2016）。ClyA是一种伸长的全α-螺旋蛋白，其α-螺旋在没入脂质层后大量重排，寡聚并环绕成膜孔，穿过脂质层（Mueller et al.，2009；Vaidyanathan et al.，2014；Wallace et al.，2000）；actinoporin家族由“螺旋－折叠－螺旋三明治”结构构成，成孔时，其氨基端α-螺旋脱离蛋白核心插入富含鞘磷脂的质膜，actinoporin膜孔的构象至今尚不清楚（Athanasiadis et al.，2001；Hinds et al.，2002；Kristan et al.，2009；Mancheno et al.，2003）。

β-PFT形成β-折叠桶样膜孔结构，主要由溶血素、气溶素（aerolysin）和胆固醇依赖性细胞溶素（CDC）三个家族组成。溶血素家族成员在寡聚成孔前体环状结构时，重排成反平行β-折叠桶，插入脂质层（Song et al.，1996）；单体气溶素组装成寡聚的孔前体结构，与膜表面糖蛋白受体结合，引发前体柔性环状结构重排，最终折叠成跨膜的β-桶（Degiacomi et al.，2013；Parker et al.，1994）；CDC是β-PFT中最为庞大的家族，大部分CDCs由革兰氏阳性菌分泌，它们以不同的方式促进细菌感染（Hotze et al.，2013；Hotze and Tweten，2012）。PFO是该家族蛋白的典型代表，其特点是一旦发生寡聚化，膜孔前体结构域发生朊病毒样螺旋－折叠转换。构象变化产生的β-发夹随后排列成微微弯曲的β-片层，形成一个包含80～200个β-折叠股的巨大β-桶（Ramachandran et al.，2004；Rossjohn et al.，1997；Sato et al.，2013；Shatursky et al.，1999）。此外，真核细胞产生的补体系统MACPF结构域也属于CDC，具有相同的折叠和组装方式（Hadders et al.，2007；Law et al.，2010；Rosado et al.，2007）。

膜裂解相关的程序性坏死主要有细胞裂亡和细胞焦亡两大类，分别由

MLKL和焦孔素（gasdermin，GSDM）家族介导。其中，我国科学家在细胞焦亡领域处于领先地位，首先发现焦孔素D（GSDMD）被上游胱天蛋白酶（caspase）切割激活，解除其羧基端对氨基端孔形成结构域的自抑制，通过形成跨膜孔复合物，诱导细胞裂解（Ding et al.，2016；Kayagaki et al.，2015；Liu et al.，2016b；Shi et al.，2015）。GSDMD属于gasdermin家族，包括GSDM-A、GSDM-B、GSDM-C、GSDM-D、GSDM-E和DFNB59几个成员，除DFNB59外，其他的GSDM都包含两个相似的结构域，能够诱导细胞焦亡，因此从分子水平看，细胞焦亡被重新定义为由gasdermin蛋白介导的程序性细胞坏死（Shi et al.，2017）。目前鼠源GSDMA3-NT“双环孔”近原子分辨率结构已经被解析，揭示了该家族PFP与CDC极其相似的孔组织形式，即由两个β-发夹构成结构单元，组装成由108个β-折叠股构成的大型反平行β-桶（Ruan et al.，2018）。未来对于该家族成员，尤其是细胞焦亡的主要执行蛋白GSDMD的膜孔组装动力学过程研究以及相关药物的研发设计将成为本领域发展的主要趋势。

另外一个重要的由膜裂解导致的死亡方式是细胞裂亡。当坏死信号激活MLKL上游激酶的催化活性，MLKL羧基端357/358丝氨酸残基被磷酸化修饰，进一步释放其氨基端功能结构域。MLKL随即形成寡聚体并转运到质膜，引起质膜裂解（Cai et al.，2014；Murphy et al.，2013；Sun et al.，2012；Wang et al.，2014）。近年来，越来越多的证据表明，MLKL介导的细胞坏死在炎症反应激活、微生物感染和防御、肿瘤免疫、胚胎发育、神经退行性疾病、生殖系统退化、组织修复以及癌症的发生、发展、转移等生理病理过程中发挥了重要的作用（Galluzzi et al.，2017；Jin et al.，2019；Li et al.，2017a;Weinlich et al.，2017；Yuan et al.，2019；Zhou et al.，2020）。因此，理解MLKL的作用机制，将帮助我们解决各领域的重大问题。尽管细胞裂亡领域的建立和发展早于细胞焦亡，该领域已经累积了大量的研究成果，但是由于MLKL蛋白分子极为不稳定，膜结合状态的结构难以解析，其破膜机制一直存在争议，目前主要提出了膜孔形成（Ros et al.，2017；Wang et al.，2014）、离子通道激活（Cai et al.，2014；Xia et al.，2016；Zhang and Han，2016）、淀粉样蛋白聚集（Liu et al.，2017b）等假说。然而，至今没有任何理论能够完整解释MLKL诱导质膜崩解的动态过程，阻碍了我们对细胞裂亡机制的深入理解和临床应用研究。

与膜的裂解相反的生物学过程即膜的修复，能够帮助细胞清理、重建损伤的质膜，维持质膜的完整和细胞内环境稳态，其中以 ESCRT 介导的出芽研究最为丰富。转运必须的内体分选复合物（ESCRT）系统由包括 30 多个蛋白质组装成的 4 个复合物组成，主要负责膜的融合－裂分、膜系统货物分选、囊泡运输等过程（Christ et al.，2017）。研究人员发现，细胞裂亡和细胞焦亡的膜裂解都能够被 ESCRT 复合物缓解。程序性坏死的早期，ESCRT-Ⅲ复合物在膜损伤部位富集组装，诱导细胞向外出芽并释放囊泡，从而清除损伤的质膜。ESCRT-Ⅲ复合物相关组分敲减能够明显促进细胞裂亡和细胞焦亡的进程（Gong et al.，2017；Ruhl et al.，2018）。

除了能够被 ESCRT 依赖的胞外囊泡清除，PFT 引起的膜损伤还能通过胞吐－内吞循环、膜张力调节和修复斑三种方式修复。带有膜孔的质膜通过内吞作用，形成内吞小泡将损伤的质膜运输进细胞内降解，再通过胞吐作用补充质膜损失的膜脂。当胞吐作用进行程度强于内吞时，膜脂含量增多，导致膜张力降低，也有利于膜的修复。此外，膜损伤引起 Ca^{2+} 内流，刺激胞内囊泡聚集融合，在损伤部位募集形成补丁样修复斑，介导损伤质膜的封闭（Andrews and Corrotte，2018；Cooper and McNeil，2015）。

由于膜裂解机制具有多样性，其对应的膜修复机制是否也有对应的选择性是完全未知的。总之有关膜系统的裂解、修复动态变化过程复杂，关于其调控机制和相关疾病的研究依然相当不成熟。

九、亚细胞结构在体高时空分辨解析技术

1. 研究背景及现状

光学显微镜技术的独特性来自两个方面：①光了对生物样品的伤害性小，从而可以对样品进行长时间的活体观测；②通过结合荧光蛋白或染料的分子特异性标记技术，科学家可以“点亮”特定的生物分子或细胞器，以观测其在生物过程中的行为和功能。超分辨成像技术在这两点优势的基础上突破了衍射极限分辨率，成为描绘解析亚细胞结构、追踪其动态变化、研究其相关

功能的重要工具。

现有的超分辨显微镜技术面对多样性的生命科学研究需求，依然面临诸多局限，当前超分辨显微技术的短板和发展趋势是：①传统超分辨成像技术所需荧光信号量较高、激发光较强、原始图像采集时间较长。因此，如何以尽可能低的激发光强高和尽可能短的采集时间突破衍射极限分辨率，实现活体、长时程、高速超分辨成像是当前超分辨技术开发的前沿。②虽然理论上大多数超分辨显微镜技术可从二维扩展至三维成像，但如果对全细胞进行成像的话，往往需要数分钟甚至小时量级的采集时间。因此，如何实现高速三维超分辨成像也是当前的研究热点。③现代生命科学研究越来越倾向于利用多细胞生物和类器官等体系来研究对人体更具普适性的科学问题。然而，多细胞生物内部的光学非均匀性大大增加，引起激发光和信号光波前畸变，导致超分辨成像技术所依赖的物理效应在多细胞生物中不能完美实现。因此，研发对生物样品的实时像差矫正技术，实现复杂样本的大视场超分辨成像是当前光学成像领域的难点和热点。

另外，光片显微成像技术在模式动物、类器官等组织层次的成像实验中越来越受青睐。光片成像的核心优势在于，采用激发物镜和探测物镜相互垂直的放置方式，避免了探测物镜焦平面上下的荧光分子被激发而产生大量背景荧光信号，同时显著降低了光漂白。因而，相对于其他三维成像技术（如共聚焦显微镜），光片显微镜对大尺寸样品进行长时间连续成像有着独特的优势。但是，光片显微镜技术也有局限，主要表现在两个方面：①为了在较大范围内保持激发光的厚度均匀从而在整个模式动物成像范围内获得相对均匀的照明，因此用于光片成像的激发物镜和探测物镜的数值孔径都较小，导致光薄片较厚、光片成像的分辨率不高，通常分辨率为横向 0.3～0.6 μm、纵向 1～3 μm。②激发物镜和探测物镜垂直放置的方式使得现有超分辨成像技术所需的受激辐射耗尽、照明条纹平移、旋转等操作都难以实现，导致光片成像与超分辨成像难以融合并应用于更广的生物学研究应用场景。因此，如何将超分辨成像技术，特别是适于活细胞成像的结构光超分辨成像技术与光片成像技术融合，开发超分辨光片显微镜是显微成像领域的前沿技术问题。这一技术突破可使两者的优缺点互补，将超分辨成像由单细胞层次提升到多细胞模式生物层次，为当前生命科学研究的诸多前沿问题带来新的观测技术方法。

对于更大尺寸、更深层的活体组织成像（如透过玻璃窗口观察鼠或猴脑的神经元结构及其对特定刺激的响应），多光子（双光子或三光子）荧光显微镜是目前采用的主流成像技术。虽然多光子荧光成像采用近红外Ⅱ区较长波长的激发光，从而大大增加了组织样本的成像深度，但是其往往需要点扫描成像，导致成像速度较慢。此外，组织样品引入的散射和像差会导致成像质量随着成像深度增加而下降。另外，在组织样品较深范围内获得较高分辨率甚至超分辨率成像性能目前依然是空白。因此，实现在体高时空分辨解析亚细胞结构既是今后生命科学研究的重要需求，同时也是显微成像技术领域的发展趋势。

大尺度、高通量电子显微三维重构技术可对固定样品达到 10 nm 以下的分辨率，结合连续超薄切片或原位离子束切片等方法，为研究大尺度生物样品的亚细胞三维结构提供了新的技术方案。由于多细胞生物或组织样品尺寸较大，通常要求连续收集上千张超薄切片，且要求切片不能有大的变形、次序不能乱；或者利用离子束减薄方法连续工作数天甚至数周时间才能获得纳米级分辨率下的大尺度精细三维重构。国际上哈佛大学、珍妮莉亚研究园区（Janelia Research Campus）等机构已初步实现了对商业化扫描电镜、连续切片机、离子束减薄设备等的二次开发，使这些设备能够高度自动化处理长时间连续采集过程中的种种不稳定性因素，实现一次性高通量采集整个果蝇脑等较大尺寸的样品（Hayworth et al.，2020）。随着透射电子显微镜和大尺寸高灵敏电子探测相机的发展，利用透射电子平行光进行一次性成像可大大提高以往扫描电镜成像的通量，然而传统透射电镜需要手动更换样品和位移台，导致实际成像通量较低。今后对透射电镜进行工程化改造，并结合连续切片等技术，预期可实现更高通量的体电子显微成像系统。

2. 我国研究现状与水平

上述高时空分辨亚细胞结构解析技术的发展一直是国际显微成像领域的研究前沿和热点。近年来，我国在光学显微成像技术的多个细分领域取得了一些重要进展：在基于结构光照明、单分子定位、受激辐射耗尽的三类主流超分辨成像技术上都成功开发了实验室样机（Gu et al.，2019；Yang et al.，2016）；特别是在结构光照明超分辨显微镜方面，发展了掠入射结构光（Guo et al.，2018）、非线性结构光（Li et al.，2015b）以及海森去噪算法结构光（Huang et al.，2018b）等适于高速长时程超分辨活细胞成像的新技术、新方法，形成了一定的特色技术体系。这为突破上述超分辨成像技术的瓶颈奠定了良好基础。但是在光片显微镜技术上，我国能独立搭建光片显微镜系统的实验室依然很少，且大多是重复国外的光学技术方案，只是应用到不同的生物课题中，尚未突破超分辨光片显微镜的技术瓶颈。对于多光子显微镜，目前全世界大部分神经科学实验室都能够根据特定的研究课题进行一定程度的改造。但是，涉及融合自适应光学、光场调控和探测、结合超分辨成像技术等前沿多光子技术开发，我国仍然需要加强这些技术细分领域的投入。对于大尺度电子显微三维重构技术，我国在高通量连续切片（Li et al.，2017c）、大束流和摇摆物镜等高通量扫描电镜技术上具有特色，但需要指出的是我国在透射电镜特别是高压透射电镜上尚未突破，且国内团队也缺乏对电镜成像流程的诸多设备进行二次工程化开发以实现自动化高通量的探索，未来需要加强投入。最后，无论是超分辨活体显微成像或者是高通量三维电子显微成像，其共同的特点是数据量随着分辨率提高呈立方甚至四次方关系增加，如何在海量数据中挖掘、提取、分析其中的生物学信息是目前整个领域面临的严峻挑战。近年来发展的人工智能深度学习等技术有望成为解决这一挑战的解决方案。因此，未来在解析亚细胞结构这一科学问题中，多学科交叉融合的研究范式将成为必然。

第四节　我国发展战略与重点方向

一、总体思路与发展目标

1. 内膜系统细胞器的结构功能与相互作用

内膜系统细胞器在我国面临的问题和挑战在于如何在经典领域挖掘新的

创新点、提出新的理论、开拓新的思路，以及如何摆脱“跟跑”国际前沿热点的状态。目前，我国研究人员在膜转运的精细分选调控、内膜细胞器互作的生理功能和内膜完整性的维持等方向上已经独辟蹊径，有望形成新的独立的研究特色。未来的内膜系统研究将由单个的细胞器的研究转向多个细胞器的集成研究，将更多聚焦内膜系统协同互作的精细调控机制和生理功能及意义。同时，内膜系统膜结构的完整性维持、修复与降解也将成为新的研究热点和前沿。最后，内膜系统紊乱的病理相关性，尤其是在发育缺陷和神经退行性疾病的发生发展等方面，将受到持续关注。

2. 无膜细胞器形成和动态平衡的调控机制

该学科还属于起步阶段，在国际上仍面临着包括理化机制不清楚、研究方法和手段匮乏等严峻的挑战。国内在该领域的研究中也存在研究方向覆盖不够全面、学科间交流合作不足等问题。这些问题都需要在未来得到合理的解决。国家针对这一新兴学科的布局对于揭示生物大分子凝聚体的生理功能、组装动态调控的理化机制，发展新的研究相分离和相变的方法与手段，解析相分离和相变调控和与干预相关疾病的分子机制，具有重要的科学意义和潜在的医药应用价值，也是国家重大需求。未来研究以生物大分子“相分离”和“相变”的生物学功能与可塑性调控，及其与人类重大疾病（如神经退行性疾病、癌症、糖尿病等）发生发展的关联性为核心研究内容，推动生物、物理、化学等多学科间的交叉融合，综合利用各学科的优势、理论和方法手段，揭示生物大分子“相分离”和“相变”的生物学功能、动态调控机制及其理化机理，阐明生物大分子凝聚体与疾病发生发展的关系，发现疾病相关的全新治疗靶点，并开发相应的化学、基因干预等手段，同时开发一系列用于生物大分子“相分离”和“相变”研究的新方法和新技术。在该领域取得一系列国际领先的基础研究成果和应用转化成果，确保在该领域形成多个具有国际领跑地位的研究方向。在国际上率先揭示一批重要的生物大分子凝聚体的生物学功能和动态调控机制；发现一批全新的生物大分子凝聚体，并鉴定其生物学功能，阐述其组成成分、组装方式、动态调控方式等分子机制；在生物大分子“相分离”和“相变”的物理化学机理上取得突破性的理论成果；阐明生物大分子凝聚体“相分离”和“相变”的异常调控与疾病发生发展的关联，发现一批潜在的治疗靶点，并发展相应的干预手段，用于相关疾病的治疗；开发一系列用于生物大分子“相分离”和“相变”研究的原创性新技术与新方法。

3. 核内无膜细胞器的结构、组装和功能调控

细胞核无膜细胞器失调与多种人类重要疾病息息相关。不同的无膜细胞器发挥功能与它们的结构和组装密切耦联。因此，理解细胞核无膜细胞器超微结构、动态组装以及耦联的功能调控对认识细胞核无膜细胞器发挥功能的分子机制，多种疾病的病理机制及相关诊疗有着重要的意义。系统研究细胞核无膜细胞器的组分，从分子层面理解细胞核无膜细胞器组装与调控的机制；在微观尺度研究细胞核无膜细胞器的精细结构、组装过程和发挥作用的机制。此外，在细胞、组织、器官和模式生物个体的层面，理解细胞核无膜细胞器的分布、结构、调控，研究不同组织器官内细胞核无膜细胞器的功能异同和调控通路，以及细胞核无膜细胞器相关疾病所耦联机制的特性与共性。在更高的时空分辨率尺度下，解析细胞核无膜细胞器的超微结构和分子组成；认识细胞核无膜细胞器及其所蕴含的RNA和蛋白质在相关疾病发生发展中的作用；建立新的研究体系，解析细胞核无膜细胞器异常与相关疾病发生的关联，并开发细胞核无膜细胞器相关的新靶点，拓展疾病诊疗的理论框架。

4. 不同类型蛋白酶体及其亚基间的相互作用与靶向干预

蛋白酶体抑制剂已成功应用于治疗多发性骨髓瘤和套细胞淋巴癌，但对蛋白酶体通路的干预将可能成为其他肿瘤和众多其他疾病的治疗方法。我国应优先研究不同类型蛋白酶体及其亚基间的相互作用机制与生理病理意义，解析蛋白酶体亚基翻译后修饰及其转录调控机制，阐释不同类型蛋白酶体选择性地识别蛋白质底物的机制，研发更有选择性的靶向蛋白酶体药物，为治疗肿瘤和新冠肺炎、肺结核、神经退行性疾病、自身免疫疾病等其他疾病做出实质性贡献。

5. 线粒体稳态调控机制及其在衰老和相关疾病中的作用

线粒体生物学一直是生命科学研究的前沿。从细胞代谢、线粒体蛋白结构与功能看，线粒体与很多疾病的发生都有密切关系。我国在线粒体领域的

研究发展迅速，取得了一大批具世界前沿水平的成果，但是在引领世界的科研成果方面仍然有所欠缺。目前还没有形成线粒体研究的科学高地。应继续壮大线粒体生物学研究队伍，鼓励原创性基础研究，建立线粒体生物学研究的研究高地；发展与疾病诊断和治疗相关的应用研究，开发线粒体疾病的诊断和治疗的新策略与新方法。

6. 细胞骨架的动态调控及其与生理功能和疾病的相关性

细胞骨架是细胞的基本结构组成，细胞骨架调节异常与多种疾病发生发展密切相关，因而细胞骨架动态调控的基础、生理和病理作用研究不仅对细胞生物学基础理论有贡献，而且对于理解致病机制进而有针对性地开展预防和治疗具有应用价值。由于细胞骨架研究内涵广泛，目前我国多数研究方向较为分散，缺乏系统、深度和协同研究。要进一步加强对细胞骨架本身调控机制的基础研究，开发相关研究技术和方法，系统地开展细胞骨架生理和病理作用的研究，针对前沿研究方向统筹规划和资源配置，推动相关药物开发；遵循从基础、生理到病理、药物开发的研究思路，不断深化细胞骨架基础理论的研究，探索其与生理功能和疾病的联系，指导相关疾病的致病机制和治疗策略的研究；系统地阐述细胞骨架动态调控的机制，建立相关研究技术和方法，揭示细胞骨架的生理和病理作用及其机制，推动药物开发。

7. 细胞形态形成和维持的调控机制

阐明细胞形态可塑性的调控机制不仅可揭示生命建造的相关蓝图，而且对功能细胞和人工生命的理性设计与运用将发挥重要的作用。我国目前的研究较为零散，缺乏生物学内部，尤其是与物理、数学、信息科学等方面的交叉和多领域团队的协同合作。我国应建立包括物理、数学和信息等学科的交叉团队，系统阐释重要细胞类型的形态发生、维持和调节因子及其作用机理，建立和完善其时空依赖的调控网络与其构效、调控关系，建立理论模型，模拟、预测细胞表面的形态变化，并逐步实现细胞形态、表面结构和相关功能的人工理性改造。

8. 细胞膜结构动态维持、裂解及修复

目前我国有关细胞膜的研究主要集中在膜受体蛋白的结构或信号转导，

非常具有局限性。有关膜动态完整性，如质膜破裂和修复的机制的研究与相关药物研发方面的投入是远远不够的。我国首先需要加大介导膜裂解、修复的蛋白质机器的研究，揭示一系列重要的膜破裂和修复分子机器的组装方式与其动力学过程等分子特性，同时发展与临床相关的靶向药物研发和临床应用。

9. 亚细胞结构在体高时空分辨解析技术

需要指出的是，我国在透射电镜特别是高压透射电镜上尚未突破，且国内团队也缺乏对电镜成像流程的诸多设备进行二次工程化开发以实现自动化高通量的探索，未来需要加强投入。

二、优先发展领域（或重要研究方向）

1. 内膜系统细胞器的结构功能与相互作用

内膜系统细胞器的结构功能与相互作用领域需要优先发展以下重要研究方向：基于纯化组分的内质网特征膜结构的体外重构，以及内质网等细胞器塑形的生理功能研究；内质网膜转运中蛋白质的运动特征和输出机制；内质网应激下信号的跨细胞器、跨细胞传递，以及内质网应激的病理相关性；内质网和高尔基体协同调控蛋白质稳态的机制；内膜系统调控自噬体发生的机制和生理病理功能；溶酶体稳态平衡的调控机制、生理功能，特化溶酶体的功能，以及溶酶体相关疾病的细胞学机制；以溶酶体为出发点的内膜细胞器膜完整性研究；不同内吞途径的分子机制，细胞质膜组成和性质的改变的调控，内吞体与其他内膜结构的相互作用，以及在机体水平进行内吞过程的功能研究；膜性细胞器之间的物质、能量和信息交换互作是目前研究的热点，参与细胞分泌、代谢、分裂、分化、死亡等过程，因此鉴定参与调控此类过程的关键因子并阐明其作用的分子机制是重要的研究方向。发现新的膜性细胞器以及新的互作行为和功能也是研究热点。以内膜系统细胞器为集成研究对象，抓住前瞻性研究如膜完整性等在国际上方兴未艾的契机，提前布局，重点支持，包括组建并支持内膜系统研究优势团队。加大高水平人才的引进，不断增强国内的研究实力，特别是要将具有交叉学科研究背景的、高水平的海外研究者吸引到国内；设立交叉攻关项目，组建多学科跨领域交叉融合的科研团队，搭建细胞器互作研究大型仪器平台，满足特色化用户成像需求，为致力于细胞器互作的研究者提供空间和条件。

2. 无膜细胞器形成和动态平衡的调控机制

针对相分离和相变这一新兴学科的特点，结合前期国内科学家在该领域的研究进展，建议从以下几个方面开展“相分离”和“相变”的突破性研究。这些方向的研究开展将推动我国在该领域取得国际领跑的地位，具有重大的战略意义。

（1）深化国内生物大分子“相分离”和“相变”优势研究方向。继续深化我国在神经突触形成和可塑性调控、细胞自噬、细胞不对称分裂、表观遗传调控等“相分离”和“相变”优势领域的发展，鼓励这些优势领域开展连续性、系统性和深入性的研究。在已有工作的基础上，进一步鉴定新的凝聚体组分和调控因子，阐述这些新组分和调控因子的生物学功能与作用机理，并将现有的基础研究成果向应用转化过渡。

（2）生物大分子“相分离”和“相变”的生物学功能研究。进一步探索并拓展生物大分子凝聚体在生物体内发挥的功能，鉴定新型的生物大分子凝聚体，着重研究生物大分子利用“相分离”和/或“相变”在基因表达、物质运输、信号转导、细胞命运决定、生殖发育、胁迫环境适应等方面发挥的生理功能，阐明生物大分子凝聚体之间和与有膜细胞器之间的关系及协同参与多种生理功能的机理。探讨生物大分子“相分离”在生命起源中可能发挥的作用。

（3）生物大分子凝聚体的动态组装和分子调控机制研究。鉴定生物大分子凝聚体的组成成分，阐明各组分在生物大分子凝聚体组装过程中的具体作用。研究多层相分离凝聚体的动态调控方式和组装机制，以及不同的相分离凝聚体之间发生相互作用的分子机制和生物学功能。发现在生物大分子凝聚体的组装、性质变化、功能维持和动态变化等过程中发挥调控功能的分子，探究细胞周期、化学修饰、温度、渗透压、机械力、氧化还原环境等内外部条件变化对生物大分子凝聚体的调节作用，以全面揭示生物大分子凝聚体稳态和/或动态精细调控的具体分子机制。

（4）生物大分子“相分离”和“相变”与疾病的关系研究及干预手段的开发。研究生物大分子“相分离”“相变”异常与各种疾病如神经退行性疾病、癌症、糖尿病、心血管疾病等之间的关联，揭示生物大分子凝聚体异常调控在各类疾病发生发展过程中的作用和分子机制。寻找新型疾病治疗靶点，开发全新的疾病预测和治疗策略，利用生物大分子“相分离”“相变”的模型和潜在疾病靶标设计或进行新药筛选，或开发相应的治疗策略，以实现对相关疾病进行干预治疗的目的。

（5）生物大分子“相分离”和“相变”研究新技术、新方法的开发。改进物理、化学、材料学等学科中已有的用于“相分离”和“相变”研究的实验方法与技术手段，使其适用于生物大分子凝聚体的研究。发展高分辨/单分子荧光、液体/固体核磁共振波谱、单分子磁共振、亚细胞结构显微切割技术、单分子声镊、单分子磁镊、单分子光镊、粒子跟踪技术、红外及拉曼光谱、原子力显微镜、冷冻电镜、冷冻电子断层扫描技术、荧光-电镜联用、微流控等技术，开发新型小分子化合物探针，以用于生物大分子凝聚体的研究，尤其是用于在细胞内的生物大分子凝聚体的研究。根据生物大分子“相分离”“相变”的特点，有针对性地开发其他高效的原创性新技术、新方法。

（6）生物大分子“相分离”和“相变”的理化机制研究。拓展物理、化学、高分子材料等科学领域中有关“相分离”和“相变”的理论，发展适用于研究生物大分子凝聚体的理论模型和计算模拟方法。体外重构生物大分子凝聚体，综合利用生物、物理、化学、材料科学等多学科的理论模型和方法手段研究生物大分子凝聚体的动态组装与分子调控机制，在原子/分子的水平上剖析生物大分子凝聚体内部各组分之间的相互作用方式，从物理化学的角度进一步加深对生物大分子“相分离”和“相变”的理解，并在理论层次预测一系列生物大分子凝聚体的生物学行为。

（7）细胞核无膜细胞器的结构、组装和功能调控。解析细胞核无膜细胞器的超微结构、分子构成及其分布特异性是理解细胞核无膜细胞器功能发挥的前提，系统开展不同细胞核无膜细胞器的精细结构和组装过程的观察有助于理解它们发挥功能的特点，并可为人工干预细胞核无膜细胞器的特定组装过程提供理论基础。因此，建议优先发展细胞核无膜细胞器结构、组装、转录调控的相关研究。细胞核无膜细胞器的研究也依赖于多学科的有机整合，建议同步开发高精度的单细胞、亚细胞器、单分子、实时动态的技术手段与创新研究方法。对细胞核无膜细胞器的研究，建议采用多学科多交叉的方式，从不同角度研究细胞核无膜细胞器的结构、组装、动态过程以及在生理和病理条件下的功能调控；着力发展细胞核无膜细胞器研究所需要的物理、化学以及生物技术手段，提高仪器设备和技术的支持。

3. 线粒体稳态调控机制及其在衰老和相关疾病中的作用

（1）线粒体生物发生和线粒体自噬调控新机制。线粒体是细胞能量的主要来源，是细胞代谢的核心枢纽。线粒体数量的多寡与好坏对细胞的功能有着十分重要的影响。由于线粒体是被氧自由基攻击的主要位点，旧的、受损伤的线粒体需要通过线粒体自噬的方式被及时清除，新的、健康的线粒体通过生物发生的方式重新生成以满足细胞的正常需求（Ploumi et al.，2017）。细胞内线粒体保持新旧更替的动态平衡对细胞维持稳态十分关键，线粒体生物发生和线粒体自噬异常与肿瘤、神经退行性疾病、糖尿病等多种疾病以及衰老密切相关（Li et al.，2019b；Sliter et al.，2018；Wu et al.，2019a）。深入研究线粒体生物发生和线粒体自噬，探索其调控新机制，对于理解线粒体的功能和治疗线粒体相关疾病都有重要意义。

（2）线粒体蛋白质稳态和线粒体与细胞核互作。与其他细胞器不同，线粒体是半自主细胞器，线粒体蛋白质由细胞核基因和线粒体基因共同编码。细胞核编码蛋白和线粒体编码蛋白在基因转录、翻译，蛋白质转运、降解等水平上的协同是维持线粒体正常运转的基础。线粒体蛋白质稳态的维持不仅与线粒体内部蛋白质稳态维持机器息息相关（Li et al.，2019c），而且细胞核编码了线粒体的大部分蛋白质，因此需要线粒体与细胞核的密切协作（Bragoszewski et al.，2017；Pfanner et al.，2019；Priesnitz and Becker，2018）。线粒体如何通过逆向调控的方式向细胞核发出信号，以满足自身所需，细胞核又如何通过正向调控的方式向线粒体发出指示，以应对挑战，对于理解不同环境下的线粒体功能调控十分关键。

（3）线粒体与细胞代谢和细胞死亡。在机体的生殖、发育、生长、衰老、疾病等过程中，细胞的功能和环境不断发生变化，细胞的代谢活动也随之改变，甚至发生细胞死亡，以应对不同需求。大量研究表明，线粒体是细胞代谢和细胞凋亡的调控中心。然而，随着研究的深入，越来越多的细胞死亡方式（如坏死、铁死亡、焦亡等）被发现，线粒体是如何在这些死亡方式中起作用的，仍然有很多问题待解决（Bock and Tait，2020）。另外，在不同环境下，线粒体如何调控细胞代谢，以实现细胞代谢重编程，是理解疾病和发育过程中细胞功能变化的前提（Ma et al.，2015）。深入理解线粒体在细胞代谢和细胞死亡中的作用与机制对于治疗肿瘤等疾病有关键性的影响。

（4）线粒体蛋白质结构与功能。解析蛋白质结构是理解其功能的基础。结构生物学的发展极大地促进了人们对线粒体功能的理解。未来仍需进一步解析线粒体蛋白质结构，尤其是线粒体中大或者超大蛋白复合体的结构（Letts and Sazanov，2017）。通过结构生物学的手段揭秘线粒体大蛋白复合体的组装机制，是进一步理解线粒体功能的基础，也将为线粒体相关疾病的治疗提供新的药物靶点。

（5）线粒体与免疫和衰老及相关疾病。线粒体在免疫调节中有重要作用，线粒体产生的自由基、线粒体 DNA 等物质是重要的免疫激活剂。长期以来的研究表明，线粒体功能异常是衰老及相关疾病的重要致病因素。在衰老的过程中往往伴随着严重的炎症反应。然而，线粒体在免疫和衰老中的具体分子机制仍然有待进一步的研究（Kauppila et al.，2017；Sliter et al.，2018；West et al.，2015）。目前，我国面临着极大的老龄化压力，其严重威胁社会经济发展、人民生命健康和家庭幸福。因此，对线粒体在免疫和衰老中作用的研究显得尤为重要。

（6）线粒体 DNA 突变如何导致疾病。线粒体有自己的 DNA，可以编码 13 种多肽、2 种核糖体 RNA、22 种转运 RNA，对于线粒体的功能十分关键。线粒体 DNA 突变是导致衰老以及肿瘤等多种疾病的重要原因（Schon et al.，2012）。线粒体 DNA 突变可通过遗传的方式获得，另外，因其易被氧自由基攻击，在生理或者病理性条件下，线粒体 DNA 也容易发生突变，导致线粒体异质性（Sharpley et al.，2012）。但是，线粒体 DNA 突变的致病机制仍然不清楚，研究线粒体 DNA 突变相关疾病的具体分子机制将为这些疾病的治疗带来曙光。

（7）靶向线粒体的小分子化合物筛选和线粒体疾病治疗的新策略。线粒体功能异常可能会影响机体的各个组织和器官，包括大脑、心脏、肝脏、眼睛、肌肉、神经等。线粒体疾病影响人群广泛，往往症状严重，但是缺乏有效的治疗方法，迫切需要新的靶向线粒体的药物和疾病治疗策略（Russell et al.，2020）。近年来，随着线粒体研究领域的不断深入，对靶向线粒体的小分子化合物筛选提出了新的要求，而技术的快速进步也为发展出线粒体疾病治疗的新策略提供了深厚的土壤。

线粒体是细胞中十分重要的细胞器，为机体提供了90%以上的能量，也是物质代谢的核心枢纽，是细胞凋亡、天然免疫等途径的调控中心。关于线粒体的研究曾多次获得诺贝尔奖。目前，线粒体研究仍然有很多未解之谜，线粒体功能异常依然严重威胁人类的健康，却缺乏有效的治疗手段。关于线粒体的研究在未来相当长的时间里仍然有巨大的科学价值和社会经济价值。为了促进我国线粒体生物学的研究和社会经济发展，保障人民身体健康，十分有必要建立线粒体生物学科学研究中心，并适时发布重大计划，推动线粒体研究向纵深发展，并开发出新的线粒体研究技术。

4. 细胞骨架的动态调控及其与生理功能和疾病的相关性

（1）微管、微丝和中间丝的动态调控机制。这些基础性问题的回答将为细胞骨架领域的发展打下坚实的基础。

（2）微管、微丝和中间丝互作调控。此方面研究有利于对细胞骨架整体功能的认识。

（3）微管、微丝和中间丝的功能调控与细胞功能实现和疾病发生的关系。这些研究将揭示细胞骨架的生理和病理作用与机制。

（4）靶向细胞骨架的药物研发。这一方向将推动细胞骨架领域的应用研究。

5. 细胞膜结构动态维持、裂解及修复

需要加大介导膜裂解、修复的蛋白质机器的研究，揭示一系列重要膜破裂和修复分子机器的组装方式和其动力学过程等分子特性，同时发展与临床相关的靶向药物研发和临床应用。关注膜脂分子在膜裂解、修复过程中的角色。除了传统的细胞生物学、生物化学、结构生物学的实验手段，我们还需要结合超高分辨率成像、表面化学、生物力学、蛋白质－膜系统的体外重构、计算机模拟等生物物理、计算生物学的技术方法对膜破裂和修复机制展开系

统研究。加大投入膜裂解、修复的疾病模型和临床应用的研究，吸引整合相关科研人员和建设发展相应基础设施，促进该领域的人才培养、科学研究和临床转化。

6. 亚细胞结构在体高时空分辨解析技术

当今生命科学研究越来越倾向于利用多细胞生物来研究对人类健康更具普适性的科学问题。与传统的单细胞样品相比，多细胞模式生物的尺寸更大且内部结构更复杂，使得光学非均匀性大大增加。这需要发展毫米级成像视野、实时自适应像差矫正、高时空分辨率、低光毒性的显微镜技术。这既是国际显微成像领域的技术发展前沿，同时也是生命科学研究的普遍且迫切需求。

生物显微成像技术的发展需要光学工程、计算机、自动化、物理、生物等多学科的交叉融合。近年来人工智能深度学习算法有了重大突破，多个开放的深度学习平台使得人工智能开始对许多行业的发展产生变革性影响。然而，将深度学习等人工智能方法应用于生物显微成像技术研究领域尚处于起步阶段。在显微成像系统硬件开发的基础上，探索开发基于深度学习网络的生物图像去噪和图像重建算法，发展新型智能显微成像系统，有望突破对多细胞生物样品进行高时空分辨、长时程成像的瓶颈，进一步降低获得高质量图像所需的采集时间和照明光强，开拓显微成像技术发展的新技术路径。

本章参考文献

Adams J. 2004. The development of proteasome inhibitors as anticancer drugs. Cancer Cell, 5（5）: 417-421.

Adriaens C, Standaert L, Barra J, et al. 2016. p53 induces formation of NEAT1 lncRNA-containing paraspeckles that modulate replication stress response and chemosensitivity. Nat Med, 22（8）: 861-868.

Ahat E, Li J, Wang Y. 2019. New insights into the golgi stacking proteins. Front Cell Dev Biol, 7: 131.

Akhmanova A, Steinmetz M O. 2015. Control of microtubule organization and dynamics: two ends in the limelight. Nature Reviews Molecular Cell Biology, 16（12）: 711-726.

Al-Shami A, Jhaver K G, Vogel P, et al. 2010. Regulators of the proteasome pathway, Uch37 and Rpn13, play distinct roles in mouse development. PLoS One, 5（10）: e13654.

Ali A, Wang Z, Fu J, et al. 2013. Differential regulation of the REGgamma-proteasome pathway by p53/TGF-beta signalling and mutant p53 in cancer cells. Nat Commun, 4: 2667.

Anchoori R K, Karanam B, Peng S, et al. 2013. A bis-benzylidine piperidone targeting proteasome ubiquitin receptor RPN13/ADRM1 as a therapy for cancer. Cancer Cell, 24（6）: 791-805.

Andrews N W, Corrotte M. 2018. Plasma membrane repair. Curr Biol, 28（8）: R392-R397.

Athanasiadis A, Anderluh G, Macek P, et al. 2001. Crystal structure of the soluble form of equinatoxin Ⅱ, a pore-forming toxin from the sea anemone Actinia equina. Structure, 9（4）: 341-346.

Audas T E, Jacob M D, Lee S. 2012. Immobilization of proteins in the nucleolus by ribosomal intergenic spacer noncoding RNA. Mol Cell, 45（2）: 147-157.

Badros A, Hyjek E, Ma N, et al. 2017. Pembrolizumab, pomalidomide, and low-dose dexamethasone for relapsed/refractory multiple myeloma. Blood, 130（10）: 1189-1197.

Ballabio A, Bonifacino J S. 2020. Lysosomes as dynamic regulators of cell and organismal homeostasis. Nat Rev Mol Cell Biol, 21（2）: 101-118.

Banani S F, Lee H O, Hyman A A, et al. 2017. Biomolecular condensates: organizers of cellular biochemistry. Nat Rev Mol Cell Biol, 18（5）: 285-298.

Bartel B. 2015. Proteaphagy-selective autophagy of inactive proteasomes. Mol Cell, 58（6）: 970-971.

Boccon-Gibod C, Talbot A, Le Bras F, et al. 2020. Carfilzomib, venetoclax and dexamethasone for relapsed/refractory multiple myeloma. Br J Haematol, 189（3）: e73-e76.

Bock F J, Tait S W G. 2020. Mitochondria as multifaceted regulators of cell death. Nat Rev Mol Cell Biol, 21（2）: 85-100.

Bodakuntla S, Jijumon A S, Villablanca C, et al. 2019. Microtubule-associated proteins: structuring the cytoskeleton. Trends Cell Biol, 29（10）: 804-819.

Boeynaems S, Alberti S, Fawzi N L, et al. 2018. Protein phase separation: a new phase in cell biology. Trends Cell Biol, 28（6）: 420-435.

Boisvert F M, van Koningsbruggen S, Navascues J, et al. 2007. The multifunctional nucleolus. Nat Rev Mol Cell Biol, 8（7）: 574-585.

Bragoszewski P, Turek M, Chacinska A. 2017. Control of mitochondrial biogenesis and function by the ubiquitin-proteasome system. Open Biol, 7（4）: 170007.

Brangwynne C P, Mitchison T J, Hyman A A. 2011. Active liquid-like behavior of nucleoli determines their size and shape in Xenopus laevis oocytes. Proc Natl Acad Sci U S A, 108（11）: 4334-4339.

Cai Z, Jitkaew S, Zhao J, et al. 2014. Plasma membrane translocation of trimerized MLKL protein is required for TNF-induced necroptosis. Nat Cell Biol, 16（1）: 55-65.

Calo E, Gu B, Bowen M E, et al. 2018. Tissue-selective effects of nucleolar stress and rDNA damage in developmental disorders. Nature, 554（7690）: 112-117.

Campellone K G, Welch M D. 2010. A nucleator arms race: cellular control of actin assembly. Nat Rev Mol Cell Biol, 11（4）: 237-251.

Cao J, Shen Y, Zhu L, et al. 2012. miR-129-3p controls cilia assembly by regulating CP110 and actin dynamics. Nat Cell Biol, 14（7）: 697-706.

Cao Y, Zhu H, He R, et al. 2020. Proteasome, a promising therapeutic target for multiple diseases beyond cancer. Drug Des Devel Ther, 14: 4327-4342.

Cardinale S, Cisterna B, Bonetti P, et al. 2007. Subnuclear localization and dynamics of the Pre-mRNA 3' end processing factor mammalian cleavage factor I 68-kDa subunit. Mol Biol Cell, 18（4）: 1282-1292.

Chen B, Deng S, Ge T, et al. 2020a. Live cell imaging and proteomic profiling of endogenous NEAT1 lncRNA by CRISPR/Cas9-mediated knock-in. Protein Cell, 11（9）: 641-660.

Chen J, Wang Y, Xu C, et al. 2021. Cryo-EM of mammalian PA28alphabeta-iCP immunoproteasome reveals a distinct mechanism of proteasome activation by PA28alphabeta. Nat Commun, 12（1）: 739.

Chen L B, Ma S, Jiang T X, et al. 2020b. Transcriptional upregulation of proteasome activator Blm10 antagonizes cellular aging. Biochem Biophys Res Commun, 532（2）: 211-218.

Chen S, Wang R, Zheng D, et al. 2019. The mRNA export receptor NXF1 coordinates transcriptional dynamics, alternative polyadenylation, and mRNA export. Mol Cell, 74（1）: 118-131 e117.

Cheng S, Wang K, Zou W, et al. 2015. PtdIns（4,5）P（2）and PtdIns3P coordinate to regulate phagosomal sealing for apoptotic cell clearance. J Cell Biol, 210（3）: 485-502.

Christ L, Raiborg C, Wenzel E M, et al. 2017. Cellular functions and molecular mechanisms of the ESCRT membrane-scission machinery. Trends Biochem Sci, 42（1）: 42-56.

Chu B B, Liao Y C, Qi W, et al. 2015. Cholesterol transport through lysosome-peroxisome membrane contacts. Cell, 161（2）: 291-306.

Clemson C M, Hutchinson J N, Sara S A, et al. 2009. An architectural role for a nuclear noncoding RNA: NEAT1 RNA is essential for the structure of paraspeckles. Mol Cell, 33（6）: 717-726.

Commisso C, Davidson S M, Soydaner-Azeloglu R G, et al. 2013. Macropinocytosis of protein is an amino acid supply route in Ras-transformed cells. Nature, 497（7451）: 633-637.

Cooper S T, McNeil P L. 2015. Membrane repair: mechanisms and pathophysiology. Physiol Rev, 95（4）: 1205-1240.

Cosentino K, Ros U, Garcia-Saez A J. 2016. Assembling the puzzle: oligomerization of alpha-pore forming proteins in membranes. Biochim Biophys Acta, 1858（3）: 457-466.

Costa-Mattioli M, Walter P. 2020. The integrated stress response: from mechanism to disease. Science, 368（6489）: eaat5314.

Cramer L P. 2010. Forming the cell rear first: breaking cell symmetry to trigger directed cell migration. Nat Cell Biol, 12（7）: 628-632.

D'Arcy P, Brnjic S, Olofsson M H, et al. 2011. Inhibition of proteasome deubiquitinating activity as a new cancer therapy. Nat Med, 17（12）: 1636-1640.

Dal Peraro M, van der Goot F G. 2016. Pore-forming toxins: ancient, but never really out of fashion. Nat Rev Microbiol, 14（2）: 77-92.

Daussy C F, Wodrich H. 2020. "Repair me if you can": membrane damage, response, and control from the viral perspective. Cells, 9（9）: 2042.

Degiacomi M T, Iacovache I, Pernot L, et al. 2013. Molecular assembly of the aerolysin pore reveals a swirling membrane-insertion mechanism. Nat Chem Biol, 9（10）: 623-629.

Diao J, Liu R, Rong Y, et al. 2015. ATG14 promotes membrane tethering and fusion of autophagosomes to endolysosomes. Nature, 520（7548）: 563-566.

Diao L, Liu M, Bao L. 2019. Tubulin isotype and its function. Chin J Cell Biol, 41（3）: 322-332.

Ding H, Lu H, Jiang D, et al. 1999. The relationship between polymorphism of LMP2 and LMP7 genes and the phenotype of ankylosing spondylitis. Chinese Journal of Medical Genetics, 16（4）: 242-245.

Ding J, Wang K, Liu W, et al. 2016. Pore-forming activity and structural autoinhibition of the gasdermin family. Nature, 535（7610）: 111-116.

Dong Y, Zhang S, Wu Z, et al. 2019. Cryo-EM structures and dynamics of substrate-engaged human 26S proteasome. Nature, 565（7737）: 49-55.

Doyon J B, Zeitler B, Cheng J, et al. 2011. Rapid and efficient clathrin-mediated endocytosis revealed in genome-edited mammalian cells. Nat Cell Biol, 13（3）: 331-337.

Du W, Su Q P, Chen Y, et al. 2016. Kinesin 1 drives autolysosome tubulation. Dev Cell, 37（4）: 326-336.

Duan Y, Du A, Gu J, et al. 2019. PARylation regulates stress granule dynamics, phase separation, and neurotoxicity of disease-related RNA-binding proteins. Cell Res, 29（3）: 233-247.

Elbaz-Alon Y, Guo Y, Segev N, et al. 2020. PDZD8 interacts with Protrudin and Rab7 at ER-late endosome membrane contact sites associated with mitochondria. Nat Commun, 11（1）: 3645.

Etienne-Manneville S. 2018. Cytoplasmic intermediate filaments in cell biology. Annu Rev Cell Dev Biol, 34: 1-28.

Fan J, Wang K, Du X, et al. 2019. ALYREF links 3′-end processing to nuclear export of non-polyadenylated mRNAs. EMBO J, 38（9）: e99910.

Farley-Barnes K I, Ogawa L M, Baserga S J. 2019. Ribosomopathies: old concepts, new controversies. Trends Genet, 35（10）: 754-767.

Farquhar M G, Palade G E. 1998. The Golgi apparatus: 100 years of progress and controversy. Trends Cell Biol, 8（1）: 2-10.

Fededa J P, Gerlich D W. 2012. Molecular control of animal cell cytokinesis. Nat Cell Biol, 14（5）: 440-447.

Ferguson S M, Brasnjo G, Hayashi M, et al. 2007. A selective activity-dependent requirement for dynamin 1 in synaptic vesicle endocytosis. Science, 316（5824）: 570-574.

Feric M, Vaidya N, Harmon T S, et al. 2016. Coexisting liquid phases underlie nucleolar subcompartments. Cell, 165（7）: 1686-1697.

Fletcher D A, Mullins R D. 2010. Cell mechanics and the cytoskeleton. Nature, 463（7280）: 485-492.

Fox A H. 2018. A mitochondria-paraspeckle crosstalk. Nat Cell Biol, 20（10）: 1108-1109.

Fox A H, Bond C S, Lamond A I. 2005. P54nrb forms a heterodimer with PSP1 that localizes to paraspeckles in an RNA-dependent manner. Mol Biol Cell, 16（11）: 5304-5315.

Fox A H, Lam Y W, Leung A K, et al. 2002. Paraspeckles: a novel nuclear domain. Curr Biol, 12（1）: 13-25.

Frakes A E, Metcalf M G, Tronnes S U, et al. 2020. Four glial cells regulate ER stress resistance and longevity via neuropeptide signaling in *C. elegans*. Science, 367（6476）: 436-440.

Freeman S A, Uderhardt S, Saric A, et al. 2020. Lipid-gated monovalent ion fluxes regulate endocytic traffic and support immune surveillance. Science, 367（6475）: 301-305.

Fulka H, Aoki F. 2016. Nucleolus precursor bodies and ribosome biogenesis in early mammalian embryos: old theories and new discoveries. Biol Reprod, 94（6）: 143.

Galluzzi L, Kepp O, Chan F K, et al. 2017. Necroptosis: mechanisms and relevance to disease. Annu Rev Pathol, 12: 103-130.

Gao Y, Pei G, Li D, et al. 2019. Multivalent m（6）A motifs promote phase separation of YTHDF proteins. Cell Res, 29（9）: 767-769.

Ghosh R, Wang L, Wang E S, et al. 2014. Allosteric inhibition of the IRE1alpha RNase preserves cell viability and function during endoplasmic reticulum stress. Cell, 158（3）: 534-548.

Goldberg A L. 2005. Nobel committee tags ubiquitin for distinction. Neuron, 45（3）: 339-344.

Gong Y N, Guy C, Olauson H, et al. 2017. ESCRT-Ⅲ acts downstream of MLKL to Regulate necroptotic cell death and its consequences. Cell, 169（2）: 286-300 e216.

Gu L, Li Y, Zhang S, et al. 2019. Molecular resolution imaging by repetitive optical selective exposure. Nat Methods, 16（11）: 1114-1118.

Guan H, Wang Y, Yu T, et al. 2020. Cryo-EM structures of the human PA200 and PA200-20S complex reveal regulation of proteasome gate opening and two PA200 apertures. PLoS Biol, 18（3）: e3000654.

Gui X, Luo F, Li Y, et al. 2019. Structural basis for reversible amyloids of hnRNPA1 elucidates their role in stress granule assembly. Nat Commun, 10（1）: 2006.

Guo B, Liang Q, Li L, et al. 2014a. *O*-GlcNAc-modification of SNAP-29 regulates autophagosome maturation. Nat Cell Biol, 16（12）: 1215-1226.

Guo Y, Li D, Zhang S, et al. 2018. Visualizing intracellular organelle and cytoskeletal interactions at nanoscale resolution on millisecond timescales. Cell, 175（5）: 1430-1442 e1417.

Guo Y, Sirkis D W, Schekman R. 2014b. Protein sorting at the trans-Golgi network. Annu Rev Cell Dev Biol, 30: 169-206.

Hadders M A, Beringer D X, Gros P. 2007. Structure of C8alpha-MACPF reveals mechanism of membrane attack in complement immune defense. Science, 317（5844）: 1552-1554.

Hamazaki J, Hirayama S, Murata S. 2015. Redundant roles of Rpn10 and Rpn13 in recognition of ubiquitinated proteins and cellular homeostasis. PLoS Genet, 11（7）: e1005401.

Hamazaki J, Sasaki K, Kawahara H, et al. 2007. Rpn10-mediated degradation of ubiquitinated proteins is essential for mouse development. Mol Cell Biol, 27（19）: 6629-6638.

Hatch A L, Gurel P S, Higgs H N. 2014. Novel roles for actin in mitochondrial fission. J Cell Sci, 127（Pt 21）: 4549-4560.

Hauer M H, Seeber A, Singh V, et al. 2017. Histone degradation in response to DNA damage enhances chromatin dynamics and recombination rates. Nat Struct Mol Biol, 24（2）: 99-107.

Hayworth K J, Peale D, Januszewski M, et al. 2020. Gas cluster ion beam SEM for imaging of large tissue samples with 10 nm isotropic resolution. Nat Methods, 17（1）: 68-71.

Hinds M G, Zhang W, Anderluh G, et al. 2002. Solution structure of the eukaryotic pore-forming cytolysin equinatoxin Ⅱ: implications for pore formation. J Mol Biol, 315（5）: 1219-1229.

Hirose T, Virnicchi G, Tanigawa A, et al. 2014. NEAT1 long noncoding RNA regulates transcription via protein sequestration within subnuclear bodies. Mol Biol Cell, 25（1）: 169-183.

Hotze E M, Le H M, Sieber J R, et al. 2013. Identification and characterization of the first cholesterol-dependent cytolysins from Gram-negative bacteria. Infect Immun, 81（1）: 216-225.

Hotze E M, Tweten R K. 2012. Membrane assembly of the cholesterol-dependent cytolysin pore complex. Biochim Biophys Acta, 1818（4）: 1028-1038.

Hu J, Rapoport T A. 2016. Fusion of the endoplasmic reticulum by membrane-bound GTPases. Semin Cell Dev Biol, 60: 105-111.

Hu S B, Xiang J F, Li X, et al. 2015. Protein arginine methyltransferase CARM1 attenuates the paraspeckle-mediated nuclear retention of mRNAs containing IRAlus. Genes Dev, 29（6）: 630-645.

Huang L, Haratake K, Miyahara H, et al. 2016a. Proteasome activators, PA28gamma and PA200, play indispensable roles in male fertility. Sci Rep, 6: 23171.

Huang N, Zhang D, Li F, et al. 2018a. M-Phase Phosphoprotein 9 regulates ciliogenesis by modulating CP110-CEP97 complex localization at the mother centriole. Nat Commun, 9（1）: 4511.

Huang S, Xing Y, Liu Y. 2019. Emerging roles for the ER stress sensor IRE1alpha in metabolic regulation and disease. J Biol Chem, 294（49）: 18726-18741.

Huang X, Fan J, Li L, et al. 2018b. Fast, long-term, super-resolution imaging with Hessian structured illumination microscopy. Nat Biotechnol, 36（5）: 451-459.

Huang X, Luan B, Wu J, et al. 2016b. An atomic structure of the human 26S proteasome. Nat Struct Mol Biol, 23（9）: 778-785.

Huo X, Ji L, Zhang Y, et al. 2020. The nuclear matrix protein SAFB cooperates with major satellite RNAs to stabilize heterochromatin architecture partially through phase separation. Mol Cell, 77（2）: 368-383 e367.

Iarovaia O V, Minina E P, Sheval E V, et al. 2019. Nucleolus: a central hub for nuclear functions. Trends Cell Biol, 29（8）: 647-659.

Imamura K, Imamachi N, Akizuki G, et al. 2014. Long noncoding RNA NEAT1-dependent SFPQ relocation from promoter region to paraspeckle mediates IL8 expression upon immune stimuli. Mol Cell, 53（3）: 393-406.

Jacob M D, Audas T E, Uniacke J, et al. 2013. Environmental cues induce a long noncoding RNA-dependent remodeling of the nucleolus. Mol Biol Cell, 24（18）: 2943-2953.

Janke C, Magiera M M. 2020. The tubulin code and its role in controlling microtubule properties and functions. Nat Rev Mol Cell Biol, 21（6）: 307-326.

Jiang T X, Ma S, Han X, et al. 2021a. Proteasome activator PA200 maintains stability of histone marks during transcription and aging. Theranostics, 11（3）: 1458-1472.

Jiang T X, Zou J B, Zhu Q Q, et al. 2019. SIP/CacyBP promotes autophagy by regulating levels of BRUCE/Apollon, which stimulates LC3-I degradation. Proc Natl Acad Sci U S A, 116（27）: 13404-13413.

Jiang T Y, Feng X F, Fang Z, et al. 2021b. PTEN deficiency facilitates the therapeutic vulnerability to proteasome inhibitor bortezomib in gallbladder cancer. Cancer Lett, 501: 187-199.

Jiang T Y, Pan Y F, Wan Z H, et al. 2020. PTEN status determines chemosensitivity to proteasome inhibition in cholangiocarcinoma. Sci Transl Med, 12（562）: eaay0152.

Jiao Z, Cai H, Long Y, et al. 2020. Statin-induced GGPP depletion blocks macropinocytosis and starves cells with oncogenic defects. Proc Natl Acad Sci U S A, 117（8）: 4158-4168.

Jin S, Zong Y, Gao Q, et al. 2019. Cytosine, but not adenine, base editors induce genome-wide off-target mutations in rice. Science, 364（6437）: 292-295.

Kaganovich D. 2017. There is an inclusion for that: material properties of protein granules provide a platform for building diverse cellular functions. Trends Biochem Sci, 42（10）: 765-776.

Kang L, He Z, Xu P, et al. 2006. Munc13-1 is required for the sustained release of insulin from pancreatic beta cells. Cell Metab, 3（6）: 463-468.

Kauppila T E S, Kauppila J H K, Larsson N G. 2017. Mammalian mitochondria and aging: an update. Cell Metab, 25（1）: 57-71.

Kayagaki N, Stowe I B, Lee B L, et al. 2015. Caspase-11 cleaves gasdermin D for non-canonical inflammasome signalling. Nature, 526（7575）: 666-671.

Kincaid E Z, Che J W, York I, et al. 2011. Mice completely lacking immunoproteasomes show major changes in antigen presentation. Nat Immunol, 13（2）: 129-135.

Kors S, Geijtenbeek K, Reits E, et al. 2019. Regulation of proteasome activity by（post-）transcriptional mechanisms. Front Mol Biosci, 6: 48.

Kristan K C, Viero G, Dalla Serra M, et al. 2009. Molecular mechanism of pore formation by actinoporins. Toxicon, 54（8）: 1125-1134.

Law R H, Lukoyanova N, Voskoboinik I, et al. 2010. The structural basis for membrane binding and pore formation by lymphocyte perforin. Nature, 468（7322）: 447-451.

Lawrence R E, Zoncu R. 2019. The lysosome as a cellular centre for signalling, metabolism and quality control. Nat Cell Biol, 21（2）: 133-142.

Lee E B, Lee V M, Trojanowski J Q. 2011. Gains or losses: molecular mechanisms of TDP43-mediated neurodegeneration. Nat Rev Neurosci, 13（1）: 38-50.

Lee M C, Miller E A, Goldberg J, et al. 2004. Bi-directional protein transport between the ER and Golgi. Annu Rev Cell Dev Biol, 20: 87-123.

Letts J A, Sazanov L A. 2017. Clarifying the supercomplex: the higher-order organization of the mitochondrial electron transport chain. Nat Struct Mol Biol, 24（10）: 800-808.

Li D, Dong Q, Tao Q, et al. 2015a. c-Abl regulates proteasome abundance by controlling the ubiquitin-proteasomal degradation of PSMA7 subunit. Cell Rep, 10（4）: 484-496.

Li D, Meng L, Xu T, et al. 2017a. RIPK1-RIPK3-MLKL-dependent necrosis promotes the aging of mouse male reproductive system. Elife, 6: e27692.

Li D, Shao L, Chen B C, et al. 2015b. ADVANCED IMAGING. Extended-resolution structured illumination imaging of endocytic and cytoskeletal dynamics. Science, 349（6251）: aab3500.

Li J, Wang S, Zhang Y L, et al. 2019a. Immunoproteasome subunit beta5i promotes Ang Ⅱ（angiotensin Ⅱ）-induced atrial fibrillation by targeting ATRAP（Ang Ⅱ type Ⅰ receptor-associated protein）degradation in mice. Hypertension, 73（1）: 92-101.

Li R, Harvey A R, Hodgetts S I, et al. 2017b. Functional dissection of NEAT1 using genome editing reveals substantial localization of the NEAT1_1 isoform outside paraspeckles. RNA, 23（6）: 872-881.

Li W, Li Y, Siraj S, et al. 2019b. FUN14 Domain-containing 1-mediated mitophagy suppresses hepatocarcinogenesis by inhibition of inflammasome activation in mice. Hepatology, 69（2）: 604-621.

Li X, Amazit L, Long W, et al. 2007. Ubiquitin- and ATP-independent proteolytic turnover of p21 by the REGgamma-proteasome pathway. Mol Cell, 26（6）: 831-842.

Li X, Ji G, Chen X, et al. 2017c. Large scale three-dimensional reconstruction of an entire *Caenorhabditis elegans* larva using AutoCUTS-SEM. J Struct Biol, 200（2）: 87-96.

Li Y, Chen B, Zou W, et al. 2016a. The lysosomal membrane protein SCAV-3 maintains lysosome integrity and adult longevity. J Cell Biol, 215（2）: 167-185.

Li Y, Xu M, Ding X, et al. 2016b. Protein kinase C controls lysosome biogenesis independently of mTORC1. Nat Cell Biol, 18（10）: 1065-1077.

Li Y, Xue Y, Xu X, et al. 2019c. A mitochondrial FUNDC1/HSC70 interaction organizes the proteostatic stress response at the risk of cell morbidity. EMBO J, 38（3）: e98786.

Li Z, Wu J, Chavez L, et al. 2019d. Reiterative Enrichment and Authentication of CRISPRi Targets（REACT）identifies the proteasome as a key contributor to HIV-1 latency. PLoS Pathog, 15（1）: e1007498.

Lian G, Sheen V L. 2015. Cytoskeletal proteins in cortical development and disease: actin associated proteins in periventricular heterotopia. Front Cell Neurosci, 9: 99.

Liang Y, Pang Y, Wu Q, et al. 2014. FLA8/KIF3B phosphorylation regulates kinesin-Ⅱ interaction with IFT-B to control IFT entry and turnaround. Dev Cell, 30（5）: 585-597.

Liang Y, Zhu X, Wu Q, et al. 2018. Ciliary length sensing regulates IFT entry via changes in FLA8/KIF3B phosphorylation to control ciliary assembly. Curr Biol, 28（15）: 2429-2435 e2423.

Lin C, Li H, Liu J, et al. 2020. Arginine hypomethylation-mediated proteasomal degradation of histone H4-an early biomarker of cellular senescence. Cell Death Differ, 27（9）: 2697-2709.

Liu K, Jian Y, Sun X, et al. 2016a. Negative regulation of phosphatidylinositol 3-phosphate levels in early-to-late endosome conversion. J Cell Biol, 212（2）: 181-198.

Liu K, Xing R, Jian Y, et al. 2017a. WDR91 is a Rab7 effector required for neuronal development. J Cell Biol, 216（10）: 3307-3321.

Liu L, Cai J, Wang H, et al. 2019. Coupling of COPII vesicle trafficking to nutrient availability by the IRE1alpha-XBP1s axis. Proc Natl Acad Sci U S A, 116（24）: 11776-11785.

Liu S, Liu H, Johnston A, et al. 2017b. MLKL forms disulfide bond-dependent amyloid-like polymers to induce necroptosis. Proc Natl Acad Sci U S A, 114（36）: E7450-E7459.

Liu X, Huang W, Li C, et al. 2006. Interaction between c-Abl and Arg tyrosine kinases and proteasome subunit PSMA7 regulates proteasome degradation. Mol Cell, 22（3）: 317-327.

Liu X, Kim C N, Yang J, et al. 1996. Induction of apoptotic program in cell-free extracts: requirement for dATP and cytochrome c. Cell, 86（1）: 147-157.

Liu X, Xiao W, Zhang Y, et al. 2020a. Reversible phosphorylation of Rpn1 regulates 26S proteasome assembly and function. Proc Natl Acad Sci U S A, 117（1）: 328-336.

Liu X, Zhang Z, Ruan J, et al. 2016b. Inflammasome-activated gasdermin D causes pyroptosis by forming membrane pores. Nature, 535（7610）: 153-158.

Liu Z, Yang Y, Gu A, et al. 2020b. Par complex cluster formation mediated by phase separation. Nat Commun, 11（1）: 2266.

Liu Z, Zhang S, Gu J, et al. 2020c. Hsp27 chaperones FUS phase separation under the modulation of stress-induced phosphorylation. Nat Struct Mol Biol, 27（4）: 363-372.

Long H, Zhang F, Xu N, et al. 2016. Comparative analysis of ciliary membranes and ectosomes. Curr Biol, 26（24）: 3327-3335.

Lu J Y, Shao W, Chang L, et al. 2020. Genomic repeats categorize genes with distinct functions for orchestrated regulation. Cell Rep, 30（10）: 3296-3311 e3295.

Luan B, Huang X, Wu J, et al. 2016. Structure of an endogenous yeast 26S proteasome reveals two major conformational states. Proc Natl Acad Sci U S A, 113（10）: 2642-2647.

Ma M, Stoyanova M, Rademacher G, et al. 2019. Structure of the decorated ciliary doublet microtubule. Cell, 179（4）: 909-922 e912.

Ma T, Li J, Xu Y, et al. 2015. Atg5-independent autophagy regulates mitochondrial clearance and is essential for iPSC reprogramming. Nat Cell Biol, 17（11）: 1379-1387.

Mais C, Wright J E, Prieto J L, et al. 2005. UBF-binding site arrays form pseudo-NORs and sequester the RNA polymerase Ⅰ transcription machinery. Genes Dev, 19（1）: 50-64.

Malhotra V, Erlmann P. 2015. The pathway of collagen secretion. Annu Rev Cell Dev Biol, 31: 109-124.

Mancheno J M, Martin-Benito J, Martinez-Ripoll M, et al. 2003. Crystal and electron microscopy structures of sticholysin Ⅱ actinoporin reveal insights into the mechanism of membrane pore formation. Structure, 11（11）: 1319-1328.

Mandemaker I K, Geijer M E, Kik I, et al. 2018. DNA damage-induced replication stress results in PA200-proteasome-mediated degradation of acetylated histones. EMBO Rep, 19（10）: e45566.

Mao Y S, Zhang B, Spector D L. 2011. Biogenesis and function of nuclear bodies. Trends Genet, 27（8）: 295-306.

Mayor S, Pagano R E. 2007. Pathways of clathrin-independent endocytosis. Nat Rev Mol Cell Biol, 8（8）: 603-612.

Mekhail K, Gunaratnam L, Bonicalzi M E, et al. 2004. HIF activation by pH-dependent nucleolar sequestration of VHL. Nat Cell Biol, 6（7）: 642-647.

Morimoto M, Boerkoel C F. 2013. The role of nuclear bodies in gene expression and disease. Biology（Basel）, 2（3）: 976-1033.

Mueller M, Grauschopf U, Maier T, et al. 2009. The structure of a cytolytic alpha-helical toxin pore reveals its assembly mechanism. Nature, 459（7247）: 726-730.

Murphy J M, Czabotar P E, Hildebrand J M, et al. 2013. The pseudokinase MLKL mediates necroptosis via a molecular switch mechanism. Immunity, 39（3）: 443-453.

Nemeth A, Grummt I. 2018. Dynamic regulation of nucleolar architecture. Curr Opin Cell Biol, 52: 105-111.

Nicastro D, Schwartz C, Pierson J, et al. 2006. The molecular architecture of axonemes revealed by cryoelectron tomography. Science, 313（5789）: 944-948.

Niu L, Ma T, Yang F, et al. 2019. Atlastin-mediated membrane tethering is critical for cargo mobility and exit from the endoplasmic reticulum. Proc Natl Acad Sci U S A, 116（28）: 14029-14038.

Palm W, Park Y, Wright K, et al. 2015. The utilization of extracellular proteins as nutrients is suppressed by mTORC1. Cell, 162（2）: 259-270.

Parker M W, Buckley J T, Postma J P, et al. 1994. Structure of the *Aeromonas* toxin proaerolysin in its water-soluble and membrane-channel states. Nature, 367（6460）: 292-295.

Pfanner N, Warscheid B, Wiedemann N. 2019. Mitochondrial proteins: from biogenesis to functional networks. Nat Rev Mol Cell Biol, 20（5）: 267-284.

Ploumi C, Daskalaki I, Tavernarakis N. 2017. Mitochondrial biogenesis and clearance: a balancing act. FEBS J, 284（2）: 183-195.

Powers R E, Wang S, Liu T Y, et al. 2017. Reconstitution of the tubular endoplasmic reticulum network with purified components. Nature, 543（7644）: 257-260.

Prasanth K V, Prasanth S G, Xuan Z, et al. 2005. Regulating gene expression through RNA nuclear retention. Cell, 123（2）: 249-263.

Preissler S, Ron D. 2019. Early events in the endoplasmic reticulum unfolded protein response. Cold Spring Harb Perspect Biol, 11（4）: a033894.

Priesnitz C, Becker T. 2018. Pathways to balance mitochondrial translation and protein import.

Genes Dev, 32（19-20）: 1285-1296.

Prieto J L, McStay B. 2007. Recruitment of factors linking transcription and processing of pre-rRNA to NOR chromatin is UBF-dependent and occurs independent of transcription in human cells. Genes Dev, 21（16）: 2041-2054.

Qian M X, Pang Y, Liu C H, et al. 2013. Acetylation-mediated proteasomal degradation of core histones during DNA repair and spermatogenesis. Cell, 153（5）: 1012-1024.

Qiu X B, Ouyang S Y, Li C J, et al. 2006. hRpn13/ADRM1/GP110 is a novel proteasome subunit that binds the deubiquitinating enzyme, UCH37. EMBO J, 25（24）: 5742-5753.

Ramachandran R, Tweten R K, Johnson A E. 2004. Membrane-dependent conformational changes initiate cholesterol-dependent cytolysin oligomerization and intersubunit beta-strand alignment. Nat Struct Mol Biol, 11（8）: 697-705.

Ramos I, Stamatakis K, Oeste C L, et al. 2020. Vimentin as a multifaceted player and potential therapeutic target in viral infections. Int J Mol Sci, 21（13）: 4675.

Ran J, Liu M, Feng J, et al. 2020. ASK1-mediated phosphorylation blocks HDAC6 ubiquitination and degradation to drive the disassembly of photoreceptor connecting cilia. Dev Cell, 53（3）: 287-299 e285.

Revenu C, Athman R, Robine S, et al. 2004. The co-workers of actin filaments: from cell structures to signals. Nat Rev Mol Cell Biol, 5（8）: 635-646.

Rock K L, Goldberg A L. 1999. Degradation of cell proteins and the generation of MHC class I-presented peptides. Annu Rev Immunol, 17: 739-779.

Rong Y, Liu M, Ma L, et al. 2012. Clathrin and phosphatidylinositol-4,5-bisphosphate regulate autophagic lysosome reformation. Nat Cell Biol, 14（9）: 924-934.

Ros U, Pena-Blanco A, Hanggi K, et al. 2017. Necroptosis execution is mediated by plasma membrane nanopores independent of calcium. Cell Rep, 19（1）: 175-187.

Rosado C J, Buckle A M, Law R H, et al. 2007. A common fold mediates vertebrate defense and bacterial attack. Science, 317（5844）: 1548-1551.

Rossjohn J, Feil S C, McKinstry W J, et al. 1997. Structure of a cholesterol-binding, thiol-activated cytolysin and a model of its membrane form. Cell, 89（5）: 685-692.

Ruan J, Xia S, Liu X, et al. 2018. Cryo-EM structure of the gasdermin A3 membrane pore. Nature, 557（7703）: 62-67.

Ruhl S, Shkarina K, Demarco B, et al. 2018. ESCRT-dependent membrane repair negatively regulates pyroptosis downstream of GSDMD activation. Science, 362（6417）: 956-960.

Russell O M, Gorman G S, Lightowlers R N, et al. 2020. Mitochondrial diseases: hope for the future. Cell, 181（1）: 168-188.

Saheki Y, De Camilli P. 2017. Endoplasmic reticulum-plasma membrane contact sites. Annu Rev Biochem, 86: 659-684.

Sancak Y, Peterson T R, Shaul Y D, et al. 2008. The rag GTPases bind raptor and mediate amino acid signaling to mTORC1. Science, 320（5882）: 1496-1501.

Sardiello M, Palmieri M, di Ronza A, et al. 2009. A gene network regulating lysosomal biogenesis and function. Science, 325（5939）: 473-477.

Sato T K, Tweten R K, Johnson A E. 2013. Disulfide-bond scanning reveals assembly state and beta-strand tilt angle of the PFO beta-barrel. Nat Chem Biol, 9（6）: 383-389.

Schon E A, DiMauro S, Hirano M. 2012. Human mitochondrial DNA: roles of inherited and somatic mutations. Nat Rev Genet, 13（12）: 878-890.

Scott C C, Vacca F, Gruenberg J. 2014. Endosome maturation, transport and functions. Semin Cell Dev Biol, 31: 2-10.

Shan Z, Tu Y, Yang Y, et al. 2018. Basal condensation of Numb and Pon complex via phase transition during Drosophila neuroblast asymmetric division. Nat Commun, 9（1）: 737.

Shao L W, Peng Q, Dong M, et al. 2020. Histone deacetylase HDA-1 modulates mitochondrial stress response and longevity. Nat Commun, 11（1）: 4639.

Sharpley M S, Marciniak C, Eckel-Mahan K, et al. 2012. Heteroplasmy of mouse mtDNA is genetically unstable and results in altered behavior and cognition. Cell, 151（2）: 333-343.

Shatursky O, Heuck A P, Shepard L A, et al. 1999. The mechanism of membrane insertion for a cholesterol-dependent cytolysin: a novel paradigm for pore-forming toxins. Cell, 99（3）: 293-299.

Shi J, Gao W, Shao F. 2017. Pyroptosis: gasdermin-mediated programmed necrotic cell death. Trends Biochem Sci, 42（4）: 245-254.

Shi J, Zhao Y, Wang K, et al. 2015. Cleavage of GSDMD by inflammatory caspases determines pyroptotic cell death. Nature, 526（7575）: 660-665.

Shibata Y, Voeltz G K, Rapoport T A. 2006. Rough sheets and smooth tubules. Cell, 126（3）: 435-439.

Shin Y, Brangwynne C P. 2017. Liquid phase condensation in cell physiology and disease. Science, 357（6357）: eaaf4382.

Sliter D A, Martinez J, Hao L, et al. 2018. Parkin and PINK1 mitigate STING-induced inflammation. Nature, 561（7722）: 258-262.

Snider N T, Omary M B. 2014. Post-translational modifications of intermediate filament proteins: mechanisms and functions. Nat Rev Mol Cell Biol, 15（3）: 163-177.

Song L, Hobaugh M R, Shustak C, et al. 1996. Structure of staphylococcal alpha-hemolysin, a heptameric transmembrane pore. Science, 274（5294）: 1859-1866.

Souquere S, Beauclair G, Harper F, et al. 2010. Highly ordered spatial organization of the structural long noncoding NEAT1 RNAs within paraspeckle nuclear bodies. Mol Biol Cell, 21（22）: 4020-4027.

Stanek D, Fox A H. 2017. Nuclear bodies: news insights into structure and function. Curr Opin Cell Biol, 46: 94-101.

Steinmetz M O, Prota A E. 2018. Microtubule-targeting agents: strategies to hijack the cytoskeleton. Trends Cell Biol, 28（10）: 776-792.

Sturgill T W, Cohen A, Diefenbacher M, et al. 2008. TOR1 and TOR2 have distinct locations in live cells. Eukaryot Cell, 7（10）: 1819-1830.

Sun D, Wu R, Zheng J, et al. 2018. Polyubiquitin chain-induced p62 phase separation drives autophagic cargo segregation. Cell Res, 28（4）: 405-415.

Sun L, Gao Y, He J, et al. 2019. Ultrastructural organization of NompC in the mechanoreceptive organelle of Drosophila campaniform mechanoreceptors. Proc Natl Acad Sci U S A, 116（15）: 7343-7352.

Sun L, Wang H, Wang Z, et al. 2012. Mixed lineage kinase domain-like protein mediates necrosis signaling downstream of RIP3 kinase. Cell, 148（1-2）: 213-227.

Tanaka K, Ii K, Ichihara A, et al. 1986. A high molecular weight protease in the cytosol of rat liver. I. Purification, enzymological properties, and tissue distribution. J Biol Chem, 261（32）: 15197-15203.

Thiry M, Lafontaine D L. 2005. Birth of a nucleolus: the evolution of nucleolar compartments. Trends Cell Biol, 15（4）: 194-199.

Thottacherry J J, Sathe M, Prabhakara C, et al. 2019. Spoiled for choice: diverse endocytic pathways function at the cell surface. Annu Rev Cell Dev Biol, 35: 55-84.

Tian Y, Li Z, Hu W, et al. 2010. *C. elegans* screen identifies autophagy genes specific to multicellular organisms. Cell, 141（6）: 1042-1055.

Tong L, Shen S, Huang Q, et al. 2020. Proteasome-dependent degradation of Smad7 is critical for lung cancer metastasis. Cell Death Differ, 27（6）: 1795-1806.

Vaidyanathan M S, Sathyanarayana P, Maiti P K, et al. 2014. Lysis dynamics and membrane oligomerization pathways for cytolysin A（ClyA）pore-forming toxin. RSC Advances, 4（10）: 4930-4942.

Wallace A J, Stillman T J, Atkins A, et al. 2000. *E. coli* hemolysin E（HlyE, ClyA, SheA）: X-ray crystal structure of the toxin and observation of membrane pores by electron microscopy. Cell, 100（2）: 265-276.

Walter P, Ron D. 2011. The unfolded protein response: from stress pathway to homeostatic regulation. Science, 334（6059）: 1081-1086.

Wang C, Duan Y, Duan G, et al. 2020a. Stress induces dynamic, cytotoxicity-antagonizing TDP-43 nuclear bodies via paraspeckle LncRNA NEAT1-mediated liquid-liquid phase separation. Mol Cell, 79（3）: 443-458 e447.

Wang H, Sun L, Su L, et al. 2014. Mixed lineage kinase domain-like protein MLKL causes necrotic membrane disruption upon phosphorylation by RIP3. Mol Cell, 54（1）: 133-146.

Wang J, Chen J, Wu G, et al. 2019a. NRDE2 negatively regulates exosome functions by inhibiting MTR4 recruitment and exosome interaction. Genes Dev, 33（9-10）: 536-549.

Wang L, Gao Y, Zheng X, et al. 2019b. Histone modifications regulate chromatin compartmentalization by contributing to a phase separation mechanism. Mol Cell, 76（4）: 646-659 e646.

Wang L, Hu M, Zuo M Q, et al. 2020b. Rett syndrome-causing mutations compromise MeCP2-mediated liquid-liquid phase separation of chromatin. Cell Res, 30（5）: 393-407.

Wang L, Li X, Xue W, et al. 2017a. Enhanced base editing by co-expression of free uracil DNA glycosylase inhibitor. Cell Research, 27（10）: 1289-1292.

Wang X, Hu X, Song W, et al. 2021. Mutual dependency between lncRNA LETN and protein NPM1 in controlling the nucleolar structure and functions sustaining cell proliferation. Cell Res, 31（6）: 664-683.

Wang X, Li S, Wang H, et al. 2017b. Quantitative proteomics reveal proteins enriched in tubular endoplasmic reticulum of Saccharomyces cerevisiae. Elife, 6: e23816.

Wang X, Wu Y C, Fadok V A, et al. 2003. Cell corpse engulfment mediated by *C. elegans* phosphatidylserine receptor through CED-5 and CED-12. Science, 302（5650）: 1563-1566.

Wang Y, Hu S B, Wang M R, et al. 2018. Genome-wide screening of NEAT1 regulators reveals cross-regulation between paraspeckles and mitochondria. Nat Cell Biol, 20（10）: 1145-1158.

Wang Z, Miao G, Xue X, et al. 2016. The Vici syndrome protein EPG5 is a Rab7 effector that determines the fusion specificity of autophagosomes with late endosomes/lysosomes. Mol Cell, 63（5）: 781-795.

Weinlich R, Oberst A, Beere H M, et al. 2017. Necroptosis in development, inflammation and disease. Nat Rev Mol Cell Biol, 18（2）: 127-136.

West A P, Khoury-Hanold W, Staron M, et al. 2015. Mitochondrial DNA stress primes the antiviral innate immune response. Nature, 520（7548）: 553-557.

West J A, Mito M, Kurosaka S, et al. 2016. Structural, super-resolution microscopy analysis of paraspeckle nuclear body organization. J Cell Biol, 214（7）: 817-830.

Wu H, Carvalho P, Voeltz G K. 2018. Here, there, and everywhere: the importance of ER membrane contact sites. Science, 361（6401）: eaan5835.

Wu H, Wang Y, Li W, et al. 2019a. Deficiency of mitophagy receptor FUNDC1 impairs mitochondrial quality and aggravates dietary-induced obesity and metabolic syndrome. Autophagy, 15（11）: 1882-1898.

Wu H, Yin Q F, Luo Z, et al. 2016. Unusual processing generates SPA LncRNAs that sequester multiple RNA binding proteins. Mol Cell, 64（3）: 534-548.

Wu X, Cai Q, Shen Z, et al. 2019b. RIM and RIM-BP form presynaptic active-zone-like condensates via phase separation. Mol Cell, 73（5）: 971-984 e975.

Xia B, Fang S, Chen X, et al. 2016. MLKL forms cation channels. Cell Res, 26（5）: 517-528.

Xiao J, Luo J, Hu A, et al. 2019. Cholesterol transport through the peroxisome-ER membrane contacts tethered by PI（4,5）P2 and extended synaptotagmins. Sci China Life Sci, 62（9）: 1117-1135.

Xie C, Li L, Li M, et al. 2020. Optimal sidestepping of intraflagellar transport kinesins regulates structure and function of sensory cilia. EMBO J, 39（12）: e103955.

Xie S, Zhou J. 2019. Regulation of microtubule dynamics and functions by post-translational modifications and mcrotubule-binding proteins. Chin J Cell Biol, 41（3）: 333-341.

Xie X, Bi H L, Lai S, et al. 2019. The immunoproteasome catalytic beta5i subunit regulates cardiac hypertrophy by targeting the autophagy protein ATG5 for degradation. Sci Adv, 5（5）: eaau0495.

Xing Y H, Yao R W, Zhang Y, et al. 2017. SLERT regulates DDX21 rings associated with Pol I transcription. Cell, 169（4）: 664-678 e616.

Xu J, Zhou L, Ji L, et al. 2016. The REGgamma-proteasome forms a regulatory circuit with IkappaBvarepsilon and NFkappaB in experimental colitis. Nat Commun, 7: 10761.

Yamazaki T, Souquere S, Chujo T, et al. 2018. Functional domains of NEAT1 architectural lncRNA induce paraspeckle assembly through phase separation. Mol Cell, 70（6）: 1038-1053 e1037.

Yan H, Chen C, Chen H, et al. 2020. TALPID3 and ANKRD26 selectively orchestrate FBF1 localization and cilia gating. Nat Commun, 11（1）: 2196.

Yang C, Wang X. 2017. Cell biology in China: focusing on the lysosome. Traffic, 18（6）: 348-357.

Yang L Z, Wang Y, Li S Q, et al. 2019. Dynamic imaging of RNA in living cells by CRISPR-Cas13 systems. Mol Cell, 76（6）: 981-997 e987.

Yang X, Xie H, Alonas E, et al. 2016. Mirror-enhanced super-resolution microscopy. Light Sci Appl, 5（6）: e16134.

Yao R W, Xu G, Wang Y, et al. 2019. Nascent pre-rRNA sorting via phase separation drives the assembly of dense fibrillar components in the human nucleolus. Mol Cell, 76（5）: 767-783 e711.

Yin Q F, Yang L, Zhang Y, et al. 2012. Long noncoding RNAs with snoRNA ends. Mol Cell, 48（2）: 219-230.

Yin Y, Zhu Q, Jiang T, et al. 2019. Targeting histones for degradation in cancer cells as a novel strategy in cancer treatment. Sci China Life Sci, 62（8）: 1078-1086.

Yoshida S, Pacitto R, Yao Y, et al. 2015. Growth factor signaling to mTORC1 by amino acid-laden macropinosomes. J Cell Biol, 211（1）: 159-172.

Youle R J. 2019. Mitochondria-Striking a balance between host and endosymbiont. Science, 365（6454）: eaaw9855.

Yu F, Guo S, Li T, et al. 2019. Ciliary defects caused by dysregulation of *O*-GlcNAc modification are associated with diabetic complications. Cell Res, 29（2）: 171-173.

Yu J, Li T, Liu Y, et al. 2020. Phosphorylation switches protein disulfide isomerase activity to maintain proteostasis and attenuate ER stress. EMBO J, 39（10）: e103841.

Yu L, McPhee C K, Zheng L, et al. 2010. Termination of autophagy and reformation of lysosomes regulated by mTOR. Nature, 465（7300）: 942-946.

Yuan J, Amin P, Ofengeim D. 2019. Necroptosis and RIPK1-mediated neuroinflammation in CNS diseases. Nat Rev Neurosci, 20（1）: 19-33.

Zeng M, Chen X, Guan D, et al. 2018. Reconstituted postsynaptic density as a molecular platform for understanding synapse formation and plasticity. Cell, 174（5）: 1172-1187 e1116.

Zeng M, Diaz-Alonso J, Ye F, et al. 2019. Phase separation-mediated TARP/MAGUK complex condensation and AMPA receptor synaptic transmission. Neuron, 104（3）: 529-543 e526.

Zeng M, Shang Y, Araki Y, et al. 2016. Phase transition in postsynaptic densities underlies formation of synaptic complexes and synaptic plasticity. Cell, 166（5）: 1163-1175 e1112.

Zhang B, Zhuang T, Lin Q, et al. 2019. Patched1-ArhGAP36-PKA-inversin axis determines the ciliary translocation of smoothened for sonic hedgehog pathway activation. Proc Natl Acad Sci U S A, 116（3）: 874-879.

Zhang C, Zhang W, Lu Y, et al. 2016a. NudC regulates actin dynamics and ciliogenesis by stabilizing cofilin 1. Cell Res, 26（2）: 239-253.

Zhang C S, Jiang B, Li M, et al. 2014. The lysosomal v-ATPase-ragulator complex is a common activator for AMPK and mTORC1, acting as a switch between catabolism and anabolism. Cell Metab, 20（3）: 526-540.

Zhang C S, Li M, Ma T, et al. 2016b. Metformin activates AMPK through the lysosomal pathway. Cell Metab, 24（4）: 521-522.

Zhang G, Wang Z, Du Z, et al. 2018a. mTOR regulates phase separation of PGL granules to modulate their autophagic degradation. Cell, 174（6）: 1492-1506 e1422.

Zhang H, Hu J. 2016. Shaping the endoplasmic reticulum into a social network. Trends Cell Biol, 26（12）: 934-943.

Zhang H, Ji X, Li P, et al. 2020a. Liquid-liquid phase separation in biology: mechanisms, physiological functions and human diseases. Sci China Life Sci, 63（7）: 953-985.

Zhang H, Wu Z, Lu J Y, et al. 2020b. DEAD-box helicase 18 counteracts PRC2 to safeguard ribosomal DNA in pluripotency regulation. Cell Rep, 30（1）: 81-97 e87.

Zhang J, Zhu Q, Wang X, et al. 2018b. Secretory kinase Fam20C tunes endoplasmic reticulum redox state via phosphorylation of Ero1alpha. EMBO J, 37（14）: e98699.

Zhang X, Jia S, Chen Z, et al. 2018c. Cilia-driven cerebrospinal fluid flow directs expression of urotensin neuropeptides to straighten the vertebrate body axis. Nat Genet, 50（12）: 1666-1673.

Zhang X W, Yan X J, Zhou Z R, et al. 2010. Arsenic trioxide controls the fate of the PML-RARalpha oncoprotein by directly binding PML. Science, 328（5975）: 240-243.

Zhang Y, Han J. 2016. Electrophysiologist shows a cation channel function of MLKL. Cell Res, 26（6）: 643-644.

Zhang Y Y, Fu Z Y, Wei J, et al. 2018d. A LIMA1 variant promotes low plasma LDL cholesterol and decreases intestinal cholesterol absorption. Science, 360（6393）: 1087-1092.

Zhang Z, Zhang R. 2008. Proteasome activator PA28 gamma regulates p53 by enhancing its MDM2-mediated degradation. EMBO J, 27（6）: 852-864.

Zhang Z H, Jiang T X, Chen L B, et al. 2020c. Proteasome subunit alpha4s is essential for formation of spermatoproteasomes and histone degradation during meiotic DNA repair in spermatocytes. J Biol Chem, 296: 100130.

Zhao H, Zhu L, Zhu Y, et al. 2013. The Cep63 paralogue Deup1 enables massive *de novo* centriole biogenesis for vertebrate multiciliogenesis. Nat Cell Biol, 15（12）: 1434-1444.

Zhou K M, Dong Y M, Ge Q, et al. 2007. PKA activation bypasses the requirement for UNC-31 in the docking of dense core vesicles from *C. elegans* neurons. Neuron, 56（4）: 657-669.

Zhou S, Zhang W, Cai G, et al. 2020. Myofiber necroptosis promotes muscle stem cell proliferation via releasing tenascin-C during regeneration. Cell Res, 30（12）: 1063-1077.

Zhu J, Zhou Q, Xia Y, et al. 2020. GIT/PIX Condensates are modular and ideal for distinct compartmentalized cell signaling. Mol Cell, 79（5）: 782-796 e786.

Zong N, Ping P, Lau E, et al. 2014. Lysine ubiquitination and acetylation of human cardiac 20S proteasomes. Proteomics Clin Appl, 8（7-8）: 590-594.

Zorov D B, Juhaszova M, Sollott S J. 2014. Mitochondrial reactive oxygen species（ROS）and ROS-induced ROS release. Physiol Rev, 94（3）: 909-950.

Zuo W, Huang F, Chiang Y J, et al. 2013. c-Cbl-mediated neddylation antagonizes ubiquitination and degradation of the TGF-beta type Ⅱ receptor. Mol Cell, 49（3）: 499-510.

第七章

细胞类型的区分与确定

第一节　概　　述

细胞是生命的基本组成单元，据此可把生物体简单地划分为两大类型：单细胞生物和多细胞生物。单细胞生物，如细菌和原生生物等；多细胞生物，如大多数的动、植物。显然，多细胞生物通常是由一定数量的细胞组成，如一种简单的无脊椎动物“秀丽隐杆线虫”（*Caenorhabditis elegans*）的成体有1000多个细胞，而一个成年人体则有40万亿～60万亿个细胞。不论多细胞生物拥有多少个细胞，最初都源自一个细胞。对采用有性生殖方式繁衍后代的二倍体生物而言，最初的细胞就是受精卵，它由一个精子和一个卵子结合而成。在形成个体的发育过程中，单个受精卵首先以细胞分裂的方式形成2个细胞，这2个细胞再分裂形成4个细胞，就这样一次次通过细胞分裂的方式不断地进行细胞数量的扩增，直至达到一个完整个体所需要的细胞数，这个过程称为细胞增殖。

更重要的是，在细胞数量增加的同时，细胞的种类也在通过细胞分化的方式不断增加。二倍体生物的细胞类型首先可以分为两大类，具有单倍体基因组的生殖细胞和具有二倍体基因组的体细胞；前者负责在有性生殖过程中形成受精卵，后者负责执行个体的生命活动。过去人们认为，一种体细胞通常具有一种特定的形态和功能，因此，体细胞按照其形态和功能的特点可以分为相应的细胞类型。例如，在发育成完整个体的人体中，估计有200多种不同的体细胞类型，如神经细胞、肌肉细胞、白细胞和红细胞等。显然，要想揭示多细胞生物不同的生命活动与功能，首先需要认识其相应的细胞类型。也就是说，细胞类型的区分和确定是生命科学最重要的基础之一。随着后基因组时代的到来，研究者开始从分子水平对细胞进行分类。2016年10月，在英国伦敦召开了一个国际会议，提出了一个宏大的“人类细胞图谱”（Human Cell Atlas，HCA）的大科学计划；该计划堪比“人类基因组计划”，其组织委员会由来自10个国家的不同研究机构的27名科学家组成。2017年5月，HCA的主要领导人Regev A和Teichmann SA及其合作者发表了关于HCA的报告，系统地论述了该计划的意义、目标，任务和实施路径；其基本目标是：采用特定的分子表达谱来确定人体的所有细胞类型，并将此类信息与经典的细胞空间位置和形态的描述连接起来（Regev et al.，2017）。2019年，美国国立卫生研究院（NIH）也启动了一个名为“人类生物分子图谱计划”（Human Biomolecular Atlas Program，HuBMAP）（Hu，2019）；该计划拟用7年的时间，发展一个能够在单细胞水平上全面分析人体细胞图谱的研究框架和研究技术（Hu，2019）。可以看到，从分子水平上来区分和确定细胞类型已经成为当前生命科学的一个重要的前沿领域。

第二节　关键科学问题

一、细胞类型的遗传学基础

生物在从受精卵分裂到个体发育成熟的生长过程中，一代又一代地产生着各种新的细胞类型，其中子代细胞获得了亲代细胞的遗传信息，细胞间形成了具有特定遗传关系的细胞谱系。20 世纪 70 年代，英国科学家 Sulston 利用微分干涉显微镜对线虫整个生长过程的持续观察，描绘出了第一个完整的动物细胞谱系图，清楚地揭示了线虫的1090个体细胞中每一个细胞的“身世”和“命运”。可以说，要区分和鉴定多细胞生物的细胞类型的一个重要任务就是要确定细胞谱系。如何构建复杂生物体的细胞谱系，尤其是人类的细胞谱系，显然是当前研究者面对的一个重大挑战。

过去研究者认为，在多细胞生物个体中，产生体细胞的基本过程是细胞增殖，即亲代细胞通过 DNA 复制机制将自身的基因组完整地复制为两份拷贝并传给两个子代细胞，因此不同的体细胞拥有的基因组应该都是一样的。近年来，研究者利用单细胞全基因组测序技术规模化地分析了个体内的各种细胞，发现不同体细胞之间的基因组存在着大量的差别，这被称为体细胞的基因组“镶嵌现象”（mosaicism），它们主要是复制过程中核酸序列随机突变和染色体结构变异导致的。因此，如何揭示细胞的镶嵌性基因组也成为研究细胞类型和细胞谱系的一个重大科学问题。

二、细胞类型与环境的关系

多细胞生物的个体发育与生长离不开环境。遗传学的基本定律是基因型与环境之间相互作用形成表型。因此，细胞类型的形成也受到环境的影响。当前鉴定细胞类型的一个重要手段是单细胞转录组测序，进而根据不同的基因表达谱来区分不同类型的细胞。显然，环境因素会参与到基因的转录调控中。此外，不同的环境条件能够引起基因组不同的表观遗传修饰，进而影响到基因的表达调控。因此，一个重要的科学问题是，如何在区分和确定细胞类型的研究中认识环境因素在基因转录表达调控中的作用，以及排除偶然的环境因素对基因转录表达的干扰。

环境还能够导致体细胞基因组序列产生随机变异。有研究表明，抽烟、紫外线照射等环境刺激会引发体细胞基因组的碱基突变或者染色体结构变异。此外，机体内部的病理变化如癌变以及衰老等过程也会引发体细胞的基因组异常。显然，环境与基因组突变之间的关系是复杂的、随机的，并且是不可忽略的。如何在区分和确定细胞类型的研究中鉴定出环境引发的基因组异常是一个研究者面临的重大挑战。

三、细胞类型的分子分型技术与分型标准

过去研究者按照体细胞的形态和功能特点来进行细胞类型的划分，而今天则按照分子特点来进行细胞分型。目前在鉴定细胞类型方面应用最为广泛的分子分型技术是单细胞转录组测序，主要是依据基因表达谱的特征来区分不同类型的细胞。单细胞转录组测序技术的广泛应用源于该技术的高灵敏度和高通量，并且测序成本很低。单细胞水平的全基因组测序技术尽管在细胞类型鉴定方面已经有一些成果，但目前还存在着通量低、成本高、数据分析难度大的问题。蛋白质由于不具备核酸可以扩增的特性，因此单细胞蛋白质组技术目前还没有具备单细胞测序技术的灵敏度，要用于细胞分型面临着很大的技术挑战。

如果用不同的分子分型技术进行细胞类型的区分和确定，需要解决这样一个科学问题：这些不同类型的生物大分子之间是什么关系？首先要回答的是基因表达水平与蛋白质表达水平的关系。过去人们认为，基因转录水平与蛋白质合成水平是线性关系，如果基因表达水平高，则蛋白质表达水平高；反之亦然。随着近年来研究工作的深入，研究者发现，总体而言基因表达水平与蛋白质表达水平之间相关性并不高。也就是说，在同时测量一个细胞内二者的表达水平时有可能得到不同的甚至是相反的结果。如果把基因组的表观遗传修饰、非编码 RNA 的转录和蛋白质翻译后修饰等也用作细胞分型的分子标准，那就更容易导致细胞分类的复杂性和不确定性，因为这些类型的分子显然与基因表达或者蛋白质表达之间不存在线性关系。显然，细胞分型的分子标准是未来细胞类型区分和确定的关键科学问题。

第三节　重要研究方向的发展现状与趋势分析

一、细胞图谱

“人类细胞图谱”计划（以下简称HCA）目前在国际上正引领细胞类型的分子分型研究工作。2017年10月16日，该计划首批资助的38个项目正式公布，分别来自欧洲、亚洲、非洲和北美洲的8个国家，研究内容涉及神经系统和免疫系统等不同的组织器官以及分析技术。2018年3月8日，英国Sanger研究所在其网站上宣布，从一系列捐赠的发育中的人体组织（包括肝脏、皮肤、肾脏和胎盘）中提取的25万个细胞已经完成单细胞转录组测序工作；美国Broad研究所紧随其后也发布了50多万个人体免疫细胞的单细胞转录组测序数据（https://data.humancellatlas.org/）。据不完全统计，至今已经有上千篇与人类细胞图谱相关的研究论文发表。

目前人类细胞图谱分析的主要技术是单细胞转录组测序。英美科学家利用该项技术，改写了成年人血液中特定的免疫细胞图谱，其中树突状细胞（dendritic cell，DC）由4种类型扩大为6种，单核细胞（monocyte）由原来的两种类型变成了4种，还发现了一种新的DC祖细胞（DC progenitor）（Villani et al.，2017）。此外，英国Sanger研究所的研究人员分析了人类妊娠早期（6～14周）胎盘的约7万个细胞的转录组，绘制了人类妊娠早期的胎盘细胞图谱，并鉴定出了蜕膜自然杀伤细胞（decidual natural killer cell，dNK）的三个新细胞亚群（Vento-Tormo et al.，2018）。不久前，人类肝细胞图谱研究取得重大进展：通过对肝实质细胞、血管内皮细胞、胆管细胞等不同肝组织的10 000多个细胞进行单细胞转录组测序分析，鉴定出了39种细胞类型，其中包括了内皮细胞、库普弗细胞和肝细胞中许多新的细胞亚群，并在胆管中发现了一类新的上皮祖细胞——它们不仅能够分化成胆管细胞，还可以在合适的条件下分化成肝实质细胞（Aizarani et al.，2019）。

其他动物种类的细胞图谱研究也正在进行中。小鼠的细胞图谱研究文献仅次于人类，已经报道了许多新的细胞类型。例如，加拿大科学家利用单细胞转录组测序技术分析了小鼠小肠上皮的成体干细胞群，发现了一类新的小肠干细胞——revSCs；这是一种缓慢循环的细胞类型，会在小肠受损伤后受YAP1 信号调控激活，进而维持干细胞池稳态并促进小肠上皮再生（Ayyaz et al.，2019）。*Science* 杂志在 2018 年 6 月 1 日同一期登载了 3 篇有关动物胚胎发育早期细胞图谱的研究论文，研究者利用单细胞转录组测序技术分析了爪蟾和斑马鱼这两种模式生物胚胎发育早期不同的时间段的细胞基因表达谱，发现了许多动物胚胎发育过程中细胞类型变化的特征，如许多胚胎细胞类型在发育早期就出现，其发生的时间早于过去的预期（Briggs et al.，2018）；又如，来自一种细胞类型并以克隆方式扩增的细胞谱系通常是严格受限于胚胎的“全局状态”（state landscape）（Wagner et al.，2018）。研究者通过单细胞转录组测序技术系统地分析了一种脊索动物——海鞘完整的个体发育过程，构建了包括神经系统在内的全部组织类型的细胞谱系，确定了60种细胞类型，并详细分析了幼虫神经系统的 41 种细胞亚群（Cao et al.，2019）。

我国科学家在细胞图谱方面的研究工作也很有成效，总体与国际研究保持同步，并取得了一系列国际领先的研究成果。浙江大学郭国骥教授团队利用自主研发的低成本、高效率、完全国产化的高通量单细胞测序平台，对小鼠不同生命阶段的近 50 种器官组织的 40 余万个细胞进行了系统性的单细胞转录组分析，构建了全球首个哺乳动物细胞图谱（Han et al.，2018）。该研究团队又利用其技术平台对来自胎儿和成人的 8 个主要器官的 60 种组织样品进行了单细胞转录组分析，构建了跨越胚胎和成年两个时期的人体细胞图谱，包括 100 余种细胞大类和 800 余种细胞亚类（Han et al.，2020）。除生物个体全局性的细胞图谱研究以外，我国科学家在具体组织器官的细胞图谱研究方面也取得了很多进展，如陆军军医大学研究者发表的一项研究工作揭示，小鼠的体液免疫系统中存在一类新亚群细胞——$SOSTDC1^{+}$ 滤泡辅助性 T 细胞（follicular helper T cell，Tfh），它们能够通过分泌 SOSTDC1 促进滤泡调节性 T 细胞（follicular regulatory T cell，Tfr）生成，进而维持体液免疫稳态（Wu et al.，2020）。

我国科学家在人体发育过程中的细胞谱系研究方面处于国际领先水平。

早在2013年，北京大学的汤富酬课题组和乔杰课题组就合作发表了124个来自人类着床前胚胎细胞的高精度单细胞转录组图谱（Yan et al.，2013）。他们发展了单细胞DNA甲基化组测序技术并用于人类着床前胚胎细胞的细胞图谱研究，提出“DNA甲基化分析可以用于确定分裂球早期的细胞遗传谱系”（Zhu et al.，2018）；此外，他们利用单细胞转录组测序技术分析了人类精子发生中的细胞命运决定过程（Wang et al.，2018a）；并利用该技术分析了人类胚胎7～25周的肾脏组织3000多个细胞的细胞图谱，揭示了起始肾单位形成的前体细胞之间的异质性，以及肾单位前体细胞逐步分化产生不同类型的肾小管上皮细胞的过程（Wang et al.，2018b）；分析了人类胚胎4000多个心脏细胞的细胞图谱，确定了心肌细胞、心脏成纤维细胞、内皮细胞和瓣膜间质细胞等4种主要类型，发现心房和心室的心肌细胞在心脏早期发育阶段分别具有不同的特征（Cui et al.，2019）。他们还利用单细胞转录组测序技术绘制了人脑前额叶在胚胎发育过程的细胞图谱（Zhong et al.，2018）以及人脑皮层发育中的细胞谱系（Fan et al.，2020）。2020年，中国科学家研发了一种三维（3D）人囊胚培养体系，并在此体系的基础上绘制了人原肠前胚胎的发育全景图（Xiang et al.，2020）。

由于高等生物如哺乳动物个体的细胞数量巨大，要分析个体内的每一个细胞是很难做到的。正如HCA的组织者所说，为了得到人体细胞的精确图谱而对人体中所有细胞进行研究是不可能的，也没有这个必要（Regev et al.，2017）。HCA的组织者提出的研究策略可以称为“局部采样”——在细胞图谱的研究过程中，可以先进行少量稀疏的细胞采样，经过分析后再决定采用更深度的采样方式（Regev et al.，2017）。如果细胞分类仍然采用传统的表型特征，如形态辨识方法，那么这种“局部采样”的研究策略是有一定的合理性，一方面可以按照明确的可操作的标准从一群具有稳定表型特征的细胞中采集样本，另一方面由样本分析得到的结果也是具有相应的代表性的。但是，目前的细胞图谱研究是建立在分子水平之上，根据什么标准来“局部采样”以及样本的代表性就成了问题。对于如脑、心脏和肾脏等实体组织，需要通过蛋白酶等方法进行解离和制备单细胞；这些方法倾向于将易于解离的细胞类型解离下来，导致了不易解离的细胞丢失；与此同时，一些较为敏感的细胞可能会因为解离过度而破碎。这样来看，现在的解离方法难以有效获取实体组织中的所有细胞类型，结果的准确性大为降低。

从海鞘个体发育的细胞谱系研究可以看到，只有对个体的所有细胞进行分析，研究者才能得到比较确定的结果（Cao et al.，2019）。海鞘这种脊索动物在胚胎发育到原肠期时的细胞总数在 100～200 个，即使发育成可以游动的蝌蚪时也只不过有 2500 个左右的细胞；研究者用单细胞转录组测序技术分析了 9 万多个细胞，相当于把每一个发育阶段的每一个细胞重复分析了 12 次（Cao et al.，2019）。正因为有如此高“覆盖率”的单细胞分析，研究者才能够发现许多“低频”出现的细胞类型，包括在海鞘幼虫仅仅拥有的 177 个神经细胞中鉴定出近 40 个亚型（Cao et al.，2019）。由此可以看到，通过单细胞“局部采样”策略开展的高等动物细胞图谱研究在分析结果的精度和可重复性方面目前还远不能满足人们的预期。

二、衰老和病理情况下细胞图谱变化

多细胞生物不仅有胚胎发育到个体成熟过程，而且存在着衰老过程，因此需要研究个体在衰老阶段的细胞图谱。已经有多项研究表明，衰老期的细胞与成年期的细胞有很大的差别。早在 20 世纪末，研究者已经知道衰老的哺乳动物细胞具有很特殊的表型，包括停止细胞分裂、抗细胞凋亡、分泌大量的炎性细胞因子和其他种类的蛋白质。一项转录组研究工作指出，年轻人的成纤维细胞与老年人的成纤维细胞有 600 多个基因表达的差异，研究者据此基因表达差异确定了衰老细胞特有的转录组“指纹图”（fingerprint）（Zhang et al.，2003）。2020 年，中美科学家合作研究了食蟹猴卵巢衰老的单细胞转录组图谱，揭示了卵巢中的 7 种主要细胞类型：卵母细胞、颗粒细胞、基质细胞、平滑肌细胞、内皮细胞、自然杀伤 T 细胞和巨噬细胞，这些细胞类型的分布在不同年龄的猴子中没有明显差异；但是与衰老 / 长寿相关的基因和与卵巢疾病相关的基因主要在卵母细胞、颗粒细胞及基质细胞中表达，这些细胞类型对卵巢内环境平衡和衰老有重要作用（Wang et al.，2020）。此外，我国科学家还利用单细胞转录组测序技术绘制了食蟹猴动脉血管衰老的单细胞基因表达图谱，包括了主动脉弓及冠状动脉的内皮细胞、平滑肌细胞和成纤维细胞等血管细胞类型的基因表达图谱，并鉴定出 8 种区分主动脉和冠状动脉的新型分子标志物（Zhang et al.，2020）。

在个体的衰老过程中，不同的器官和组织的衰老程度与过程是很不一样的。斯坦福大学的研究者对100多个健康人进行了两年的研究，从蛋白质组学、代谢组学等分子层面解析了不同人衰老过程中特定组织器官的不同衰老变化，划分出了4种不同的衰老类型：免疫型、代谢型、肝型和肾型，不同的个体具有不同的衰老类型（Ahadi et al.，2020）。美国科学家通过对小鼠中17个器官和组织在10个不同时间节点的总RNA测序和血浆蛋白质组分析（Schaum et al.，2020），同时对小鼠全生命周期中20个器官的529 823个细胞进行了单细胞RNA测序，生成了一个小鼠衰老细胞图谱（Tabula Muris，2020）；对细胞转录组数据的分析发现，大多数器官早期和晚期的差异表达基因存在着明显的相关性，但是也有少数基因的表达变化的器官特异性大于年龄特异性；此外，与免疫应答途径相关的差异表达基因存在明显的富集（Schaum et al.，2020；Tabula Muris，2020）。

一般来说，衰老会导致器官和组织正常功能减退，扰乱其中复杂的组织和细胞间的相互作用，而这种互动与交流对于维持机体的健康状态至关重要。目前对于器官之间的衰老影响缺乏了解。随着单细胞分析技术的发展，研究者逐渐有了一些新的发现。美国研究人员利用单细胞转录组测序技术，系统地比较了老年小鼠与年轻小鼠免疫系统内各种细胞类型的基因表达谱，结果发现年轻小鼠的同类型免疫细胞之间的基因表达谱没有明显差异，各个细胞之间能够很好地协同工作。但是，在老年小鼠体内的同类型免疫细胞中，各种细胞间的基因表达差异明显增加，导致这些细胞彼此的协调性明显降低，从而使得老年小鼠对免疫刺激的响应变差（Martinez-Jimenez et al.，2017）。一项单细胞染色质修饰谱分析工作也发现了同样的现象：老人体内不同免疫细胞之间的染色质上的组蛋白修饰差异要远多于年轻人的（Cheung et al.，2018）。斯坦福大学的研究者研发了一种名为“EpiTOF”的质谱分析技术，可以在单细胞水平研究组蛋白的表观遗传修饰；他们利用该技术研究了年轻人和老年人的外周血单核细胞，发现不同年龄的免疫细胞具有不同的表观遗传学特征，并发现个体间与细胞间的染色质信号的异质性在衰老过程中明显增加；通过对这些样本中的双胞胎数据分析发现，衰老过程中的染色质信号异质性主要由环境因素所导致（Cheung et al.，2018）。

在分子水平上区分和确定细胞类型引出一个重要的问题——分子特征，尤其是基因表达或者表观遗传修饰，可以随着外部环境变化或者机体的病理变化而发生改变，进而导致细胞类型发生改变。这一点在肿瘤的单细胞研究方面已经得到了证实。经靶向疗法后再次生长的肿瘤往往表现出耐药性，意味着获得耐药性的肿瘤细胞发展成为了新的细胞类型。研究者利用单细胞转录组测序技术对肺腺癌患者在靶向治疗前和病灶残留阶段等不同病程的病理样本的 2 万多个细胞进行了分析，将病灶残留阶段的癌细胞与靶向治疗前的癌细胞相比，前者的增殖相关基因表达降低、肺泡细胞特征基因表达升高（Maynard et al.，2020）。

肿瘤的分子分型实际上就是对病态细胞的类型进行区分和确定，而单细胞水平的分子分型显然能够明显提高肿瘤分子分型的精确性，这对于具有高度异质性的肿瘤细胞研究和诊治大有帮助。2012 年，国际上根据测序技术和芯片技术将髓母细胞瘤分为 4 种分子亚型：WNT 型、SHH 型、三型和四型，其中前两者是基于 WNT 信号通路和 SHH 信号通路的变异而确定的，但后两者的分子特征却并不清楚。研究者利用单细胞转录组测序技术检测了 25 个儿童髓母细胞瘤样品，发现三型肿瘤存在未分化和已分化两种细胞类型，而四型肿瘤却只有已分化的细胞类型，即这两种被视为不同类型的肿瘤亚型很可能是起源于同一系谱的细胞，具有相似的致癌机理，但为发展到不同阶段的肿瘤（Hovestadt et al.，2019）。

单细胞分子分型技术还有助于揭示正常细胞类型和病理类型细胞之间可能存在的内在联系。加拿大科学家利用单细胞转录组测序技术首先检测了小鼠小脑发育过程中的 6 万多个细胞，分析了小脑发育细胞亚群的类型以及这些细胞亚群之间的演化途径；然后再将这些数据与人类小儿小脑肿瘤的转录组数据进行比对，确定了与各个肿瘤细胞亚型对应的正常小鼠小脑发育亚细胞群，发现肿瘤细胞亚群与处于某个特定时间节点内的特定发育亚细胞谱系的基因表达谱较为相似，如 PFA 型室管膜瘤的基因表达特征与小鼠胚胎期第 16 天的胶质前体细胞群的最为接近；这些结果提示，小儿小脑肿瘤的发生是源自早期小脑发育的紊乱（Vladoiu et al.，2019）。新加坡研究人员分析了采自人类胎肝、肝癌和小鼠肝脏的 21 万多个单细胞转录组序列，揭示了肿瘤微环境可以进行内皮细胞的胚胎性重编程，使得胎肝细胞和肝癌细胞共享一个抑制免疫的肿瘤-胚胎生态系统（onco-fetal ecosystem）（Sharma et al., 2020）。

综上所述，多细胞生物的细胞谱系研究并不是简单地局限在动物胚胎发育时期，要想获得完整的个体细胞图谱，需要研究者关注从胚胎期到成年期再到衰老期的全生长过程，而且还需要关注在变化的环境下和病理状态下的细胞类型与谱系演变过程。这些需求对研究者来说显然是巨大的挑战。从时间的角度来看，简单生物如线虫的细胞图谱分析比较容易做到，因为线虫生活周期短，从受精卵发育到可以产卵的成虫不超过三天；但对小鼠等哺乳动物来说就比较复杂了，其生长期长达数月。而对人来说，要研究个体在数十年生长期里可能出现的细胞谱系显然不是一件易事。从生理病理状态变化的角度来看，个体内外多种因素的相互作用以及生物大分子的随机性变化等会导致细胞类型变化的复杂性和不确定性，从而增大在分子层面研究细胞类型的难度。

三、细胞类型分子分型技术

目前用于研究细胞类型的分子分型技术主要是单细胞转录组测序。但是，研究者显然不满足只用这样一种技术从分子水平来对细胞进行分类。HCA 的实施方案是通过确定一系列不同的分子标志物来描述每种人类细胞，如非编码基因的表达水平，转录物可变剪接的水平，启动子和增强子的染色质状态，蛋白质表达水平以及翻译后修饰状态等（Regev et al., 2017）。NIH 不久前启动的 HuBMAP 则提出了更为庞大的研究计划，除了测序技术和染色质分析等技术，还将通过各种成像方法从组织切片的核酸、蛋白质和脂质等生物分子的分析中获得单细胞分辨率的组织器官的空间信息，并建立一个跨尺度的整合组织图谱（Hu, 2019）。

测定细胞的 DNA 及其染色质结构和修饰是当前单细胞研究的一个重点。从单细胞水平测定 DNA/ 染色质可接近性（DNA/chromatin accessibility）的方法包括 scNOME-seq、scATAC-seq、scDNAse-seq；测定染色体组装的方法为 scHIC；测定 DNA 修饰的方法包括 scBS-seq、scAba-seq、CLEVER-seq；测定组蛋白修饰的方法为 scChIP-seq。美国科学家采用 ATAC-seq 技术描绘了人胚胎脑发育过程中 6 个脑区的染色质可接近性动态变化特征图谱，识别出具有统计上的显著性的13万个开放染色质区域，其中仅有23%的区域在6个区域中共有；此外，研究者还利用部分脑区样本进行了单细胞转录组测序，并发现不同脑区之间的差异表达基因对应的染色质可接近性往往也具有非常显著的差异（Markenscoff-Papadimitriou et al.，2020）。

北京大学汤富酬研究组在2014年研发了微量细胞DNA甲基化组高通量测序技术，并首次实现了对人类早期胚胎发育过程中DNA甲基化组重编程的系统研究（Guo et al.，2014）。不久前，该研究组进一步研发了一种单细胞多组学测序技术seCOOL-seq（single-cell chromatin overall omic-scale landscape sequencing），可以对一个单细胞同时进行染色质状态、DNA甲基化、基因组拷贝数变异以及染色体倍性的全基因组测序，并利用该技术分析了小鼠早期胚胎和胚胎干细胞的染色质重编程与甲基化状态（Guo et al.，2017）。该研究组随后与乔杰研究组合作，利用该技术在单细胞分辨率水平上系统地描绘了着床前人类胚胎发育过程中DNA甲基化组和染色质状态组的重编程过程（Li et al.，2018）。此外，我国其他科学家团队也分别利用DNase-seq技术（Gao et al.，2018）和miniATAC-seq技术（Wu et al.，2018）系统地描绘了人类胚胎早期发育过程染色质调控的动态图谱。

为了将染色质结构研究与基因转录研究整合，同时能够实现全基因组范围的检测，哈佛大学庄小威研究组开发了全基因组范围染色质三维结构影像检测技术，并实现了在单细胞中同时进行DNA、RNA和蛋白质的标记。他们发展了一种名为“MERFISH”的荧光原位杂交技术，通过在一条染色体上连续标记50 kb大小的DNA片段，能够以50 kb的分辨率观测全染色体结构；研究者进一步将散布在全基因组的1000个区域进行编码标记，实现了全基因组三维结构的直接观测（DNA-MERFISH）；随后把DNA-MERFISH与标记的1000多个新生RNA（nascent RNA）以及核仁等细胞核结构特征蛋白相结合，揭示了转录水平与细胞核内环境的关系（Su et al.，2020）。由此可见，DNA-MERFISH可以与RNA的荧光原位杂交以及蛋白质的免疫荧光相结合，为将来各种基于细胞影像分析的应用场景提供了可能。

HCA在细胞图谱研究方面主要关注细胞类型的区分与确定，但现在的HuBMAP则开始关注细胞在组织器官中的空间分布情况。瑞典科学家开发了一种新研究技术，通过将单细胞转录组测序技术、空间转录组技术（spatial transcriptomics）和原位测序技术（*in situ* sequencing）相结合，可以在器官范围内以单细胞分辨率分析基因表达的时空差异；他们针对人类胚胎心脏的三个不同发育阶段，分别收集了4个、9个和6个组织切片，然后利用该研究策略检测了这些样本上的细胞类型和空间位置，进而构建了首个具有单细胞空间分辨率的人类器官发育转录图谱（Asp et al.，2019）。2020年，他们又开发了一种基于概率论的新算法，可以将组织样本中的空间转录组数据转换为组织中不同类型细胞的空间排布信息（Andersson et al.，2020）。

细胞谱系示踪是细胞图谱研究的一个重要内容，但单细胞转录组测序技术往往只能确定细胞类型而很难用于追踪细胞分化过程中细胞类型的演化，单细胞转录组数据的高维特性导致同一数据可以构建出不同的细胞分化轨迹。此外，在将组织器官进行单细胞化的处理过程中，细胞谱系之间的转换关系通常会丢失。有研究者研发了一种名为CellTagging的组合细胞索引技术，通过连续多轮细胞标记，能够同时捕获细胞克隆历史和细胞类型，进而重建多层细胞谱系树；他们利用这种技术对成纤维细胞的重编程和谱系进行追踪，揭示出在细胞谱系转换的最早阶段就确定了两种不同的转换路径：一种转换路径导致成功的重编程细胞产生，而另一种转换路径则导致细胞进入一条“死路”（dead-end）（Biddy et al.，2018）。哈佛大学医学院研究人员研发了一种能够同时检测细胞状态和细胞命运的新技术，即首先在细胞基因组中插入具有特定序列的“DNA条码”（DNA barcode），然后在随后的分化过程中通过单细胞基因组测序追踪细胞谱系，并通过转录组测序分析基因表达变化；他们利用该技术对小鼠造血过程中的细胞命运决定进行了研究，揭示出两条单核细胞分化路线（Weinreb et al.，2020）。研究者利用该技术进一步追踪了造血干细胞（HSC）的细胞谱系，发现转录因子TCF15可以使HSC进入静止（quiescence）状态并保持长时程的自我更新（long-term self-renewal）（Rodriguez-Fraticelli et al.，2020）。

由此可以看到，细胞类型鉴定和分子分型技术正在快速发展，研究者正在从分子水平研究细胞图谱跨越到将解剖部位和组织等不同尺度整合的微观－宏观多层次，正在从注重细胞的基因表达谱特征进入到同时关注细胞的基因组、染色质和蛋白质等各种生物分子信息，正在从强调离体的、静态的单细胞分析发展到在体的、动态的细胞类型研究，从而可以使人们以前所未有的时空分辨率观察组织的三维结构。显然，要实现这些转变和远大目标还需要开发出更多更好的研究技术。NIH 推动的 HuBMAP 的核心目标就是要进行变革性技术的开发，并搭建一个开放的、在细胞分辨率水平绘制人体组织图谱的数据库和协同研究框架。HuBMAP 为此设计了三大任务模块：①组织测绘中心（Tissue Mapping Centres）；② HuBMAP 整合、可视化与交互组（HuBMAP Integration, Visualization and Engagement）；③创新技术组（Innovative Technologies Groups）（Hu，2019）。HuBMAP 旨在开发一些创新方法，从而突破现存的技术的局限性。例如，为了对多种组织中的 RNA 和蛋白质进行成像，变革性技术，诸如交换反应信号扩增法（SABER）和序贯荧光原位杂交（seqFISH）等，将被进一步改进以提升放大能力、敏感性及通量。此外，新型质谱成像技术将以高空间分辨率和灵敏度对同一组织切片中的数百种脂肪、代谢物和蛋白质进行定量成像。这些新技术的开发都离不开新的计算工具和机器学习算法的发展（Hu，2019）。

在开发用来判别细胞类型的各种生物大分子技术的同时，有一个需要研究者高度关注的问题：这些不同类型的生物大分子之间是什么关系？要首先是要回答基因表达水平与蛋白质表达水平的关系。过去人们认为，基因转录水平与蛋白质合成水平是线性关系，如果基因表达水平高，即作为模板的 mRNA 的拷贝数多，则蛋白质表达水平高，即合成出来的蛋白质就应该多。但是，越来越多的证据表明，蛋白质表达水平和其相应的 mRNA 表达水平之间的相关性并非人们想象得那样高。一篇综述文章系统地分析了蛋白质表达水平与 mRNA 表达水平的关系，指出这种关系受到细胞内外环境变化、细胞稳态和状态变化以及 mRNA 的时空分布等各种影响，因此，mRNA 表达水平本身在许多情况下不足以用来预测蛋白质的表达水平（Liu et al.，2016）。美国科学家用质谱技术定量分析了 32 个人类组织中 1.2 万多个基因的蛋白质表达情况，然后对“基因型 - 组织表达谱”（genotype-tissue expression，GTEx）研究联合体获取的基因表达数据进行了比较，其结论是：组织特有的蛋白质信息能够解释遗传疾病的表型，而仅仅采用转录组信息则做不到这一点（Jiang et al.，2020）。

显然，蛋白质表达水平与 mRNA 表达水平的非线性关系对确定细胞的类型和进行分子分型是一个巨大的挑战。瑞典科学家的一项研究已经涉及了这个问题。研究者首先根据56株人细胞系的基因表达谱构建了一个细胞图谱，在此基础上选择了22株细胞系，然后用近13 993种抗体检测了这些细胞株上的12 003种蛋白质的亚细胞器的空间分布，并通过单细胞分析技术发现了1855种蛋白质存在着表达水平或者空间分布的差异。显然，在这项工作中，研究者是把基因表达谱而非蛋白质表达谱作为细胞分类的标准（Thul et al.，2017）。与此观点相反，德国科学家通过小鼠大脑皮层神经组织发育的研究工作对该问题给出了另外一种答案：他们发现少突胶质细胞（oligodendrocyte）、星形胶质细胞（astrocyte）、小胶质细胞（microglia）和神经细胞（neuron）的转录组与蛋白质组的相关性系数均只有0.4～0.45；在此情况下，研究者提出，蛋白质组的数据能够更好地反映大脑皮层的细胞类型和差异（Sharma et al.，2015）。

从生物学现有的知识来看，基因组的表观遗传修饰、非编码RNA的转录和蛋白质翻译后修饰等显然与基因表达或者蛋白质表达之间不存在线性关系。因此，如果研究者把这些类型的信息全都视为可用于细胞类型确定、细胞谱系示踪和细胞的分子分型等的分子标志物，那么将进一步导致细胞图谱研究的复杂性和不确定性。这还只是在分子层面上反映出来的问题，如果再到更高的层次，如组织形态或生理功能等，这些涉及细胞分类的各种标准之间的关系将变得更为模糊。HCA的组织者显然也意识到了这一问题，“关键在于现在始终不清楚基于形态的、分子的和生理性质的三者各自得到的分类特征是否相互兼容”（Regev et al.，2017）。

第四节　我国发展战略与重点方向

一、发展总体思路与发展目标

1. 发展总体思路

建立和开发针对细胞分子分型相关的新技术与新方法，注重学科交叉和多学科协同攻关；在细胞类型鉴定与分子分型领域鼓励探索和综合运用新概念、新理论、新技术、新方法以开展创新性研究活动，重点关注在个体发育过程和复杂性疾病中的细胞图谱辨识及其演化研究；加强建设和完善相关的数据库与技术支撑平台，实行充分的开放和共享。

2. 发展目标

为解决制约细胞类型鉴定与分子分型领域发展的关键科学问题做出重要的贡献；在细胞类型鉴定与分子分型重要方向取得突破性成果；进一步提升我国科学家在该研究领域的国际影响力，在多数研究方向处于与国际先进水平并行，在部分研究方向处于领跑地位。

二、优先发展领域或重要研究方向

1. 基于组学技术的细胞类型鉴定与分子分型研究

利用各类单细胞分析技术、染色质结构分析技术和分子影像技术等，系统地开展多种模式生物和人体的细胞类型鉴定与分子分型研究；开展具有个体时空信息的动态细胞图谱的构建；建立和完善细胞分类与分子分型的标准；构建全国统一的模式生物和人体的细胞图谱数据库。

2. 个体发育过程中细胞谱系的确定与追踪研究

发展和建立新型的细胞谱系示踪与追踪技术及方法，重点发展能够在细胞分化过程中同时进行细胞谱系示踪和细胞状态鉴定的技术；开展多种模式生物和人体胚胎发育过程中单细胞分辨水平的细胞命运研究，研究主要组织器官过程中的细胞谱系与细胞状态演化；建立从分子到细胞到组织的跨尺度的整合组织图谱。

3. 复杂性疾病中的细胞图谱辨识及其演化研究

利用各类单细胞分析技术和分子影像技术等，系统地开展各种肿瘤细胞的分子分型研究，以及在肿瘤发生、发展、转移和复发过程中的细胞类型鉴定与演化关系分析；研究免疫系统在生理和病理状态下的细胞图谱与变化规律；研究衰老过程以及衰老相关疾病中的细胞图谱及其演化机制。

4. 细胞的分子分型新技术和新方法的建立

建立和完善各种组织的单细胞制备技术与流程；建立和开发高精度、高灵敏度及高通量的单细胞蛋白质组技术、单细胞表观遗传信息分析技术；建立和开发适用于分子影像的新型化学小分子探针与高分辨率的实时动态分子影像技术，并集成创新多尺度多模态的影像技术平台。开发细胞图谱相关的多组学大数据分析和系统整合技术，并建立和开发适用于细胞分子影像分析的人工智能技术与可视化技术。

本章参考文献

Ahadi S, Zhou W, Rose S M S F, et al. 2020. Personal aging markers and ageotypes revealed by deep longitudinal profiling. Nat Med, 26（1）: 83-90.

Aizarani N, Saviano A, Sagar, et al. 2019. A human liver cell atlas reveals heterogeneity and epithelial progenitors. Nature, 572（7768）: 199-204.

Andersson A, Bergenstrahle J, Asp M, et al. 2020. Single-cell and spatial transcriptomics enables probabilistic inference of cell type topography. Commun Biol, 3（1）: 565.

Asp M, Giacomello S, Larsson L, et al. 2019. A spatiotemporal organ-wide gene expression and cell atlas of the developing human heart. Cell, 179（7）: 1647-1660 e1619.

Ayyaz A, Kumar S, Sangiorgi B, et al. 2019. Single-cell transcriptomes of the regenerating intestine reveal a revival stem cell. Nature, 569（7754）: 121-125.

Biddy B A, Kong W, Kamimoto K, et al. 2018. Single-cell mapping of lineage and identity in direct reprogramming. Nature, 564（7735）: 219-224.

Briggs J A, Weinreb C, Wagner D E, et al. 2018. The dynamics of gene expression in vertebrate embryogenesis at single-cell resolution. Science, 360（6392）: eaar5780.

Cao C, Lemaire L A, Wang W, et al. 2019. Comprehensive single-cell transcriptome lineages of a proto-vertebrate. Nature, 571（7765）: 349-354.

Cheung P, Vallania F, Warsinske H C, et al. 2018. Single-Cell chromatin modification profiling reveals increased epigenetic variations with aging. Cell, 173（6）: 1385-1397 e1314.

Cui Y, Zheng Y, Liu X, et al. 2019. Single-cell transcriptome analysis maps the developmental track of the human heart. Cell Rep, 26（7）: 1934-1950 e1935.

Fan X, Fu Y, Zhou X, et al. 2020. Single-cell transcriptome analysis reveals cell lineage specification in temporal-spatial patterns in human cortical development. Sci Adv, 6（34）: eaaz2978.

Gao L, Wu K, Liu Z, et al. 2018. Chromatin accessibility landscape in human early embryos and its association with evolution. Cell, 173（1）: 248-259 e215.

Guo F, Li L, Li J, et al. 2017. Single-cell multi-omics sequencing of mouse early embryos and embryonic stem cells. Cell Res, 27（8）: 967-988.

Guo H, Zhu P, Yan L, et al. 2014. The DNA methylation landscape of human early embryos. Nature, 511（7511）: 606-610.

Han X, Wang R, Zhou Y, et al. 2018. Mapping the mouse cell atlas by microwell-seq. Cell, 172（5）: 1091-1107 e1017.

Han X, Zhou Z, Fei L, et al. 2020. Construction of a human cell landscape at single-cell level. Nature, 581（7808）: 303-309.

Hovestadt V, Smith K S, Bihannic L, et al. 2019. Resolving medulloblastoma cellular architecture by single-cell genomics. Nature, 572（7767）: 74-79.

Hu B C. 2019. The human body at cellular resolution: the NIH Human Biomolecular Atlas Program. Nature, 574（7777）: 187-192.

Jiang L, Wang M, Lin S, et al. 2020. A quantitative proteome map of the human body. Cell, 183（1）: 269-283 e219.

Li L, Guo F, Gao Y, et al. 2018. Single-cell multi-omics sequencing of human early embryos. Nat Cell Biol, 20（7）: 847-858.

Liu Y, Beyer A, Aebersold R. 2016. On the dependency of cellular protein levels on mRNA abundance. Cell, 165（3）: 535-550.

Markenscoff-Papadimitriou E, Whalen S, Przytycki P, et al. 2020. A chromatin accessibility atlas of the developing human telencephalon. Cell, 182（3）: 754-769 e718.

Martinez-Jimenez C P, Eling N, Chen H C, et al. 2017. Aging increases cell-to-cell transcriptional variability upon immune stimulation. Science, 355（6332）: 1433-1436.

Maynard A, McCoach C E, Rotow J K, et al. 2020. Therapy-induced evolution of human lung cancer revealed by single-cell RNA sequencing. Cell, 182（5）: 1232-1251 e1222.

Regev A, Teichmann S A, Lander E S, et al. 2017. The human cell atlas. Elife, 6: e27041.

Rodriguez-Fraticelli A E, Weinreb C, Wang S W, et al. 2020. Single-cell lineage tracing unveils a role for TCF15 in haematopoiesis. Nature, 583（7817）: 585-589.

Schaum N, Lehallier B, Hahn O, et al. 2020. Ageing hallmarks exhibit organ-specific temporal signatures. Nature, 583（7817）: 596-602.

Sharma A, Seow J J W, Dutertre C A, et al. 2020. Onco-fetal reprogramming of endothelial cells drives immunosuppressive macrophages in hepatocellular carcinoma. Cell, 183（2）: 377-394 e321.

Sharma K, Schmitt S, Bergner C G, et al. 2015. Cell type- and brain region-resolved mouse brain proteome. Nat Neurosci, 18（12）: 1819-1831.

Su J H, Zheng P, Kinrot S S, et al. 2020. Genome-scale imaging of the 3D organization and transcriptional activity of chromatin. Cell, 182（6）: 1641-1659 e1626.

Tabula Muris C. 2020. A single-cell transcriptomic atlas characterizes ageing tissues in the mouse. Nature, 583（7817）: 590-595.

Thul P J, Akesson L, Wiking M, et al. 2017. A subcellular map of the human proteome. Science, 356（6340）: eaal3321.

Vento-Tormo R, Efremova M, Botting R A, et al. 2018. Single-cell reconstruction of the early maternal-fetal interface in humans. Nature, 563（7731）: 347-353.

Villani A C, Satija R, Reynolds G, et al. 2017. Single-cell RNA-seq reveals new types of human blood dendritic cells, monocytes, and progenitors. Science, 356（6335）: eaah4573.

Vladoiu M C, El-Hamamy I, Donovan L K, et al. 2019. Childhood cerebellar tumours mirror conserved fetal transcriptional programs. Nature, 572（7767）: 67-73.

Wagner D E, Weinreb C, Collins Z M, et al. 2018. Single-cell mapping of gene expression landscapes and lineage in the zebrafish embryo. Science, 360（6392）: 981-987.

Wang M, Liu X, Chang G, et al. 2018a. Single-cell RNA sequencing analysis reveals sequential cell fate transition during human spermatogenesis. Cell Stem Cell, 23（4）: 599-614 e594.

Wang P, Chen Y, Yong J, et al. 2018b. Dissecting the global dynamic molecular profiles of human fetal kidney development by single-cell RNA sequencing. Cell Rep, 24（13）: 3554-3567 e3553.

Wang S, Zheng Y, Li J, et al. 2020. Single-cell transcriptomic atlas of primate ovarian aging. Cell, 180（3）: 585-600 e519.

Weinreb C, Rodriguez-Fraticelli A, Camargo F D, et al. 2020. Lineage tracing on transcriptional landscapes links state to fate during differentiation. Science, 367（6479）: eaaw3381.

Wu J, Xu J, Liu B, et al. 2018. Chromatin analysis in human early development reveals epigenetic transition during ZGA. Nature, 557（7704）: 256-260.

Wu X, Wang Y, Huang R, et al. 2020. SOSTDC1-producing follicular helper T cells promote regulatory follicular T cell differentiation. Science, 369（6506）: 984-988.

Xiang L, Yin Y, Zheng Y, et al. 2020. A developmental landscape of 3D-cultured human pre-gastrulation embryos. Nature, 577（7791）: 537-542.

Yan L, Yang M, Guo H, et al. 2013. Single-cell RNA-Seq profiling of human preimplantation embryos and embryonic stem cells. Nat Struct Mol Biol, 20（9）: 1131-1139.

Zhang H, Pan K H, Cohen S N. 2003. Senescence-specific gene expression fingerprints reveal cell-type-dependent physical clustering of up-regulated chromosomal loci. Proc Natl Acad Sci U S A, 100（6）: 3251-3256.

Zhang W, Zhang S, Yan P, et al. 2020. A single-cell transcriptomic landscape of primate arterial aging. Nat Commun, 11（1）: 2202.

Zhong S, Zhang S, Fan X, et al. 2018. A single-cell RNA-seq survey of the developmental landscape of the human prefrontal cortex. Nature, 555（7697）: 524-528.

Zhu P, Guo H, Ren Y, et al. 2018. Single-cell DNA methylome sequencing of human preimplantation embryos. Nat Genet, 50（1）: 12-19.

第八章

细胞命运的决定及其可塑性

第一节　概　　述

细胞是生物体结构和功能的基本单元，承载了生物的生长、发育、繁殖与进化等基本生命活动。在机体生命过程中，细胞每时每刻都在面临着选择——增殖、分化、迁移或者死亡，这些选择使得细胞维持着现有的状态，或者转化成另一种身份（如生理的终末分化细胞或病理的癌细胞），这个过程称为细胞命运决定。细胞命运决定是多种细胞的生命活动的核心事件。如生殖细胞的产生与分化，以雄性生殖细胞为例，其中涉及精原干细胞的自我更新和分化、减数分裂以及精子生成与成熟等多个重要生物学问题。此外，成体干细胞作为各个器官中未分化的细胞，能够通过自我更新和分化为器官发育、自稳态维持及损伤修复提供新的细胞来源。成体干细胞不仅对各个组织器官的发育和损伤修复都起着十分重要的作用，而且很有可能是癌症发生的靶点。成体干细胞具有自我更新（对称分裂）和多系分化（不对称分裂）两种“互斥的”命运。自我更新维持着干细胞群体自身，而多系分化则负责着成体组织多种功能细胞的更替和损伤修复。成体干细胞的身份属性的解析及成体干细胞巢的微环境的体外重构将为器官的体外重建与制造、疾病的预防和靶向治疗提供重要的理论基础。除了干细胞外，终末分化的体细胞还可以通过转分化或去分化的方式转变成其他类型的细胞，这一过程称为细胞属性转换。近年来，研究人员通过谱系示踪技术和组织损伤再生模型，深入地研究了组织损伤再生的细胞来源，发现在多种组织中都存在以细胞属性转换为基础的组织再生。以上细胞命运决定的生物学事件由细胞内在因素和周围环境因素共同调控。一些重要的细胞内信号通路，如 Hippo 信号通路和 Wnt 信号通路，能够通过整合胞外环境和胞内信号的变化，调控细胞特异基因的表达，精细控制细胞的增殖、分化和死亡。而这些信号通路的异常将会导致细胞生长和分化异常，进一步导致组织发育异常、再生受损或肿瘤发生等疾病。此外，细胞外环境为细胞分化提供的复杂而多样的驱动信号，如细胞间的接触信号、代谢信号等，对细胞命运的调控有待探索。所以，深入研究干细胞命运决定、终末分化细胞属性转换的生命过程，构建相关信号通路调控网络，阐明细胞命运决定相关的分子机制，总结相关因子在不同生理学过程中的个性和共性特征，将有助于我们加深对特定细胞命运决定过程的理解，并为未来人工干预细胞命运奠定基础，为再生障碍和癌症等疾病的治疗提供理论指导。

第二节　关键科学问题

一、生殖干细胞与生殖细胞的发生

生殖细胞是生命体发生的基础。该方向重要的科学问题包括：①精原干细胞维持自我更新及分化的分子机制是什么？如何高效地将胚胎干细胞 / 诱导多能干细胞（ESC/iPSC）定向分化为精原干细胞以及如何高效稳定地在体外长期培养精原干细胞？如何在精原干细胞中进行高效的基因编辑并依此开发新的转基因技术？②在从精原细胞向精子的转变过程中，启动减数分裂的关键因子是什么？减数分裂同源重组过程中双链DNA断裂（double-stranded DNA break，DSB）形成交叉或非交叉是如何被决定的？哺乳动物减数分裂的体外重构是如何形成的？③减数分裂完成后，精子生成过程如何启动？精子生成中细胞核凝聚延长（组蛋白－鱼精蛋白置换）如何被精密调控，其生物学意义又是什么？顶体、鞭毛生成及细胞质丢弃等事件是如何被精确调控的？如何在体外诱导单倍体精细胞发生为成熟精子？

二、成体干细胞发现、功能调控与应用

成体干细胞是维持机体组织稳态的重要细胞类型。相关研究领域的重要科学问题包括：①是否所有的组织器官中都有成体干细胞？新发现的成体干细胞对于人类健康非常重要，如新发现的胰岛成体干细胞为胰岛功能的修复带来了可能性。②成体干细胞如何维持自身的自我更新能力？③成体干细胞的维持需要怎样的微环境条件（如免疫细胞、血管等）？④干细胞向特定功能细胞分化的过程中命运是如何被决定的？如何在体外实现成体干细胞的扩增与定向分化？比如，是否能通过胰岛成体干细胞建立体外长期大量扩增的体系、源源不断地获得胰岛β细胞？如何在体外实现造血干细胞的扩增和维持？间充质干细胞具有怎样的分化等级和谱系层次？⑤成体干细胞与疾病（器官退行性疾病、肿瘤）发生发展的关系是什么？

三、细胞属性转换的机制与应用

细胞属性转换是指终末分化的体细胞通过转分化或去分化的方式转变成其他类型的细胞，参与稳态维持或损伤修复等过程，其中有很多重要的问题有待解答。这些问题包括：①如何诱导细胞属性转换进而弥补损失或者衰老的细胞类型？②发生细胞属性转换的细胞是随机产生的，还是在特定的细胞微环境下产生的？如果是随机产生的，那么诱导信号是什么？如果是特定微环境下产生的，那么具体构成是什么？③微环境信号与细胞属性转换的互作关系是什么？④细胞属性转换的具体调控机制是什么？细胞是否具有特定的

性质，以至于可以相应信号发生转换？⑤细胞属性转换的程度及适时地关闭属性转换的过程的机制并不明确。⑥细胞属性转换过程中通常会被细胞属性检查点机制所抑制，而对正常细胞响应并抑制转录因子引发的细胞属性转变的机制也知之甚少。⑦细胞属性转换与组织稳态维持的关系，以及这种调控关系在疾病状态下的调控。

四、细胞命运决定时空信号调控网络

随时间空间轴动态改变的环境和胞内信号是决定细胞命运改变的根本物质基础。重要科学问题包括：①关键信号通路（如 Hippo、Wnt、代谢和细胞接触黏附的空间信号等）对细胞增殖分化和个体组织器官大小的调控所形成的时间与空间多维度网络。②关键信号通路调节细胞命运及可塑性的分子机制。③建立特色研究模型，结合我国人群资源，开展以我国人群代谢特征为基础的细胞命运及可塑性的研究。④开发研究各个信号通路的新方法和新技术，结合分子机制筛选鉴定具有潜在应用价值的小分子化合物或抗体，以实现细胞命运干预和相关疾病的治疗。

五、新技术的建立与开发

新技术是基础研究的重要工具。在该领域，一些关键技术需要建立与完善。例如：①单倍体干细胞系的建立与应用。哺乳动物单倍体胚胎干细胞是从单倍体囊胚中建立的细胞系，该细胞具有二倍体胚胎干细胞所有的特性，为在细胞中开展高通量正反向遗传筛选提供了新的工具。另外，单倍体干细胞可以替代精子通过卵子注射产生小鼠，因此可作为载体将基因编辑器通过“受精”带到胚胎中，为研究胚胎发育和细胞命运决定提供新的遗传学工具。单倍体干细胞是新生事物，还有待于进一步完善和发展，其应用潜力还有待于进一步的挖掘和拓展，如建立更多物种、不同分化状态、具有配子表观遗传特性的单倍体干细胞，建立替代卵子遗传物质的单倍体干细胞，如何稳定其单倍体特性，挖掘单倍体干细胞作为配子替代物在胚胎发育和细胞命运决定研究中的作用等，是该领域的关键科学问题。②类器官作为一种新型的细胞研究模型能够高度模拟原位组织的生理结构和功能（Fatehullah et al.，2016；Lancaster and Knoblich，2014）。类器官模型不仅是研究干细胞的有力工具，还是研究器官发育和细胞分化的理想模型。但是，目前类器官体系存在细胞类型不完善、不成熟、不具备生理状态的空间结构、不能有效地响应内分泌和特定外界环境刺激等缺陷。因此，建立一套可控的三维类器官培养新技术体系，体外培养出具有管道化特征、完整血管网络、功能成熟、免疫及外界刺激响应性好、能够快速评价药物毒性及疗效、具有临床应用价值的新一代类器官是类器官研究领域的关键科学问题。③在生物体内对细胞命运进行长时间标记及实时示踪是该领域内一个重大的技术需求。细胞命运的转换通常需要对有限材料和细胞数目进行分析，如何对少量的细胞进行表达谱、表观遗传修饰等检测，预测并追踪少量细胞转换的过程带来了技术上的难度。对细胞状态和组织结构修复的描述，涉及生命现象的原则描述，需要数理模型对其本质进行刻画。

第三节　重要研究方向的发展现状与趋势分析

一、雄性生殖细胞研究领域

1. 精原干细胞

精原干细胞（spermatogonia stem cell，SSC）最初由原始生殖细胞（primitive germ cell，PGC）分化而来，而PGC是一种独立于三胚层之外的特殊细胞类型。当胚胎植入子宫后，伴随着原肠运动的发生，在胚外外胚层分泌的BMP4、BMP8B和原始内胚层的Wnt信号通路协同作用下，上胚层（epiblast）中产生了PGC的前体细胞（Ohinata et al.，2009；Ying and Zhao，2001）。小鼠PGC最早发现于胚胎发育的第6.25天（E6.25），位于原条后端；

伴随胚胎的发育，PGC 数量逐渐增加，并且进入后肠（hindgut）内胚层中；在 E9.5 时离开后肠进入背侧肠系膜，向生殖嵴迁移；在 E12.5 时几乎所有 PGC 到达生殖嵴并持续增殖；在 E13.5 时雄性 PGC 逐渐建立起雄性印记并进入有丝分裂阻滞，随后分化为精原细胞前体细胞（prospermatogonia，ProSG）（Bendel-Stenzel et al.，1998；Ginsburg et al.，1990；Kanatsu-Shinohara and Shinohara，2013）。这些位于曲细精管管腔中的 ProSG 在出生前一直处于阻滞状态（ProSG-T1），并且不断建立包括父源印记的雄性表观遗传修饰。出生之后，ProSG 迁移到曲细精管的基底膜上，重新开始有丝分裂（ProSG-T2）并分化成精原细胞（Law and Oatley，2020）。ProSG 向 SSC 的转变是一个高度动态变化的过程，由于技术限制，目前对该过程的分子调控机制还不甚了解。Tan 等（2020）利用单细胞测序技术对此过程中不同细胞类型转录组水平的动态变化进行了系统性分析，并筛选出一系列具有潜在功能的基因，但这些基因在此过程中究竟分别发挥何种功能仍需进一步探究。

由于缺乏成熟的人 SSC 体外培养系统，关于 SSC 维持自我更新与分化的机制研究主要利用小鼠 SSC 进行（Makela and Hobbs，2019）。在小鼠中，胶质细胞源性神经营养因子（GDNF）和成纤维细胞 2（FGF2）是维持 SSC 自我更新最重要的两个因子。GDNF 主要由曲细精管中的支持细胞（Sertoli cell）分泌，GDNF 通过与 SSC 细胞膜表面的由人胶质细胞系来源神经营养因子受体 α 1（GFR α 1）和受体酪氨酸激酶 RET 组成的共受体结合，激活下游的 PI3K-Akt 信号通路（Lee et al.，2007；Meng et al.，2000）。FGF2 与 SSC 细胞膜表面的受体 FGFR 结合，与 GDNF 协同激活 Src 家族激酶，进而激活 Ras 蛋白，Ras 蛋白再进一步激活下游的 Akt 和 MER2K 信号通路（Saitou and Miyauchi，2016）。此外，支持细胞分泌的趋化因子 CXCL12 与精原细胞上表达的 CXCL12 的受体 CXCR4 相互作用，也在调控小鼠 SSC 命运决定中发挥作用（Loveland et al.，2017）。视黄酸（retinoic acid,RA）是诱导小鼠 SSC 分化的关键因子。RA 主要由支持细胞分泌，能诱导减数分裂起始基因 *Stra8* 表达，在精原细胞减数分裂起始的调控中发挥关键作用（Griswold et al.，2012）。此外，RA 还能够通过抑制支持细胞表达 GDNF，并同时拮抗 GDNF 调控的信号通路，促进 SSC 的分化（Barrios et al.，2012；Pellegrini et al.，2008）。但 RA 诱导 SSC 分化的具体机制尚不明确。目前，领域内对 GDNF

依赖的维持 SSC 自我更新的信号通路已经比较了解，但对其他不受 GDNF 调控的信号通路及相关因子的作用仍需进一步研究。从上述对 SSC 自我更新与分化的分子机制的总结中可以发现，以支持细胞为主的 SSC 周围的微环境（niche）对于 SSC 的命运决定起着至关重要的作用，但微环境具体是如何对 SSC 进行精确调控的，有哪些分子参与调控这一过程尚不清楚。

SSC 的分离与富集是对其进行体外培养、移植的首要步骤。小鼠中 SSC 的含量极低，仅约占小鼠睾丸细胞的 0.01%，如何高效精确地分离出 SSC 一直是领域内亟待解决的难题之一（Kanatsu-Shinohara et al.，2013）。现有的方法无法分离出纯粹的 SSC，对 SSC 筛选标志物的使用仍需进一步的探究。经过领域内不断探索，目前小鼠 SSC 的体外培养技术已经较为成熟：使用饲养层细胞培养法、无饲养层细胞培养法、组织块培养法或三维细胞培养法均能够在体外培养小鼠 SSC 达数月之久，且使用组织块培养法可以在体外完成小鼠 SSC 的精子发生（Komeya et al.，2018；Kubota and Brinster，2018）。但由于许多培养关键因子的不保守性，小鼠 SSC 的体外培养及精子发生条件不能很好地应用在其他哺乳动物上，尤其是人和大动物中，因此对于人和其他物种 SSC 的体外培养仍需进一步探索。

2. 减数分裂

减数分裂为生殖细胞过程所特有的生物学事件，是生物有性生殖的基础，也是保证物种繁衍、染色体数目稳定和物种适应环境变化而不断进化的基本前提。所以，减数分裂如何发生和调控，是探索生命奥秘的基本科学问题。

不育不孕症作为一个危害全球公共健康的大问题，影响着 10%～15% 育龄夫妇，其中男方因素约占 50%。临床研究表明，生精细胞减数分裂异常是造成男性不育及父源性出生缺陷的主要原因。然而，目前对生精细胞减数分裂异常发生的确切机制所知甚少。因此，深入解析减数分裂调控机制对临床诊治男性不育和预防父源性出生缺陷的发生等均至关重要。从单细胞酵母到哺乳动物，减数分裂的关键生物学过程包括同源染色体配对联会、重组交换与分离都高度保守，然而不同生物体之间调控减数分裂的分子机制却显著不同。如绝大多数调控酵母、果蝇、线虫减数分裂的关键基因，在哺乳动物中并无同源基因存在。减数分裂启动是有丝分裂细胞转换为减数分裂细胞的关键步骤。就细胞生物学研究现状而言，学术界对哺乳动物减数分裂启动的调控机制几乎一无所知，对减数分裂重组的调控研究也所知不多。

迄今，减数分裂还难以在体外高效实现，而在体外重建体内的减数分裂过程对阐明遗传稳定性、基因多样性的起源和表观遗传重编程发生的程度具有重大意义，对遗传疾病的防治、干预和男性不育的治疗以及对物种改良、动物保种都具有重要价值。

3. 精子生成与成熟

哺乳动物精子发生的最后阶段为精子生成与成熟。这一过程中精子细胞形态结构发生剧烈变化，最终形成高度特化的成熟精子。其间会发生一系列生物学事件，包括顶体生成、鞭毛产生、染色质凝聚、胞质丢弃等，这些事件精确有序地进行，为成功受精提供了必要基础。

雄性生殖细胞中顶体的正确生成对受精作用具有重要意义。顶体产生主要是由内质网膜合成顶体生成所需的蛋白质，高尔基体加工产生糖蛋白，并且由反面高尔基网产生前顶体颗粒，前顶体颗粒会被陆续运输到核膜外表面聚集形成顶体（Khawar et al.，2019）。目前已知 Agfg1、GOPC、PICK1、ZPBP1 等多种蛋白质参与上述形成过程（Xiao et al.，2009；Yao et al.，2002；Yu et al.，2009），但是顶体形成过程的调控机制还不甚清楚。有研究表明在较早期小鼠精子细胞中，MIWI/piRNA 通过与翻译起始因子 eIF3f 及 AU-rich 元件结合蛋白 HuR 等相互作用，激活精子细胞中的 *Agfg1* 和 *Tbpl1* 翻译，协助精子细胞顶体的形成，拓展了顶体产生过程中的转录调控机制（Dai et al.，2019）。

染色质凝聚是精子生成过程中最为重要和显著的生物学事件。在人类、小鼠等哺乳动物减数分裂完成后的球形精子细胞期，体细胞组蛋白开始逐渐被睾丸特异组蛋白变体取代。例如，人类和小鼠中发现的组蛋白变体 HILS1，在精子发生的第 9～15 步中高表达，可能参与“组蛋白→鱼精蛋白”的替换过程（Yan et al.，2003）。这些组蛋白变体形成的松散核小体结构为后续组蛋白修饰提供了基础，而组蛋白的转录后修饰对于精子染色质重塑具有重要作用。近年来研究发现，精子发生后期 piRNA 介导 PiWi 蛋白经由 APC/C 途径的泛素化降解，能够帮助泛素连接酶 RNF8 从细胞质进入细胞核内催化组蛋白 H2A、H2B 的泛素化过程，进而促进精子细胞组蛋白的替换（Gou et al., 2017；Zhao et al., 2013）。

随着精子细胞染色质凝聚，延长型精子细胞的转录活性逐渐下降，并且伴随着大量 mRNA 的降解。在后期精子细胞中，MiWi/piRNA 与其结合蛋白和脱腺苷酶 CAF1 组成 pi-RISC 复合物，指导精子细胞中 mRNA 大规模的脱腺苷酸化及降解，同时还发现 MIWI/piRNA 还可以类似干扰小 RNA（siRNA）的机制，直接介导部分睾丸 mRNA 的切割（Gou et al., 2014；Zhang et al., 2015）。上述两种由 MIWI/piRNA 介导的大规模 mRNA 降解可能与精子生成过程中的特定生物学过程密切相关。

越来越多的研究为揭示"组蛋白→鱼精蛋白"置换的调控机制及生物学功能提供了基础，但其中过渡蛋白如何替换组蛋白，鱼精蛋白又如何替换过渡蛋白，以及整个替换过程如何被精密调控仍有待进一步探究。一项研究揭示了小鼠精卵结合后精子 DNA 起始解压缩及染色质重塑的重要机制：即母源 SRPK1 激酶负责催化进入卵细胞的精子鱼精蛋白磷酸化，从而启动并完成"鱼精蛋白→组蛋白"置换，协助受精卵早期染色质重构，从而保障雄原核与雌原核逐步发育形成并随后融合（Gou et al., 2020）。这项工作将为深入研究精子生成过程中"组蛋白→鱼精蛋白"置换的精密调控机制和生物学功能提供重要的研究线索。

精子离开睾丸后并未完全成熟，还需要经历一系列加工修饰才能具有运动和受精的能力。附睾作为哺乳动物重要的生殖器官，参与了精子修饰成熟过程。Kiyozumi 等（2020）发现，附睾上皮细胞受体 ROS1 会与精细胞分泌的 NELL2 结合，进而激活产生分泌蛋白酶 OVCH2。OVCH2 进入附睾头部腔体，对精子表面蛋白 ADAM3 进行修饰，保证其能够与卵细胞透明带结合。此外，附睾中存在的其他蛋白质如 CRISP1、SPINK13、CES5A 等分别对精子的钙信号、表面蛋白以及脂质组成存在调控作用（Bjorkgren and Sipila, 2019）。已有研究报道附睾小体通过传递 Wnt 信号通路蛋白调控精子获能过程。尽管已有研究对附睾中精子的成熟做出了一定阐释，但是仍存在许多亟待解决的问题。例如，附睾中还存在什么重要蛋白因子协助精子成熟，以及附睾腔体微环境中离子、酸碱度等理化因素对精子成熟有何影响？

二、成体干细胞领域

1. 胰岛干细胞

糖尿病被列为对人类健康威胁最大的三类疾病之一，成为全球重点关注的公共卫生问题。全球有 4.6 亿的糖尿病患者，其中 1/4 在中国。胰岛 β 细胞衰竭是 1 型和 2 型糖尿病发生发展的最终结局，许多患者需要终生使用胰岛素注射治疗。对于很多未成年的 1 型糖尿病患者，胰岛素注射存在自身操作困难和应激情况下低血糖的风险；对 2 型糖尿病患者而言，长期注射胰岛素除了导致体重增加，也会导致胰岛素抵抗，结果是胰岛素注射剂量的不断增加。每日单次或多次皮下注射胰岛素带来的心理和疼痛负担导致患者的依从性较差，血糖波动较大，应激情况下低血糖风险大幅增加，长期注射导致血管、微血管事件的发生。因此，如果能够建立起一种恢复体内自动应答血糖水平、胰岛素分泌自动调节的治疗方法（Shapiro et al.，2000），是糖尿病临床治疗的根本和终极目标。自从世界上第一例胰岛移植手术成功后，这个方案就一直被视作糖尿病治疗的希望。然而，由于供体器官的严重缺乏极大限制了临床推广，目前全球仅有少数医疗中心开展胰岛移植治疗。如何打破供体的局限，大量获得可用于移植的功能性胰岛 β 细胞，一直是糖尿病治疗领域的巨大挑战。

一种安全的思路是以组织的成体干细胞为来源，这些器官特有的干细胞可以遵循“天然”的分化路径，在体外培养通过增殖和分化形成该器官的功能细胞。然而过去的十几年里，该领域内的“主流”认识是“胰岛不存在成体干细胞”，这成为走通这条道路的第一个路障。因此，发现并鉴定胰岛干细胞是培养功能性胰岛类器官的先决条件。2020 年，研究人员发现小鼠胰岛中存在成体干细胞，并利用这些成体干细胞首次建立了具有完备细胞谱系构成的胰岛类器官培养体系，并实现了在体外的长期、大量的扩增（Wang et al.，2020a）。这一发现修正了该领域在过去十几年里认为“胰岛没有成体干细胞”的认识偏差，同时，为“胰岛体外制造”的设计思路带来了前所未有的变革，引出令人激动的转化应用问题：人类是否存在胰岛成体干细胞？是否也能利用这些干细胞建立体外长期大量扩增的体系，从而源源不断地获得胰岛 β 细胞？1型糖尿病患者中胰岛成体干细胞是否得以保存（逃逸免疫攻击）？能否找到方法激活成体干细胞的分裂以促进胰岛细胞的体内再生？

2. 前列腺干细胞

前列腺癌是欧美国家发病率第一的男性恶性肿瘤，在我国男性中前列腺癌的发病率也呈逐年上升的趋势（Watson et al.，2015）。目前，前列腺癌的治疗仍然以雄激素去势治疗为主，但是大多数患者最终会发展成去势抵抗性前列腺癌。因此，寻找生理条件下去势抵抗性前列腺成体干细胞是前列腺研究领域的重要科学问题。

前列腺上皮细胞由基底细胞、管腔细胞和神经内分泌细胞组成（Li and Shen，2018；Shen et al.，2010）。前列腺细胞的这种分类方法主要基于细胞的物理位置定位和少数经典的细胞谱系标志物。因此，前列腺细胞谱系分类需要进一步解析，细胞谱系之间的关系还需全面的阐明。

值得注意的是，不同的研究方法会得到相互矛盾的结论，将来还需整合多种不同的实验方法的研究。利用细胞肾被膜移植重建的方法（Xin et al.，2003），研究发现只有基底细胞可以重构前列腺上皮的所有细胞类型（Burger et al.，2005；Goldstein et al.，2010；Leong et al.，2008；Xin et al.，2005）。利用类器官培养的方法，发现基底细胞和管腔细胞都可以在体外重构前列腺类器官，这一结果也为前列腺成体管腔干细胞的存在提供了有力证据（Chua et al.，2014；Gao et al.，2014；Karthaus et al.，2014）。在前列腺去势损伤的条件下，前列腺中有一群表达 *Nkx3.1* 的去势抵抗性前列腺成体干细胞（Goldstein et al.，2010；Wang et al.，2009）。体内的细胞谱系示踪实验表明，前列腺发育过程中前列腺中存在多能干细胞（Ousset et al.，2012），而在成体前列腺中的基底细胞和管腔细胞都是独立地自我更新（Choi et al.，2012；Ousset et al.，2012）。矛盾的是，细胞谱系示踪的方法也发现了一群具有多能性的前列腺基底细胞能够产生前列腺管腔细胞（Lu et al.，2013；Wang et al.，2013）。近期，利用单细胞测序技术、细胞谱系示踪技术、类器官培养技术联合发现前列腺中存在一群定位于前列腺内陷顶端的前列腺管腔细胞祖细胞（Guo et al.，2020；Karthaus et al.，2020）。进一步的研究证明，只有在管腔细胞全部丢失的情况下基底细胞才可以产生管腔细胞（Centonze et al.，2020）。

单细胞测序技术可以同时检测数千个细胞的大量信息，将会在成体干细胞研究领域发挥越来越重要的作用（Macosko et al.，2015；Svensson et al.，2017；Zheng et al.，2017）。

3. 造血干细胞

造血干细胞（hematopoietic stem cell）是临床应用最早、最成熟的一种多能成体干细胞。尽管如此，在体外培养过程中，如何抑制造血干细胞分化、促进其自我更新，仍是科学家尚未解决的问题。过去三十多年的研究积累为分离造血干细胞建立了一套完备的细胞表面标志物组合（Tian and Zhang，2016），利用流式分选技术分离得到的单个造血干细胞能够连续地重建受体小鼠的全血液谱系，这被认为是造血干细胞研究的黄金法则，目前这条法则尚无法推行到其他任何成体干细胞上。因此，造血干细胞被作为试金石来研究新基因、新信号通路在成体干细胞中的功能。例如，以造血干细胞为模型研究 RNA 甲基化 m6A 在成体干细胞自我更新中的作用（Yao et al.，2018；Zhang et al.，2017a）。反过来，这些研究也正不断地拓展和刷新我们对造血干细胞自我更新内在机制的认识。

造血干细胞的自我更新不仅需要自身基因的调控，也有赖于周围细胞的支持，这些周围细胞提供的环境叫作造血干细胞巢。造血干细胞巢包含两个基本内涵：第一，它在物理学上是造血干细胞自我更新的场所；第二，它在功能上能通过结构性或分泌性的机制直接调节造血活动。过去十年的研究极大地扩展了我们对造血干细胞巢组成和功能的认识。利用先进的标记技术、组织透明化技术和成像技术，科学家已经能准确地捕捉到造血干细胞在骨髓内的准确定位（Acar et al.，2015），并发现它们处于一种低氧的微环境中（Spencer et al.，2014）；利用不同的遗传突变小鼠，科学家鉴定出包括成骨细胞（Calvi et al.，2003；Zhang et al.，2003）、$Nestin^{+}$ 细胞（Mendez-Ferrer et al.，2010）、$NG2^{+}$ 细胞（Kunisaki et al.，2013）、$Lepr^{+}$ 细胞（Zhou et al.，2014a）、巨核细胞（Bruns et al.，2014）、脂肪细胞（Naveiras et al.，2009）在内的细胞是骨髓造血干细胞巢的关键组分。

4. 间充质干细胞

过去二十多年间，间充质干细胞是用于临床试验最多的成体干细胞，但人们对这群细胞的认识非常有限，多数实验只是局限于体外。近七八年的研究革新了我们对这个广泛存在于所有组织中的特殊细胞群的认识。一方面，科学家在骨髓中鉴定并分离到了负责成骨的主要间充质干细胞群体（Zhou et al., 2014a）；另一方面，研究发现，虽然骨髓间充质干细胞在体外能够分化成软骨，但在体内软骨是骨髓间充质干细胞的主要前体细胞（Ono et al., 2014；Yang et al., 2014；Zhou et al., 2014b）。在骨骼损伤后，骨外的而不是骨髓内的间充质干细胞是损伤修复的主要贡献者（Debnath et al., 2018；Ortinau et al., 2019）。这些工作成果颠覆了人们对骨髓间充质干细胞等级结构和生理功能的认识。骨髓间充质干细胞被广泛应用于非骨髓组织的损伤修复中，但有研究提示组织特异性的间充质干细胞是组织损伤修复的主要群体（Kramann et al., 2015），至于骨髓来源的间充质干细胞是否也参与其中，仍然缺乏直接的体内证据。

三、细胞属性转换领域

1. 细胞属性转换与组织器官再生

细胞属性转换是指终末分化的体细胞通过转分化或去分化的方式转变成其他类型的细胞，进而贡献到稳态维持或损伤修复等过程。近些年来通过谱系示踪技术和组织损伤再生模型，深入地研究了组织损伤再生的细胞来源，发现在多种组织中都存在以细胞属性转换为基础的组织再生。例如，在肝脏中，人们利用谱系示踪（lineage tracing）新技术发现，在药物诱导肝脏损伤再生的过程中，主要再生模式是通过细胞属性转换来实现肝细胞的再生，而并不存在肝脏干细胞所介导的损伤再生模式（Li et al., 2020）。在胰腺中，特异性诱导 β 细胞大量死亡可导致 α 细胞转分化变成 β 细胞（Kopp et al., 2016）。在小肠中，在腺窝基底的 $Lgr5^+$ 干细胞中特异性过表达白喉毒素诱导 $Lgr5^+$ 干细胞死亡，可导致小肠上皮前体细胞（enterocyte-progenitor cell）去分化成 $Lgr5^+$ 干细胞，从而促进小肠上皮细胞的再生（Tetteh et al., 2016）。在肺泡中，在部分肺切除的模型中，肺切除会诱导 I 型的肺泡上皮细胞转化成Ⅱ型的肺泡上皮细胞，促进肺上皮组织的再生（Jain et al., 2015）。由此可见，细胞属性转换存在于在多种组织中，提示其可能是组织中非依赖干细胞而存在的一种广泛的组织再生方式。

2. 细胞属性转换与疾病

异常的细胞属性转换会导致发育不良以及肿瘤等重大疾病。阿拉日耶综合征（Alagille syndrome，ALGS）是一种累及多器官系统的显性遗传性疾病，该病在肝脏中的表型主要为先天性肝内胆管缺乏。Huppert 等构建了小鼠模型，模拟了 ALGS 疾病，并发现肝细胞可转分化为胆管细胞（Schaub et al.，2018），这为 ALGS 的发病机制和治疗提供了新的视角。此外，通过对人类胆管细胞癌的样品测序数据分析，发现部分胆管细胞癌可能也来源于肝细胞。最近，通过在人源的诱导肝细胞（hiHep）类器官体内过表达 Ras，模拟了胆管细胞癌的发生发展过程（Sun et al.，2019），证明了人类肝细胞可以作为胆管细胞癌的起源细胞。在另一个上皮组织胰腺中，也发现在胰腺癌发生过程中，损伤的腺泡细胞转分化为腺管细胞，进而在 KRAS 驱动下发展为胰腺导管癌（Alonso-Curbelo et al.，2021）。在损伤情况下终末分化细胞也会通过去分化为前体细胞贡献到食道癌、胃癌、肺癌等肿瘤发生过程（Dotto and Rustgi，2016；Jiang et al.，2017；Leushacke et al.，2017）。这些工作提示对细胞属性转换的研究可能为疾病研究和治疗提供新靶点。

3. 细胞属性转换的分子调控

损伤过程的细胞内外信号通路可能在上游调控着细胞属性转换。以哺乳动物肝脏为例。肝脏是重要的内脏器官，发挥着多种生理功能，包括代谢和解毒功能（Si-Tayeb et al.，2010）。通过过表达 NICD（Notch1 的组成型激活形式）激活 Notch 信号通路，或者过表达 Yap 的活化形式激活 Yap 信号通路，可以使成熟的肝细胞属性转换形成 LPLC（Yanger et al.，2013；Yimlamai et al.，2014）；反之则能抑制损伤诱导的肝细胞属性转换（Li et al.，2019；Yanger et al.，2013）。在 DDC 等诱导的肝损伤中，Wnt 信号通路被激活（Planas-Paz et al.，2019），肝细胞特异性地过表达 β- 连环蛋白（β-catenin）可以促进肝细胞属性转换（Okabe et al.，2016）。此外，在体外肝细胞培养中，通过激活 Wnt 信号通路可以使人的原代肝细胞发生去分化（Hu et al.，2018；Okabe et al.，2016；Zhang et al.，2018a），进一步证明 Wnt 信号通路可能是体内肝细胞属性转换的重要调控机制。而在肝细胞中敲除 *Lgr4* 和 *Lgr5* 基因并不影响损伤诱导的肝细胞属性转换，提示着其中存在着其他信号通路的作用（Planas-Paz et al.，2019）。利用肝内胆管缺失的 ALGS 小鼠模型，发现 Tgf-β 信号对肝细胞向胆管细胞转分化发挥重要作用（Schaub et al.，2018）。

成体组织中细胞属性转换研究主要集中在转录因子上，近年来的研究显示转录因子直接作用于染色质水平进行细胞命运调控。新近的研究明确了染色质可及性（chromatin accessibility）对细胞属性转换的关键作用。Arid1a 作为染色体重塑复合体的核心亚基之一，可以介导肝细胞属性转换基因在静息状态下处于预打开状态，使得损伤应激下这些基因可以响应活化的 Yap 信号，完成肝细胞属性转换过程（Li et al.，2019）。这个工作从染色体水平上揭示了分化成熟的肝细胞具有细胞属性转换的潜能。

除了染色质开放程度的调控，近年来研究表明 DNA 甲基化、组蛋白修饰、非编码 RNA 调控和染色体空间结构的调控也是细胞命运调控的关键因素。其中，DNA 甲基化是最早发现、也是最稳定的基因表观修饰方式之一，在调控基因表达、染色体结构维持、X 染色体失活和基因组印迹中起着重要的作用。胚胎期发育和原始生殖细胞发育的过程中，新的细胞产生需要全基因组范围内的 DNA 甲基化模式重编程。因此，发育过程中甲基化模式的适时消除和重建对于个体的生存与健康至关重要。在体细胞中，甲基化重编程通常会发生在局部区域范围内，具有位点特异性，但其对于成体组织的 DNA 甲基化对细胞命运可塑性的研究并不清楚。

4. 外在微环境信号对细胞属性转换的调控

机体内细胞的命运与周围细胞类型及微环境成分紧密联系、相互影响。以肝脏为例，肝细胞主要由肝实质细胞和非肝实质细胞（non-parenchymal cell，NPC）组成。NPC 包括胆管细胞、肝脏星形细胞、内皮细胞、肝脏巨噬细胞和肝脏自然杀伤细胞（NK 细胞）等（MacParland et al.，2018；Miyajima et al.，2014；Racanelli et al.，2006），这些 NPC 所处环境中的细胞外基质和细胞因子构成了 LPLC 所处的微环境。组织损伤诱导的细胞属性转换常常伴随着免疫细胞的浸润和激活、基质细胞的活化与增殖等现象（Jiang et al.，2017），而组织再生的完成往往伴随着这些现象的消失。那么微环境是否调控细胞属性转换过程？如果是，微环境细胞或者因子如何调控细胞属性转换？

机体的免疫系统在抵御肿瘤、病原微生物感染、自身免疫性疾病或参与组织修复中起到重要的、不可替代的作用。值得关注的是，免疫细胞是极具可塑性的细胞群，能应对不同的生理或病理微环境和炎症因子的刺激，分化成不同的细胞亚型，继而发挥有利的——力图帮助机体回归稳态或有害的——促进疾病发生发展的免疫调控功能。其中，T 细胞的分化和可塑性非常强大，并且各种 T 细胞亚型之间复杂的相互转化也是目前肿瘤、病原微生物感染、自身免疫性疾病研究领域中的热点。T 细胞通过表达多样化的表面受体 TCR，有能力识别自然界中成千上万种病原微生物组分或肿瘤抗原，进而分化成具有不同免疫功能的亚群，如 Th1（分泌 IFN-γ）、Th2［分泌白细胞介素 -4（IL-4）］、Th17（分泌 IL-17）和调节性 T 细胞［Treg，分泌转化生长因子 -β（TGF-β）］（Bird，2019）。另外，免疫细胞的分化和可塑性在促进或缓解疾病发展中发挥重要作用（Qiu et al.，2020）。Th1、Th17 细胞亚群使多种自身免疫疾病恶化，虽然发挥免疫抑制功能的 Treg 在自身免疫性疾病中发挥保护性作用，但 Treg 有利于肿瘤或病原微生物逃逸免疫监视（DuPage and Bluestone，2016；Togashi et al.，2019）。此外，先天免疫反应是抵御病原微生物等的第一道防线，其中巨噬细胞、中性粒细胞、NK 细胞或骨髓来源的抑制性细胞（myeloid-derived suppressor cell，MDSC）等发挥重要功能。受不同疾病微环境的影响和转录因子的调控，巨噬细胞能分化成促炎的 M1 型巨噬细胞和抑炎的 M2 型巨噬细胞（Pathria et al.，2019）。

研究人员通过遗传操作技术方法建立相关动物模型和活体观察模型，发现组织损伤修复过程中会出现炎症反应期、修复期和重塑期等一系列“瀑布式”事件的多细胞协同作用。例如，肝脏在静息状态下一般处于免疫抑制状态，而大多数肝脏损伤的一个重要特征是过度的炎症反应。例如，乙型肝炎病毒（HBV）感染导致的肝炎是危害中国人身体健康最重要的肝脏疾病，原发性硬化胆管炎（PSC）等胆汁淤积性肝病也常常和免疫调节异常密切相关（Guicciardi et al.，2018；Jiang et al.，2017；Ravichandran et al.，2019）。免疫细胞和肝细胞的相互作用可能是肝脏损伤再生中的一个关键调控机制。肝脏组织中的免疫细胞由固有免疫细胞和适应性免疫细胞组成，包括巨噬细胞、中性粒细胞、NK 细胞、T 细胞和 B 细胞等（Racanelli et al.，2006）。已有研究报道指出，巨噬细胞在肝组织损伤再生过程中发挥着重要的作用（Abdullah and Knolle，2017；Eming et al.，2017）。例如，在胆碱缺乏、乙硫氨酸补充（choline-deficient ethionine-supplemented，CDE）肝损伤模型中发现，巨噬细胞在吞噬死亡的肝细胞碎片后可分泌 Wnt3a 促进肝再生（Boulter et al.，2012）。此外，在李斯特菌造成的肝损伤模型中发现巨噬细胞能促进肝脏的再生过程（Minutti et al.，2017）。这些证据提示肝脏损伤微环境中的免疫细胞，尤其是巨噬细胞，很可能通过和肝细胞相互作用参与肝细胞的细胞属性转换的调控。这些发现提示，免疫系统以及其他微环境信号，如基质成分、代谢物等，都可能对损伤再生过程的细胞属性转换起到重要的调控作用。对相关的微环境细胞和信号的研究，会大大提升我们对组织损伤再生的了解和操控能力。

四、细胞命运决定信号调控网络领域

从受精卵到成熟个体直至死亡的每一个生命活动中，细胞都面临着选择：是继续保持原有的状态，还是转化成另一种身份。细胞通过感知细胞内和细胞外信号的变化，调动并整合关键信号通路，调控特异性基因的表达，形成独有的表达谱，从而精准控制细胞的命运。

1. Hippo 信号通路

Hippo 信号通路是近年来被发现并确定的调控细胞命运的关键信号通路，其通过调控细胞生长、增殖、分化与凋亡来控制组织器官的发育、再生和损伤修复（Wang et al.，2017；Yu et al.，2015）。此外，Hippo 信号通路在组织的稳态维持与干细胞的自我更新、分化等生命活动过程中也发挥着极其重要的作用（Mo et al.，2014；Patel et al.，2017a；Yin et al.，2015）。Hippo 信号通路与多个重要的信号通路（如 Wnt、mTOR、Notch、TGF-β），以及免疫，代谢和细胞黏附等，有密切的相互调控关系，它们共同协作决定细胞的命运（Ardestani et al.，2018；Kim and Jho，2014；Kim et al.，2017；Mo et al.，2014；Tumaneng et al.，2012；Wang et al.，2020c；Zhang et al.，2017b）。

Hippo 信号通路的核心成员由上游的激酶复合物 MST/SAV/LATS/MOB 以及下游的转录复合物 YAP/TAZ-TEADs 组成（Yin et al.，2015）。Hippo 信号通路通过感知胞内外营养、机械压力、激素、细胞极性和细胞黏附等信号的变化（Yu and Guan et al.，2013），改变上游激酶复合物的活性，并协同表观遗传因子（Hillmer and Link et al.，2019），调控 YAP/TAZ-TEADs 转录复合物的活性，从而调控与细胞增殖、分化和凋亡相关的靶基因的表达（Zhou et al.，2016）。已知 Hippo 信号通路在胚胎发育阶段发挥重要的功能，其在囊胚阶段调控内细胞团和滋养外胚层的特化，Hippo 信号通路的异常将导致胚胎发育异常、器官畸形以及相关功能受损，甚至会导致个体的死亡（Downs et al.，2014；Home et al.，2012；Sasaki，2017；Wu and Guan et al.，2020；Yu et al.，2019）。此外，Hippo 信号通路在各个器官组织的增殖、分化中也扮演着至关重要的角色。Hippo 信号通路能够调控脂肪、骨、肌肉和肠等器官的分化，确保器官发育的既定大小和相关功能（Feng et al.，2019；Wen et al.，2019；Zhang et al.，2018b）。然而，当细胞失去控制无限增殖时，便会形成肿瘤。根据人的肿瘤样本和相关的动物肿瘤模型，已经证实了 Hippo 信号通路与多种肿瘤的发生发展密切相关，如肝癌、结肠癌、胃癌、头颈癌和乳腺癌等（Harvey et al.，2013；Zhou et al.，2016）。目前，以 Hippo 信号通路成员作为靶点，研究人员已经鉴定出了一些潜在的肿瘤治疗药物（He et al.，2019；Liu-Chittenden et al.，2012；Nakatani et al.，2016；Pobbati et al.，2015）。此外，在成体肝脏、心脏和肌肉等器官的损伤再生过程中，Hippo 信号通路一方面可以调控器官的再生和修复，另一方面又能够保证器官修复到合适的大小（Feng et al.，2019；Lu et al.，2018；Zhou et al.，2015）。总之，Hippo 信号通路在器官的生长分化、组织的损伤修复和肿瘤抑制等方面发挥着非常重要的功能。而深入剖析 Hippo 信号通路在不同生理学过程中的调控机制和相关功能，将为人为干预细胞命运提供强有力的分子基础，进一步为个体发育畸形、器官再生障碍以及肿瘤等疾病的治疗提供有效的治疗途径。

2. Wnt 信号通路

Wnt 信号蛋白是机体内一类重要的形态发生素，既参与调控细胞增殖，也参与细胞分化的命运决定，在细胞分化的系统控制、组织稳态的维持及干细胞的自我更新中起到了重要的作用（Clevers，2006；MacDonald et al.，2009；Steinhart and Angers，2018）。Wnt 信号通路在细胞增殖和分化的转换与维持过程中发挥多种生物学功能，具有动态选择性及特异性。然而，Wnt 信号通路的表观调控机制、转录激活复合物的差异性组成、核心蛋白的翻译后修饰等一系列生物学过程的动态选择性或特异性调控细胞增殖或分化的功能基础目前仍不清楚。因此未来将利用包括高时空分辨率显微成像、基因敲出和敲入、蛋白质组学与计算生物学等技术手段，揭示其选择性或特异性调控增殖或分化的机制以及发挥增殖与分化相互转换的功能的基础为该领域研究的中心问题。

3. 细胞间接触信号

细胞间的接触黏附是多细胞生物的一种基础生命现象，为细胞提供了重要的调控信号。一系列细胞 - 细胞间黏附分子参与了接触黏附过程，如钙黏素、黏连蛋白以及免疫球蛋白家族黏附分子等（Shimono et al.，2012）。失去细胞间接触黏附通常会导致胚胎死亡，器官、细胞功能异常等严重后果。在胚胎发育过程中，由单个受精卵通过细胞分裂形成多细胞，这些细胞间会在不同的时间和空间尺度上形成不同的接触黏附，从而可能为不同的细胞个体提供差异的分化信号，这些接触黏附帮助机体形成正常的组织结构，同时抑制发育后期组织黏合，维持机体正常的增殖信号（Fukata and Kaibuchi，2001）。此外，在成体的组织与器官的稳态维持和损伤修复过程中，多种不同类型的细胞会发生动态的接触，已产生细胞间接触的细胞会解除原先的接触黏附，迁移至新的部位并与其他细胞产生新的接触黏附，在时空水平动态调控机体的生理平衡（Curtis，1962）。例如，免疫细胞与组织中具分化潜能的干细胞的接触，可能为干细胞的更新和分化提供多种变化的调控信号而控制干细胞的行为（Naik et al.，2018）。在胚胎发育过程中细胞间的接触信号可能调控细胞的分化命运，如整合素等细胞间黏附分子通过影响细胞有丝分裂期纺锤体的方向，使得子代细胞可以快速定位于新的机体环境，通过响应环境信号从而参与调控子代细胞的命运决定（Streuli，2009）。此外，胚胎发育中脊索前板细胞间接触黏附的时间也将直接影响原肠胚发育期中内胚层细胞的分化方向（Barone et al.，2017）。另外，免疫细胞与其他类群的免疫细胞或肿瘤细胞之间的动态接触网络将直接影响其后续的增殖及分化，继而调控免疫激活或抑制功能，如：①适应性免疫细胞和抗原递呈细胞间的接触与黏附是起始细胞免疫、活化 T 细胞的必要条件；②调节性 T 细胞等可以通过竞争抗原递呈细胞的接触而影响 T 细胞的活化增殖，动态调控机体免疫平衡；③肿瘤细胞通过免疫检验点蛋白与免疫细胞的接触将直接影响 T 细胞的活化信号，调控初始 T 细胞向杀伤性亚群的分化；④肿瘤细胞之间的接触与黏附也将为肿瘤提供存活和增殖信号（Harjunpaa et al.，2019）。细胞间接触黏附信号含有在时间、空间等多维度上的信息，目前的研究还非常欠缺，一些重要的科学问题还有待回答。

4. 代谢信号

细胞命运调控及细胞可塑性与多种疾病进展相关。细胞命运调节紊乱导致了肿瘤等疾病的产生，而细胞命运的可塑性是癌症产生治疗耐受的根源。对细胞命运决定及其可塑性的机理研究，将为临床疾病治疗提供新的治疗靶点和策略。肿瘤细胞无限增殖的特点导致其对物质合成有极大的需求。正是因为肿瘤细胞的可塑性，肿瘤细胞可以调节代谢酶的活性，改变代谢通路的偏好性，进而满足自身物质、能量合成的需要。例如，肿瘤细胞可以适应外界环境，选择利用葡萄糖、谷氨酸、乙酸等为碳源合成乙酰辅酶A（Martinez-Outschoorn et al.，2017）；前列腺癌和乳腺癌细胞能够在不同治疗阶段利用不同的甾体小分子合成性激素；多种癌细胞会增强脂肪酸从头合成路径来储备能量（Bueno and Quintela-Fandino，2020；Kuhajda et al.，1994）。细胞的可塑性决定了在不同生理病理中其影响细胞命运代谢物的多样性。这些内源代谢物或者外源刺激因子除了满足细胞物质、能量需求，还能够作为信号分子调节多种信号通路，帮助细胞适应外界环境变化。例如，乙酰辅酶 A 能够影响 mTOR 信号通路、影响组蛋白乙酰化修饰；α- 酮戊二酸能够影响 HIF 和 NF-κB 信号通路（Wang et al.，2019a）；葡萄糖及其代谢物能够通过影响 SNAI mRNA 的稳定性来调节细胞的上皮 – 间充质转换（epithelial-mesenchymal transition，EMT）活性（Wang et al.，2019b）。同时膳食来源的营养因子或者临床药物也能够通过代谢影响细胞命运。前列腺癌细胞能够将治疗药物阿比特龙代谢为新的促癌代谢物，直接激活雄激素受体，导致治疗耐受（Li et al.，2016a）。细胞通过感知不同的代谢物，塑造自身特性以适应环境变化。代谢物对细胞命运决定的重要性及相关机理研究推动了临床疾病的治疗和诊断。例如，基于癌细胞大量摄取葡萄糖的特点，临床能够利用正电子发射断层成像（positron emission tomography，PET）技术结合氟代葡萄糖来发现体内肿瘤细胞（Hanahan and Weinberg，2000）；多种代谢酶是疾病治疗的靶点；基于药物不同代谢物的生理功能不同，患者体内药物代谢情况可以用来筛选药物适用人群（Li et al.，2015）。尽管目前已经发现了多种代谢物对细胞命运决定的重要性，但是相关领域还有更多的挑战。

五、研究新方法与系统领域

（一）单倍体干细胞技术

2011年，英国科学家Leeb和Wutz（2011）及奥地利科学家Elling等（2011）通过引入流式细胞分选技术首次建立了小鼠孤雌单倍体胚胎干细胞。2012年，上海和北京两个研究团队分别独立建立了小鼠孤雄单倍体胚胎干细胞系（Li et al.，2012；Yang et al.，2012）。重要的是，孤雄单倍体细胞在注射到卵子后，能够产生健康可育的“半克隆小鼠”。在过去几年间，单倍体干细胞研究取得了突飞猛进的进展，中国科学家在该领域做出了重要的贡献，建立了大鼠、猴和人的单倍体干细胞，并开展了基于单倍体干细胞的多种应用研究（Bai et al.，2016；Li et al.，2019；Wang et al.，2019；丁一夫等，2019）。

1. 不同物种单倍体胚胎干细胞的建立

小鼠单倍体胚胎干细胞建立后，大鼠孤雄和孤雌单倍体胚胎干细胞（Hirabayashi et al.，2017；Li et al.，2014），食蟹猴（Yang et al.，2013）和恒河猴（Wang et al.，2018）孤雌单倍体胚胎干细胞，人孤雌（Sagi et al.，2016；Zhong et al.，2016b）和孤雄单倍体胚胎干细胞（Zhang et al.，2020）相继建立。然而，到目前为止尚无非人灵长类孤雄单倍体胚胎干细胞成功建立的报道。

2. 单倍体胚胎干细胞的特性

单倍体胚胎干细胞是一类集单倍体性、自我更新、多能性和“受精”能力为一体的新型干细胞类型。

（1）自发二倍体化。单倍体胚胎干细胞均会自发成为二倍体，对其机制的解析，一方面可能产生新的细胞倍性维持可塑性的调控机制，另一方面有望产生促进单倍体特性稳定的新方案。

（2）单倍体细胞携带 X 染色体。细胞缺失 X 染色体无法存活，因此我们只能获得携带 X 染色体的单倍体细胞，因而产生的半克隆胚胎都是雌性的。因此，孤雄单倍体干细胞是研究胚胎发育过程中 X 染色体失活的有效工具。

（3）多能性和不同分化状态单倍体细胞的建立。单倍体胚胎干细胞具有多能性，小鼠单倍体干细胞的分化伴随着自发二倍体的加剧，而人的单倍体干细胞在分化过程中则表现出相对稳定的单倍体特性，因此，研究干性与倍性的相互关系以及建立不同分化状态的单倍体细胞是领域的重要方向。

（4）配子特性。孤雄单倍体胚胎干细胞大体上保持精子的雄性印记状态，将其注射到卵子中，可以产生健康半克隆小鼠。然而，健康半克隆小鼠的出生效率低（2% 左右），进一步通过在细胞中敲除印记调控区域 H19-DMR 和 IG-DMR 可以使健康半克隆小鼠的出生效率提高了 10 倍（称为“类精子干细胞”）（Zhong et al.，2015）。小鼠的孤雌单倍体胚胎干细胞在早期代数通过核移植方法替换卵子遗传物质后也可以支持胚胎发育，但是效率非常低（Wan et al.，2013）。进一步研究需要发现决定胚胎发育的关键的配子携带表观遗传标记，进一步提高单倍体干细胞作为配子替代物使用的效率。

3. 单倍体干细胞的应用

（1）细胞水平的正反向遗传筛选。单倍体细胞最显著的应用是利用其单倍体特性开展大规模的正反向遗传筛选，结合其具有分化为三胚层细胞的能力，可以在体外高效筛选参与细胞分化的重要因子。利用这一特点，可以构建干细胞突变文库（Elling et al.，2017）。结合体外分化和 piggBac 转座子随机插入技术，可以构建突变神经细胞文库（He et al.，2017）。

（2）异源二倍体细胞的构建和应用。通过小鼠和大鼠单倍体胚胎干细胞的融合，产生的异源二倍体胚胎干细胞可用于开展物种间性状差异的分子调控机制研究（Li et al.，2016c）。

（3）胚胎发育的印记基因调控研究。Zhong等（2016a）和Li等（2016b）在孤雌单倍体干细胞中敲除H19-DMR与IG-DMR后，获得了来自卵子的类精子干细胞，实现了小鼠孤雌胚胎的高效发育。Li等（2018b）在孤雄单倍体胚胎干细胞中敲除了7个雌性印记调控区域替代卵子遗传物质产生了两只孤雄小鼠。

（4）高效制备复杂遗传修饰小鼠。类精子干细胞可以在体外进行多轮的基因编辑并用于一步产生小鼠。在类精子干细胞中敲除多个基因，通过卵子注射一步获取多基因杂合敲除小鼠，可以用于模拟人类复杂疾病，如强直性肌营养不良征1型（携带4个基因的杂合敲除）（Yin et al.，2020）和米勒管畸形患者（携带2个杂合突变）（Wang et al.，2020b）。因此，类精子干细胞技术是验证人类复杂疾病的遗传缺陷基础的有效工具。

（5）个体水平遗传筛选。类精子干细胞与CRISPR-Cas9结合可以实现小鼠个体水平的遗传筛选，如针对骨发育开展的关键基因的遗传筛选研究（Bai et al.，2019）以及针对影响原始生殖细胞发育的重要蛋白质DND1的关键氨基酸的遗传筛选（Li et al.，2018a），填补哺乳动物遗传研究的空白。

（6）基因组标签计划（genome tagging project，GTP）。人类基因组计划揭示存在超过22 000个编码蛋白质的基因，蛋白质组功能研究进展却困难重重，其中一个原因是缺乏一个简单、高效、标准的在体研究体系。GTP是在体外编辑类精子干细胞获得蛋白质标签细胞库，进而通过“人工授精”大规模获得携带标签蛋白质的小鼠库（Jiang et al.，2018）。中国科学院生物化学与细胞生物学研究所于2017年11月成立了GTP研发中心，组建了管理和技术团队，制定了一系列标准操作流程。截止至2020年8月30日，中心已经构建了1157个标签类精子干细胞系和249个标签小鼠。GTP中心将建成为具有特色、开放共享的标签小鼠资源平台，服务于国内外科学家。

（二）类器官

20世纪90年代，Lindberg等（1993）和Pellegrini等（1997）率先将角膜缘干细胞培养在3T3滋养层细胞之上，体外成功培养出了人源3D眼角膜结构，开辟了人源细胞3D类器官培养的新领域。Li等（1987）将乳腺上皮细胞培养在一种从小鼠肉瘤中分离得到的可溶性再生基底膜上，发现相对于传统的 2D 培养方法，包括酪蛋白在内的乳蛋白分泌量有了很大的提升，后续的形态学分析也表明 3D 培养体系中的乳腺上皮细胞形成了具有导管、小导管和空腔的特殊结构，类似于乳腺中的分泌腺泡。2009 年，Hans Clevers（Barker et al.，2007）实验室将小鼠肠段中分离出来的隐窝细胞培养在含有 ENR（即含有表皮生长因子、Noggin 蛋白和 R-spondin 蛋白）的 3D Matrigel 培养体系中，发现在该培养体系下隐窝细胞能够形成类似于肠的微型结构，即隐窝－绒毛样复合体。他们还利用该培养体系对前期分离并鉴定出来的单个 $Lgr5^+$ 肠干细胞进行培养，发现也能形成具有上述特殊结构的类器官，同时他们的追踪实验表明该类器官中仍然具有 $Lgr5^+$ 干细胞的存在，所以该模型能够很好地模拟体内小肠的形态结构和功能（Sato et al.，2009）。这种肠类器官体系的成功建立开启了类器官研究的新篇章，迅速成为新的研究热点。目前 3D 类器官培养技术已经成功地培养出具有部分关键生理结构和功能的肾（Takasato et al.，2015；Xia et al.，2013）、肝（Hu et al.，2018；Huch et al.，2013b；Takebe et al.，2013；Zhang et al.，2018a）、肺（Mondrinos et al.，2014；Sachs et al.，2019）、肠（Jung et al.，2011；Sato et al.，2011；Sato et al.，2009；van de Wetering et al.，2015）、脑（Lancaster et al.，2013；Mariani et al.，2012）、前列腺（Gao et al.，2014；Karthaus et al.，2014）、胰腺（Greggio et al.，2013；Huch et al.，2013a）和视网膜（Eiraku et al.，2011）等类组织器官。与细胞系相比，类器官不仅在体外培养条件下能够无限增殖，很好地保持原位组织的异质性，还适用于大规模的药物筛选，同时可以对该模型进行包括基因敲低（knock down）、过表达和突变在内的基因编辑操作，所以类器官的成功建立对于疾病尤其是肿瘤的研究有着重要的意义。

1. 类器官的种类

（1）正常成体组织类器官。正常成体组织类器官在长期培养的过程中能够维持基因组的稳定性，并且在体外培养和移植的条件下能够重现对应部分的组织器官功能。

（2）ESC 和 iPSC 类器官，由于 ESC、iPSC 都具有体外培养能够无限增殖与多向分化的特点（Evans and Kaufman，1981；Takahashi and Yamanaka.，

2006；Thomson et al.，1998），因此两者也成为类器官培养的重要细胞来源。

（3）肿瘤组织类器官。肿瘤类器官培养只需要量非常少的肿瘤穿刺样品，具有成功率高、周期短、花费低的优势，将在肿瘤的个性化治疗领域发挥不可替代的作用，尤其在晚期肿瘤患者的个性化治疗方面有着巨大的应用前景。

2. 类器官的应用

类器官可以用于研究肿瘤的发生发展，肿瘤的发生和耐药抵抗机制，肿瘤细胞与其微环境的互作，药物筛选及个性化治疗，临床的个体化治疗等，具有非常重要的应用价值。

第四节　我国发展战略与重点方向

人口健康是国家重大战略需求。一方面，我国人口众多，老龄化加剧，癌症、糖尿病、免疫疾病等重大疾病严重威胁人民健康；另一方面，生殖健康也成为影响我国社会发展的重要问题。研究细胞命运的可塑性及其分子调节机制不仅能阐释包括人类在内的多细胞生命体生老病死的重要生物学规律，揭示细胞结构、功能及命运维持与转变（转换）的分子基础和调控规律，为人工干预细胞命运甚至改造细胞提供理论依据和模型，可望实现细胞命运的人工精准操控，从而大幅度提升人类疾病的防治能力。目前，我们对细胞命运决定与可塑性的了解还很有限，很多初步发现需要更深入的研究与应用。

一、总体思路和发展目标

1. 总体思路

瞄准雄性生殖细胞、成体干细胞、细胞属性转换、细胞命运决定的信号调控网络以及新的研究体系与方法等重要科学问题，加强投入与支持，引领解决核心科学问题，同时兼顾基础研究与转化应用，加强对人才的培养与重视，使我国在该领域的研究处于国际领先地位，并大力推进实验室的原创发现向应用转化。

2. 发展目标

通过对雄性生殖细胞、成体干细胞、细胞属性转换、细胞命运决定的信号调控网络以及新的研究体系和方法的全面研究与推进，在生殖健康以及糖尿病、肿瘤、免疫等重大疾病的诊治方面取得突破，并形成多个国际领先的研究方向，培养一批具有国际影响力的研究团队。相关研究成果的转化与应用也能促进国民经济的发展和人类医学的进步，开发干细胞治疗和肿瘤治疗的靶向药物以及器官再生与修复的人工干预手段，具有重大的现实意义。

二、优先发展领域或重要研究方向

细胞命运决定与可塑性是生命科学的根本问题，与基本的生长发育和多种重大疾病密切相关，符合人口健康的国际战略需求。该科学问题需要整合生殖细胞、成体干细胞、细胞属性转换、细胞命运决定的信号调控网络以及新的研究体系与方法的多方面研究团队，进行系统的原创性研究。不同研究方向相互关联，可以相互协同，有望在相关领域取得国际领先的原创性成果。针对该方向的持续支持和研究将有助于在器官再生与修复、重大疾病治疗及生殖健康领域带来重要突破。鉴于此应优先发展下述研究方向。

1. 基于成体干细胞的基础应用研究

组织器官的功能异常是多种人类重大疾病产生的原因，对这些疾病的治疗需要功能正常的组织与器官移植。通过对人成体干细胞命运决定的机制研究，可望开发获得多种人类正常组织器官的新途径。

（1）胰岛的体外再造。研究人员发现小鼠胰岛中存在成体干细胞，利用这些干细胞在体外建立了小鼠人工胰岛（类器官），为“胰岛体外制造”的设计思路带来了前所未有的变革，引出令人激动的转化应用问题：人类是否存在胰岛成体干细胞？是否也能利用这些干细胞建立体外长期大量扩增的体系，源源不断地获得胰岛 β 细胞？

（2）造血干细胞的体外扩增，在体外重建造血干细胞巢，抑制造血干细胞的体外分化，促进其自我更新。

（3）间充质干细胞的深入研究与应用。

2. 重大疾病与机体损伤修复中的细胞转分化

研究损伤及癌症等疾病过程中细胞属性转换的调控机制；明确一些重大的组织器官损伤修复过程、细胞移植等再生医学相关过程，理解细胞命运决定及细胞属性转换的机制；针对细胞属性转换的靶点鉴定与药物研发，发现调控细胞属性转换的小分子化合物。

3. 调控细胞命运决定的动态信号网络

细胞信号是决定细胞命运的分子基础。在生长发育与重大疾病过程中，系统研究调控细胞命运决定的信号时空动态网络（如 Hippo、Wnt 化学信号和细胞间接触黏附空间信号等），揭示其决定细胞命运的分子机制，有助于开发干细胞治疗和肿瘤治疗的靶向药物以及器官再生与修复的人工干预手段。

4. 生殖健康

当前，我国的育龄人口不育率高达 15%，由男性因素引起的不育症约占 50%，男性不育已成为影响我国人口健康的重大问题。因此，将精原干细胞自我更新与分化、减数分裂、精子生成与成熟这三大研究领域作为重点研究方向，不仅有助于我国进入生殖与发育生物学基础研究国际前沿行列，并且可以促进我国农业及畜牧业的快速发展；更为重要的是，可为我国男性不育症的精准医疗提供必要的理论依据，同时为辅助生殖技术的不断发展完善创造条件，从而综合提升我国男性生殖健康水平。

5. 创新的研究方法

研究方法是以上研究的根本和基石。单倍体干细胞系具有非常广泛的应用前景，但该系统还有待于进一步完善和发展，例如，建立更多物种、不同分化状态、具有配子表观遗传特性的单倍体干细胞；建立替代卵子遗传物质的单倍体干细胞；稳定其单倍体特性；挖掘单倍体干细胞作为配子替代物在胚胎发育和细胞命运决定研究中的作用。另外，类器官作为研究细胞命运决定与可塑性的重要工具，随着进一步的条件优化和系统性的多组学分析，类器官模型可能会在发育生物学领域发挥重要的作用，并有可能在器官制造和器官移植方面具有一定的应用价值。未来更需要结合临床实践，发展新一代的类器官培养体系。总的来说，类器官在基因水平和形态特点上能够很好地模拟患者体内相应的组织，也适用于高通量的药物筛选，同时为疾病的个性化治疗提供了研究模型。

本章参考文献

丁一夫，李劲松，周琪 . 2019. 哺乳动物单倍体胚胎干细胞的建立与应用 . 中国科学：生命科学 , 49（12）: 16.

Abdullah Z, Knolle P A. 2017. Liver macrophages in healthy and diseased liver. Pflugers Arch, 469（3-4）: 553-560.

Acar M, Kocherlakota K S, Murphy M M, et al. 2015. Deep imaging of bone marrow shows non-dividing stem cells are mainly perisinusoidal. Nature, 526（7571）: 126-130.

Alonso-Curbelo D, Ho Y J, Burdziak C, et al. 2021. A gene-environment-induced epigenetic program initiates tumorigenesis. Nature, 590（7847）: 642-648.

Ardestani A, Lupse B, Maedler K. 2018. Hippo signaling: key emerging pathway in cellular and whole-body metabolism. Trends in Endocrinology & Metabolism, 29（7）: 492-509.

Bai M, Han Y, Wu Y, et al. 2019. Targeted genetic screening in mice through haploid embryonic stem cells identifies critical genes in bone development. PLoS Biol, 17（7）: e3000350.

Bai M, Wu Y, Li J. 2016. Generation and application of mammalian haploid embryonic stem cells. J Intern Med, 280（3）: 236-245.

Barker N, Van Es J H, Kuipers J, et al. 2007. Identification of stem cells in small intestine and colon by marker gene Lgr5. Nature, 449（7165）: 1003-1007.

Barone V, Lang M, Krens S F G, et al. 2017. An effective feedback loop between cell-cell contact duration and morphogen signaling determines cell fate. Developmental Cell, 43（2）: 198-211 e12.

Barrios F, Filipponi D, Campolo F, et al. 2012. SOHLH1 and SOHLH2 control Kit expression during postnatal male germ cell development. J Cell Sci, 125（Pt 6）: 1455-1464.

Bendel-Stenzel M, Anderson R, Heasman J, et al. 1998. The origin and migration of primordial germ cells in the mouse. Semin Cell Dev Biol, 9（4）: 393-400.

Bird L. 2019. Poised plasticity of skin T cells. Nat Rev Immunol, 19（2）: 70-71.

Bjorkgren I, Sipila P. 2019. The impact of epididymal proteins on sperm function. Reproduction, 158（5）: R155-R167.

Boulter L, Govaere O, Bird T G, et al. 2012. Macrophage-derived Wnt opposes Notch signaling to specify hepatic progenitor cell fate in chronic liver disease. Nat Med, 18（4）: 572-579.

Bruns I, Lucas D, Pinho S, et al. 2014. Megakaryocytes regulate hematopoietic stem cell quiescence through CXCL4 secretion. Nat Med, 20（11）: 1315-1320.

Bueno M J, Quintela-Fandino M. 2020. Emerging role of fatty acid synthase in tumor initiation: implications for cancer prevention. Mol Cell Oncol, 7（2）: 1709389.

Burger P E, Xiong X, Coetzee S, et al. 2005. Sca-1 expression identifies stem cells in the proximal region of prostatic ducts with high capacity to reconstitute prostatic tissue. Proc Natl Acad Sci U S A, 102（20）: 7180-7185.

Calvi L M, Adams G B, Weibrecht K W, et al. 2003. Osteoblastic cells regulate the haematopoietic stem cell niche. Nature, 425（6960）: 841-846.

Centonze A, Lin S, Tika E, et al. 2020. Heterotypic cell-cell communication regulates glandular stem cell multipotency. Nature, 584（7822）: 608-613.

Choi N, Zhang B, Zhang L, et al. 2012. Adult murine prostate basal and luminal cells are self-sustained lineages that can both serve as targets for prostate cancer initiation. Cancer Cell, 21（2）: 253-265.

Chua C W, Shibata M, Lei M, et al. 2014. Single luminal epithelial progenitors can generate prostate organoids in culture. Nat Cell Biol, 16（10）: 951-961.

Clevers H. 2006. Wnt/β-catenin signaling in development and disease. Cell, 127（3）: 469-480.

Curtis A S. 1962. Cell contact and adhesion. Biol Rev Camb Philos Soc, 37: 82-129.

Dai P, Wang X, Gou L T, et al. 2019. A translation-activating function of MIWI/piRNA during mouse spermiogenesis. Cell, 179（7）: 1566-1581 e1516.

Debnath S, Yallowitz A R, McCormick J, et al. 2018. Discovery of a periosteal stem cell mediating intramembranous bone formation. Nature, 562（7725）: 133-139.

Dotto G P, Rustgi A K. 2016. Squamous cell cancers: a unified perspective on biology and genetics. Cancer Cell, 29（5）: 622-637.

Downs K M, Wicklow E, Blij S, et al. 2014. HIPPO pathway members restrict SOX2 to the inner cell mass where it promotes ICM fates in the mouse blastocyst. PLoS Genetics, 10（10）: e1004618.

DuPage M, Bluestone J A. 2016. Harnessing the plasticity of CD4（+）T cells to treat immune-mediated disease. Nat Rev Immunol, 16（3）: 149-163.

Eiraku M, Takata N, Ishibashi H, et al. 2011. Self-organizing optic-cup morphogenesis in three-dimensional culture. Nature, 472（7341）: 51-56.

Elling U, Taubenschmid J, Wirnsberger G, et al. 2011. Forward and reverse genetics through derivation of haploid mouse embryonic stem cells. Cell Stem Cell, 9（6）: 563-574.

Elling U, Wimmer R A, Leibbrandt A, et al. 2017. A reversible haploid mouse embryonic stem cell biobank resource for functional genomics. Nature, 550（7674）: 114-118.

Eming S A, Wynn T A, Martin P. 2017. Inflammation and metabolism in tissue repair and regeneration. Science, 356（6342）: 1026-1030.

Evans M J, Kaufman M H. 1981. Establishment in culture of pluripotential cells from mouse embryos. Nature, 292（5819）: 154-156.

Fatehullah A, Tan S H, Barker N. 2016. Organoids as an *in vitro* model of human development and disease. Nature cell biology, 18（3）: 246-254.

Feng X, Wang Z, Wang F, et al. 2019. Dual function of VGLL4 in muscle regeneration. EMBO J, 38（17）: e101051.

Fukata M, Kaibuchi K. 2001. Rho-family GTPases in cadherin-mediated cell-cell adhesion. Nature Reviews Molecular Cell Biology, 2（12）: 887-897.

Gao D, Vela I, Sboner A, et al. 2014. Organoid cultures derived from patients with advanced prostate cancer. Cell, 159（1）: 176-187.

Ginsburg M, Snow M H, McLaren A. 1990. Primordial germ cells in the mouse embryo during gastrulation. Development, 110（2）: 521-528.

Goldstein A S, Huang J, Guo C, et al. 2010. Identification of a cell of origin for human prostate cancer. Science, 329（5991）: 568-571.

Gou L T, Dai P, Yang J H, et al. 2014. Pachytene piRNAs instruct massive mRNA elimination during late spermiogenesis. Cell Res, 24（6）: 680-700.

Gou L T, Kang J Y, Dai P, et al. 2017. Ubiquitination-deficient mutations in human Piwi cause male infertility by impairing histone-to-protamine exchange during spermiogenesis. Cell, 169（6）: 1090-1104 e1013.

Gou L T, Lim D H, Ma W, et al. 2020. Initiation of parental genome reprogramming in fertilized oocyte by splicing kinase srpk1-catalyzed protamine phosphorylation. Cell, 180（6）: 1212-1227 e1214.

Greggio C, De Franceschi F, Figueiredo-Larsen M, et al. 2013. Artificial three-dimensional niches deconstruct pancreas development *in vitro*. Development, 140（21）: 4452-4462.

Griswold M D, Hogarth C A, Bowles J, et al. 2012. Initiating meiosis: the case for retinoic acid. Biol Reprod, 86（2）: 35.

Guicciardi M E, Trussoni C E, Krishnan A, et al. 2018. Macrophages contribute to the pathogenesis of sclerosing cholangitis in mice. J Hepatol, 69（3）: 676-686.

Guo W, Li L, He J, et al. 2020. Single-cell transcriptomics identifies a distinct luminal progenitor cell type in distal prostate invagination tips. Nat Genet, 52（9）: 908-918.

Hanahan D, Weinberg R A. 2000. The hallmarks of cancer. Cell, 100（1）: 57-70.

Harjunpaa H, Asens M L, Guenther C, et al. 2019. Cell adhesion molecules and their roles and regulation in the immune and tumor microenvironment. Frontiers in Immunology, 10: 1078.

Harvey K F, Zhang X, Thomas D M. 2013. The Hippo pathway and human cancer. Nat Rev Cancer, 13（4）: 246-257.

He L, Yuan L, Sun Y, et al. 2019. Glucocorticoid receptor signaling activates TEAD4 to promote breast cancer progression. Cancer Res, 79（17）: 4399-4411.

He Z Q, Xia B L, Wang Y K, et al. 2017. Generation of mouse haploid somatic cells by small molecules for genome-wide genetic screening. Cell Rep, 20（9）: 2227-2237.

Hillmer R E, Link B A. 2019. The roles of Hippo signaling transducers Yap and Taz in chromatin remodeling. Cells, 8（5）: 502.

Hirabayashi M, Hara H, Goto T, et al. 2017. Haploid embryonic stem cell lines derived from androgenetic and parthenogenetic rat blastocysts. J Reprod Dev, 63（6）: 611-616.

Home P, Saha B, Ray S, et al. 2012. Altered subcellular localization of transcription factor TEAD4 regulates first mammalian cell lineage commitment. Proc Natl Acad Sci U S A, 109（19）: 7362-7367.

Hu H, Gehart H, Artegiani B, et al. 2018. Long-term expansion of functional mouse and human hepatocytes as 3D organoids. Cell, 175（6）: 1591-1606 e1519.

Huch M, Bonfanti P, Boj S F, et al. 2013a. Unlimited *in vitro* expansion of adult bi-potent pancreas progenitors through the Lgr5/R-spondin axis. The EMBO Journal, 32（20）: 2708-2721.

Huch M, Dorrell C, Boj S F, et al. 2013b. *In vitro* expansion of single $Lgr5^+$ liver stem cells induced by Wnt-driven regeneration. Nature, 494（7436）: 247-250.

Jain R, Barkauskas C E, Takeda N, et al. 2015. Plasticity of Hopx（+）type I alveolar cells to regenerate type II cells in the lung. Nat Commun, 6: 6727.

Jiang J, Yan M, Li D, et al. 2018. Genome tagging project tag every protein in mice through 'artificial spermatids'. National Science Review, 6（3）: 394-396.

Jiang M, Li H, Zhang Y, et al. 2017. Transitional basal cells at the squamous-columnar junction generate Barrett's oesophagus. Nature, 550（7677）: 529-533.

Jiang X, Karlsen T H. 2017. Genetics of primary sclerosing cholangitis and pathophysiological implications. Nat Rev Gastroenterol Hepatol, 14（5）: 279-295.

Jung P, Sato T, Mcrlos-Suárez A, et al. 2011. Isolation and *in vitro* expansion of human colonic stem cells. Nature Medicine, 17（10）: 1225-1227.

Kanatsu-Shinohara M, Mori Y, Shinohara T. 2013. Enrichment of mouse spermatogonial stem cells based on aldehyde dehydrogenase activity. Biol Reprod, 89（6）: 140.

Kanatsu-Shinohara M, Shinohara T. 2013. Spermatogonial stem cell self-renewal and development. Annu Rev Cell Dev Biol, 29: 163-187.

Karthaus W R, Hofree M, Choi D, et al. 2020. Regenerative potential of prostate luminal cells revealed by single-cell analysis. Science, 368（6490）: 497-505.

Karthaus W R, Iaquinta P J, Drost J, et al. 2014. Identification of multipotent luminal progenitor cells in human prostate organoid cultures. Cell, 159（1）: 163-175.

Khawar M B, Gao H, Li W. 2019. Mechanism of acrosome biogenesis in mammals. Front Cell Dev Biol, 7: 195.

Kim M, Jho E-h. 2014. Cross-talk between Wnt/β-catenin and Hippo signaling pathways: a brief review. BMB Reports, 47（10）: 540-545.

Kim W, Khan S K, Yang Y. 2017. Interacting network of Hippo, Wnt/β-catenin and Notch signaling represses liver tumor formation. BMB Reports, 50（1）: 1-2.

Kiyozumi D, Noda T, Yamaguchi R, et al. 2020. NELL2-mediated lumicrine signaling through OVCH2 is required for male fertility. Science, 368（6495）: 1132-1135.

Komeya M, Sato T, Ogawa T. 2018. *In vitro* spermatogenesis: a century-long research journey, still half way around. Reprod Med Biol, 17（4）: 407-420.

Kopp J L, Grompe M, Sander M. 2016. Stem cells versus plasticity in liver and pancreas regeneration. Nat Cell Biol, 18（3）: 238-245.

Kramann R, Schneider R K, DiRocco D P, et al. 2015. Perivascular Gli1+ progenitors are key contributors to injury-induced organ fibrosis. Cell Stem Cell, 16（1）: 51-66.

Kubota H, Brinster R L. 2018. Spermatogonial stem cells. Biol Reprod, 99（1）: 52-74.

Kuhajda F P, Jenner K, Wood F D, et al. 1994. Fatty acid synthesis: a potential selective target for antineoplastic therapy. Proc Natl Acad Sci U S A, 91（14）: 6379-6383.

Kunisaki Y, Bruns I, Scheiermann C, et al. 2013. Arteriolar niches maintain haematopoietic stem cell quiescence. Nature, 502（7473）: 637-643.

Lancaster M A, Knoblich J A. 2014. Organogenesis in a dish: modeling development and disease using organoid technologies. Science, 345（6194）: 1247125.

Lancaster M A, Renner M, Martin C-A, et al. 2013. Cerebral organoids model human brain development and microcephaly. Nature, 501（7467）: 373-379.

Law N C, Oatley J M. 2020. Developmental underpinnings of spermatogonial stem cell establishment. Andrology, 8（4）: 852-861.

Lee J, Kanatsu-Shinohara M, Inoue K, et al. 2007. Akt mediates self-renewal division of mouse spermatogonial stem cells. Development, 134（10）: 1853-1859.

Leeb M, Wutz A. 2011. Derivation of haploid embryonic stem cells from mouse embryos. Nature, 479（7371）: 131-134.

Leong K G, Wang B E, Johnson L, et al. 2008. Generation of a prostate from a single adult stem cell. Nature, 456（7223）: 804-808.

Leushacke M, Tan S H, Wong A, et al. 2017. Lgr5-expressing chief cells drive epithelial regeneration and cancer in the oxyntic stomach. Nat Cell Biol, 19（7）: 774-786.

Li J J, Shen M M. 2018. Prostate stem cells and cancer stem cells. Cold Spring Harb Perspect Med, 9（6）: a030395.

Li M L, Aggeler J, Farson D A, et al. 1987. Influence of a reconstituted basement membrane and its components on casein gene expression and secretion in mouse mammary epithelial cells. Proceedings of the National Academy of Sciences, 84（1）: 136-140.

Li Q, Li Y, Yang S, et al. 2018a. CRISPR-Cas9-mediated base-editing screening in mice identifies DND1 amino acids that are critical for primordial germ cell development. Nat Cell Biol, 20（11）: 1315-1325.

Li W, Li L, Hui L. 2020. Cell Plasticity in liver regeneration. Trends in Cell Biology, 30（4）: 329-338.

Li W, Li X, Li T, et al. 2014. Genetic modification and screening in rat using haploid embryonic stem cells. Cell Stem Cell, 14（3）: 404-414.

Li W, Shuai L, Wan H, et al. 2012. Androgenetic haploid embryonic stem cells produce live transgenic mice. Nature, 490（7420）: 407-411.

Li W, Yang L, He Q, et al. 2019. A homeostatic arid1a-dependent permissive chromatin state licenses hepatocyte responsiveness to liver-injury-associated YAP signaling. Cell Stem Cell, 25（1）: 54-68 e55.

Li X, Cui X L, Wang J Q, et al. 2016c. Generation and application of mouse-rat allodiploid embryonic stem cells. Cell, 164（1-2）: 279-292.

Li Y, Li J. 2019. Technical advances contribute to the study of genomic imprinting. PLoS Genet, 15（6）: e1008151.

Li Z, Alyamani M, Li J, et al. 2016a. Redirecting abiraterone metabolism to fine-tune prostate cancer anti-androgen therapy. Nature, 533（7604）: 547-551.

Li Z, Bishop A C, Alyamani M, et al. 2015. Conversion of abiraterone to D4A drives anti-tumour activity in prostate cancer. Nature, 523（7560）: 347-351.

Li Z, Wan H, Feng G, et al. 2016b. Birth of fertile bimaternal offspring following intracytoplasmic injection of parthenogenetic haploid embryonic stem cells. Cell Res, 26（1）: 135-138.

Li Z K, Wang L Y, Wang L B, et al. 2018b. Generation of bimaternal and bipaternal mice from hypomethylated haploid ESCs with imprinting region deletions. Cell Stem Cell, 23（5）: 665-676 e664.

Lindberg K, Brown M E, Chaves H V, et al. 1993. *In vitro* propagation of human ocular surface epithelial cells for transplantation. Invest Ophthalmol Vis Sci, 34（9）: 2672-2679.

Liu-Chittenden Y, Huang B, Shim J S, et al. 2012. Genetic and pharmacological disruption of the TEAD-YAP complex suppresses the oncogenic activity of YAP. Genes Dev, 26（12）: 1300-1305.

Loveland K L, Klein B, Pueschl D, et al. 2017. Cytokines in male fertility and reproductive pathologies: immunoregulation and beyond. Front Endocrinol（Lausanne）, 8: 307.

Lu L, Finegold M J, Johnson R L. 2018. Hippo pathway coactivators Yap and Taz are required to coordinate mammalian liver regeneration. Exp Mol Med, 50（1）: e423.

Lu T L, Huang Y F, You L R, et al. 2013. Conditionally ablated Pten in prostate basal cells promotes basal-to-luminal differentiation and causes invasive prostate cancer in mice. Am J Pathol, 182（3）: 975-991.

MacDonald B T, Tamai K, He X. 2009. Wnt/β-catenin signaling: components, mechanisms, and diseases. Developmental Cell, 17（1）: 9-26.

Macosko E Z, Basu A, Satija R, et al. 2015. Highly parallel genome-wide expression profiling of individual cells using nanoliter droplets. Cell, 161（5）: 1202-1214.

MacParland S A, Liu J C, Ma X Z, et al. 2018. Single cell RNA sequencing of human liver reveals distinct intrahepatic macrophage populations. Nat Commun, 9（1）: 4383.

Makela J A, Hobbs R M. 2019. Molecular regulation of spermatogonial stem cell renewal and differentiation. Reproduction, 158（5）: R169-R187.

Mariani J, Simonini M V, Palejev D, et al. 2012. Modeling human cortical development *in vitro* using induced pluripotent stem cells. Proceedings of the National Academy of Sciences, 109（31）: 12770-12775.

Martinez-Outschoorn U E, Peiris-Pages M, Pestell R G, et al. 2017. Cancer metabolism: a therapeutic perspective. Nat Rev Clin Oncol, 14（2）: 113.

Mendez-Ferrer S, Michurina T V, Ferraro F, et al. 2010. Mesenchymal and haematopoietic stem cells form a unique bone marrow niche. Nature, 466（7308）: 829-834.

Meng X, Lindahl M, Hyvonen M E, et al. 2000. Regulation of cell fate decision of undifferentiated spermatogonia by GDNF. Science, 287（5457）: 1489-1493.

Minutti C M, Jackson-Jones L H, Garcia-Fojeda B, et al. 2017. Local amplifiers of IL-4Rα-mediated macrophage activation promote repair in lung and liver. Science, 356（6342）: 1076-1080.

Miyajima A, Tanaka M, Itoh T. 2014. Stem/progenitor cells in liver development, homeostasis, regeneration, and reprogramming. Cell Stem Cell, 14（5）: 561-574.

Mo J S, Park H W, Guan K L. 2014. The Hippo signaling pathway in stem cell biology and cancer. EMBO Rep, 15（6）: 642-656.

Mondrinos M J, Jones P L, Finck C M, et al. 2014. Engineering *de novo* assembly of fetal pulmonary organoids. Tissue Engineering Part A, 20（21-22）: 2892-2907.

Naik S, Larsen S B, Cowley C J, et al. 2018. Two to tango: dialog between immunity and stem cells in health and disease. Cell, 175（4）: 908-920.

Nakatani K, Maehama T, Nishio M, et al. 2016. Targeting the Hippo signalling pathway for cancer treatment. J Biochem, 161（3）: 237-244.

Naveiras O, Nardi V, Wenzel P L, et al. 2009. Bone-marrow adipocytes as negative regulators of the haematopoietic microenvironment. Nature, 460（7252）: 259-263.

Ohinata Y, Ohta H, Shigeta M, et al. 2009. A signaling principle for the specification of the germ cell lineage in mice. Cell, 137（3）: 571-584.

Okabe H, Yang J, Sylakowski K, et al. 2016. Wnt signaling regulates hepatobiliary repair following cholestatic liver injury in mice. Hepatology, 64（5）: 1652-1666.

Ono N, Ono W, Nagasawa T, et al. 2014. A subset of chondrogenic cells provides early mesenchymal progenitors in growing bones. Nat Cell Biol, 16（12）: 1157-1167.

Ortinau L C, Wang H, Lei K, et al. 2019. Identification of functionally distinct Mx1+alphaSMA+ periosteal skeletal stem cells. Cell Stem Cell, 25（6）: 784-796 e785.

Ousset M, Van Keymeulen A, Bouvencourt G, et al. 2012. Multipotent and unipotent progenitors contribute to prostate postnatal development. Nat Cell Biol, 14（11）: 1131-1138.

Patel S H, Camargo F D, Yimlamai D. 2017. Hippo signaling in the liver regulates organ size, cell fate, and carcinogenesis. Gastroenterology, 152（3）: 533-545.

Pathria P, Louis T L, Varner J A. 2019. Targeting tumor-associated macrophages in cancer. Trends Immunol, 40（4）: 310-327.

Pellegrini G, Traverso C E, Franzi A T, et al. 1997. Long-term restoration of damaged corneal surfaces with autologous cultivated corneal epithelium. Lancet, 349（9057）: 990-993.

Pellegrini M, Filipponi D, Gori M, et al. 2008. ATRA and KL promote differentiation toward the meiotic program of male germ cells. Cell Cycle, 7（24）: 3878-3888.

Planas-Paz L, Sun T, Pikiolek M, et al. 2019. YAP, but not RSPO-LGR4/5, signaling in biliary epithelial cells promotes a ductular reaction in response to liver injury. Cell Stem Cell, 25（1）: 39-53 e10.

Pobbati A V, Han X, Hung A W, et al. 2015. Targeting the central pocket in human transcription factor TEAD as a potential cancer therapeutic strategy. Structure, 23（11）: 2076-2086.

Qiu R Z, Zhou L Y, Ma Y J, et al. 2020. Regulatory T cell plasticity and stability and autoimmune diseases. Clinical Reviews in Allergy & Immunology, 58（1）: 52-70.

Racanelli V, Rehermann B. 2006. The liver as an immunological organ. Hepatology, 43（2 Suppl 1）: S54- S62.

Ravichandran G, Neumann K, Berkhout L K, et al. 2019. Interferon-gamma-dependent immune responses contribute to the pathogenesis of sclerosing cholangitis in mice. J Hepatol, 71（4）: 773-782.

Sachs N, Papaspyropoulos A, Ommen D D Z, et al. 2019. Long-term expanding human airway organoids for disease modeling. EMBO J, 38（4）: e100300.

Sagi I, Chia G, Golan-Lev T, et al. 2016. Derivation and differentiation of haploid human embryonic stem cells. Nature, 532（7597）: 107-111.

Saitou M, Miyauchi H. 2016. Gametogenesis from pluripotent stem cells. Cell Stem Cell, 18（6）:

721-735.

Sasaki H. 2017. Roles and regulations of Hippo signaling during preimplantation mouse development. Development, Growth & Differentiation, 59（1）: 12-20.

Sato T, Stange D E, Ferrante M, et al. 2011. Long-term expansion of epithelial organoids from human colon, adenoma, adenocarcinoma, and Barrett's epithelium. Gastroenterology, 141（5）: 1762-1772.

Sato T, Vries R G, Snippert H J, et al. 2009. Single Lgr5 stem cells build crypt villus structures *in vitro* without a mesenchymal niche. Nature, 459（7244）: 262-265.

Schaub J R, Huppert K A, Kurial S N T, et al. 2018. *De novo* formation of the biliary system by TGFβ-mediated hepatocyte transdifferentiation. Nature, 557（7704）: 247-251.

Shapiro J, Lakey J R T, Ryan E A, et al. 2000. Islet transplantation in seven patients with type 1 diabetes mellitus using a glucocorticoid-free immunosuppressive regimen. The New England Journal of Medicine, 27（343）: 230-238.

Shen M M, Abate-Shen C. 2010. Molecular genetics of prostate cancer: new prospects for old challenges. Genes Dev, 24（18）: 1967-2000.

Shimono Y, Rikitake Y, Mandai K, et al. 2012. Immunoglobulin superfamily receptors and adherens junctions. Subcell Biochem, 60: 137-170.

Si-Tayeb K, Lemaigre F P, Duncan S A. 2010. Organogenesis and development of the liver. Dev Cell, 18（2）: 175-189.

Spencer J A, Ferraro F, Roussakis E, et al. 2014. Direct measurement of local oxygen concentration in the bone marrow of live animals. Nature, 508（7495）: 269-273.

Steinhart Z, Angers S. 2018. Wnt signaling in development and tissue homeostasis. Development, 145（11）: dev146589.

Streuli C H. 2009. Integrins and cell-fate determination. Journal of Cell Science, 122（2）: 171-177.

Sun L, Wang Y, Cen J, et al. 2019. Modelling liver cancer initiation with organoids derived from directly reprogrammed human hepatocytes. Nat Cell Biol, 21（8）: 1015-1026.

Svensson V, Natarajan K N, Ly L H, et al. 2017. Power analysis of single-cell RNA-sequencing experiments. Nat Methods, 14（4）: 381-387.

Takahashi K, Yamanaka S. 2006. Induction of pluripotent stem cells from mouse embryonic and adult fibroblast cultures by defined factors. Cell, 126（4）: 663-676.

Takasato M, Pei X E, Chiu H S, et al. 2015. Kidney organoids from human iPS cells contain multiple lineages and model human nephrogenesis. Nature, 526（7574）: 564-568.

Takebe T, Sekine K, Enomura M, et al. 2013. Vascularized and functional human liver from an iPSC-derived organ bud transplant. Nature, 499（7459）: 481-484.

Tan K, Song H W, Wilkinson M F. 2020. Single-cell RNAseq analysis of testicular germ and somatic cell development during the perinatal period. Development, 147（3）: dev183251.

Tetteh P W, Basak O, Farin H F, et al. 2016. Replacement of lost Lgr5-positive stem cells through plasticity of their enterocyte-lineage daughters. Cell Stem Cell, 18（2）: 203-213.

Thomson J A, Itskovitz-Eldor J, Shapiro S S, et al. 1998. Embryonic stem cell lines derived from human blastocysts. Science, 282（5391）: 1145-1147.

Tian C, Zhang Y. 2016. Purification of hematopoietic stem cells from bone marrow. Ann Hematol, 95（4）: 543-547.

Togashi Y, Shitara K, Nishikawa H. 2019. Regulatory T cells in cancer immunosuppression-implications for anticancer therapy. Nat Rev Clin Oncol, 16（6）: 356-371.

Tumaneng K, Schlegelmilch K, Russell R C, et al. 2012. YAP mediates crosstalk between the Hippo and PI（3）K-TOR pathways by suppressing PTEN via miR-29. Nat Cell Biol, 14（12）: 1322-1329.

van de Wetering M, Francies H E, Francis J M, et al. 2015. Prospective derivation of a living organoid biobank of colorectal cancer patients. Cell, 161（4）: 933-945.

Wan H, He Z, Dong M, et al. 2013. Parthenogenetic haploid embryonic stem cells produce fertile mice. Cell Res, 23（11）: 1330-1333.

Wang D, Wang J, Bai L, et al. 2020a. Long-term expansion of pancreatic islet organoids from resident Procr（+）progenitors. Cell, 180（6）: 1198-1211 e1119.

Wang H, Zhang W, Yu J, et al. 2018. Genetic screening and multipotency in rhesus monkey haploid neural progenitor cells. Development, 145（11）: dev160531.

Wang L, Li J. 2019. ‘Artificial spermatid’ -mediated genome editing. Biol Reprod, 101（3）: 538-548.

Wang L, Zhang Y, Fu X, et al. 2020b. Joint utilization of genetic analysis and semi-cloning technology reveals a digenic etiology of Mullerian anomalies. Cell Res, 30（1）: 91-94.

Wang S, Zhou L, Ling L, et al. 2020c. The crosstalk between Hippo-YAP pathway and innate immunity. Front Immunol, 11: 323.

Wang X, Kruithof-de Julio M, Economides K D, et al. 2009. A luminal epithelial stem cell that is a cell of origin for prostate cancer. Nature, 461（7263）: 495-500.

Wang X, Liu R, Qu X, et al. 2019a. alpha-Ketoglutarate-activated NF-kappaB signaling promotes compensatory glucose uptake and brain tumor development. Mol Cell, 76（1）: 148-162 e147.

Wang X, Liu R, Zhu W, et al. 2019b. UDP-glucose accelerates SNAI1 mRNA decay and impairs lung cancer metastasis. Nature, 571（7763）: 127-131.

Wang Y, Yu A, Yu F X. 2017. The Hippo pathway in tissue homeostasis and regeneration. Protein Cell, 8（5）: 349-359.

Wang Z A, Mitrofanova A, Bergren S K, et al. 2013. Lineage analysis of basal epithelial cells reveals their unexpected plasticity and supports a cell-of-origin model for prostate cancer heterogeneity. Nat Cell Biol, 15（3）: 274-283.

Watson P A, Arora V K, Sawyers C L. 2015. Emerging mechanisms of resistance to androgen receptor inhibitors in prostate cancer. Nat Rev Cancer, 15（12）: 701-711.

Wen T, Liu J, He X, et al. 2019. Transcription factor TEAD1 is essential for vascular development by promoting vascular smooth muscle differentiation. Cell Death Differ, 26（12）: 2790-2806.

Wu Z, Guan K L. 2020. Hippo signaling in embryogenesis and development. Trends in Biochemical Sciences, 46（1）: 51-63.

Xia Y, Nivet E, Sancho-Martinez I, et al. 2013. Directed differentiation of human pluripotent cells to ureteric bud kidney progenitor-like cells. Nature cell biology, 15（12）: 1507-1515.

Xiao N, Kam C, Shen C, et al. 2009. PICK1 deficiency causes male infertility in mice by disrupting acrosome formation. J Clin Invest, 119（4）: 802-812.

Xin L, Ide H, Kim Y, et al. 2003. *In vivo* regeneration of murine prostate from dissociated cell populations of postnatal epithelia and urogenital sinus mesenchyme. Proc Natl Acad Sci U S A, 100 Suppl 1: 11896-11903.

Xin L, Lawson D A, Witte O N. 2005. The Sca-1 cell surface marker enriches for a prostate-regenerating cell subpopulation that can initiate prostate tumorigenesis. Proc Natl Acad Sci U S A, 102（19）: 6942-6947.

Yan W, Ma L, Burns K H, et al. 2003. HILS1 is a spermatid-specific linker histone H1-like protein implicated in chromatin remodeling during mammalian spermiogenesis. Proc Natl Acad Sci U S A, 100（18）: 10546-10551.

Yang H, Liu Z, Ma Y, et al. 2013. Generation of haploid embryonic stem cells from *Macaca fascicularis* monkey parthenotes. Cell Res, 23（10）: 1187-1200.

Yang H, Shi L, Wang B A, et al. 2012. Generation of genetically modified mice by oocyte injection of androgenetic haploid embryonic stem cells. Cell, 149（3）: 605-617.

Yang L, Tsang K Y, Tang H C, et al. 2014. Hypertrophic chondrocytes can become osteoblasts and osteocytes in endochondral bone formation. Proc Natl Acad Sci U S A, 111（33）: 12097-12102.

Yanger K, Zong Y, Maggs L R, et al. 2013. Robust cellular reprogramming occurs spontaneously during liver regeneration. Genes Dev, 27（7）: 719-724.

Yao Q J, Sang L, Lin M, et al. 2018. Mettl3-Mettl14 methyltransferase complex regulates the quiescence of adult hematopoietic stem cells. Cell Res, 28（9）: 952-954.

Yao R, Ito C, Natsume Y, et al. 2002. Lack of acrosome formation in mice lacking a Golgi protein, GOPC. Proc Natl Acad Sci U S A, 99（17）: 11211-11216.

Yimlamai D, Christodoulou C, Galli G G, et al. 2014. Hippo pathway activity influences liver cell fate. Cell, 157（6）: 1324-1338.

Yin M X, Zhang L. 2015. Hippo signaling in epithelial stem cells. Acta Biochim Biophys Sin（Shanghai）, 47（1）: 39-45.

Yin Q, Wang H, Li N, et al. 2020. Dosage effect of multiple genes accounts for multisystem disorder of myotonic dystrophy type 1. Cell Res, 30（2）: 133-145.

Ying Y, Zhao G Q. 2001. Cooperation of endoderm-derived BMP2 and extraembryonic ectoderm-derived BMP4 in primordial germ cell generation in the mouse. Dev Biol, 232（2）: 484-492.

Yu F X, Guan K L. 2013. The Hippo pathway: regulators and regulations. Genes Dev, 27（4）: 355-371.

Yu F X, Zhao B, Guan K L. 2015. Hippo pathway in organ size control, tissue homeostasis, and cancer. Cell, 163（4）: 811-828.

Yu W, Ma X, Xu J, et al. 2019. VGLL4 plays a critical role in heart valve development and homeostasis. PLoS Genet, 15（2）: e1007977.

Yu Y, Vanhorne J, Oko R. 2009. The origin and assembly of a zona pellucida binding protein, IAM38, during spermiogenesis. Microsc Res Tech, 72（8）: 558-565.

Zhang C, Chen Y, Sun B, et al. 2017a（ ）Nature, 549（7671）: 273-276.

Zhang J, Niu C, Ye L, et al. 2003. Identification of the haematopoietic stem cell niche and control of the niche size. Nature, 425（6960）: 836-841.

Zhang K, Zhang L, Liu W, et al. 2018a. *In vitro* expansion of primary human hepatocytes with efficient liver repopulation capacity. Cell Stem Cell, 23（6）: 806-819 e804.

Zhang P, Kang J Y, Gou L T, et al. 2015. MIWI and piRNA-mediated cleavage of messenger

RNAs in mouse testes. Cell Res, 25（2）: 193-207.

Zhang W, Xu J, Li J, et al. 2018b. The TEA domain family transcription factor TEAD4 represses murine adipogenesis by recruiting cofactors VGLL4 and CtBP2 into a transcriptional complex. J Biol Chem, 293（44）: 17119-17134.

Zhang X M, Wu K, Zheng Y, et al. 2020. *In vitro* expansion of human sperm through nuclear transfer. Cell Res, 30（4）: 356-359.

Zhang Y, Zhang H, Zhao B. 2017b. Hippo signaling in the immune system. Trends Biochem Sci, 43（2）: 77-80.

Zhao S, Gou L T, Zhang M, et al. 2013. piRNA-triggered MIWI ubiquitination and removal by APC/C in late spermatogenesis. Dev Cell, 24（1）: 13-25.

Zheng G X, Terry J M, Belgrader P, et al. 2017. Massively parallel digital transcriptional profiling of single cells. Nat Commun, 8: 14049.

Zhong C, Xie Z, Yin Q, et al. 2016a. Parthenogenetic haploid embryonic stem cells efficiently support mouse generation by oocyte injection. Cell Res, 26（1）: 131-134.

Zhong C, Yin Q, Xie Z, et al. 2015. CRISPR-Cas9-mediated genetic screening in mice with haploid embryonic stem cells carrying a guide RNA library. Cell Stem Cell, 17（2）: 221-232.

Zhong C, Zhang M, Yin Q, et al. 2016b. Generation of human haploid embryonic stem cells from parthenogenetic embryos obtained by microsurgical removal of male pronucleus. Cell Res, 26（6）: 743-746.

Zhou B O, Yue R, Murphy M M, et al. 2014a. Leptin-receptor-expressing mesenchymal stromal cells represent the main source of bone formed by adult bone marrow. Cell Stem Cell, 15（2）: 154-168.

Zhou Q, Li L, Zhao B, et al. 2015. The hippo pathway in heart development, regeneration, and diseases. Circ Res, 116（8）: 1431-1447.

Zhou X, von der Mark K, Henry S, et al. 2014b. Chondrocytes transdifferentiate into osteoblasts in endochondral bone during development, postnatal growth and fracture healing in mice. PLoS Genet, 10（12）: e1004820.

Zhou Y, Huang T, Cheng A S, et al. 2016. The TEAD family and its oncogenic role in promoting tumorigenesis. Int J Mol Sci, 17（1）: 138.

第九章

细胞间通信及细胞与微环境的互作与功能

第一节　概　　述

多细胞生物由不同类型的细胞以高度有序的方式形成组织、器官，乃至个体。生物体的生长发育及生理活动的协调，都离不开形态和功能各异的高度分化细胞之间的信息交流与相互协作。精细、准确、高效的细胞间通信机制是多细胞生物的基本特征，也是细胞协调它们的行为，完成一系列复杂生命活动的基础。

细胞间通信是指细胞发出的信息通过介质传递到另一个细胞产生相应反应的复杂过程。细胞通过多种多样的方式完成细胞间通信。细胞分泌水溶性、脂溶性的小分子化合物和多肽、蛋白质等大分子化合物，这些化合物作为信号分子作用于其他细胞，调节其生物学功能。细胞这种非直接接触，而是通过化学信号分子，以自分泌（autocrine）、旁分泌（paracrine）、内分泌（endocrine）的形式调控不同距离范围内其他细胞的行为，是多细胞生物最普遍采用的细胞间通信方式。细胞也可以通过直接接触与相邻细胞交换信息。细胞通过细胞膜表面蛋白特异性地识别相邻细胞的膜表面分子进行信息传递，从而达到功能上的相互协调。相邻细胞还可以通过细胞间隙连接（gap junction）等结构形成的亲水性孔道共享水溶性的小分子化合物，或分泌细胞外囊泡，快速地促进相邻细胞对外界信号的协同反应。

细胞间通信不是一个单向的简单过程。多细胞生物细胞之间形成复杂的细胞间通信网络与反馈机制，可精确、高效地发送与接收信息，协调细胞的生命活动，使之成为统一整体，对外界环境做出快速的综合反应。细胞处于复杂、动态变化的微环境之中。细胞感知微环境信号，并反过来影响其微环境，细胞与其微环境交互调控。不同的组织器官，在发育和疾病发展的不同阶段，细胞所处微环境的组成和组织结构发生时空特异性的动态改变。细胞整合微环境信号，将胞外信号转导为胞内信号，改变胞内的生理生化过程，进而调节特定基因的表达，最终决定增殖、分化、衰老、死亡的细胞命运和细胞类型特异性的生物学功能，维持多细胞生命体的生长发育和组织稳态。无论是发育畸形，还是癌症、代谢疾病，几乎所有的人类疾病都与细胞间通信、细胞与微环境相互作用异常相关。解析细胞间相互作用关系，阐明细胞－微环境的网络关系，可为深入了解各种生理和病理过程及其内在调控机制提供重要信息，势必产生新的理论和应用上的突破。

第二节　关键科学问题

尽管研究人员很早就发现了细胞间信息传递的现象，但目前该领域还存在很多既独立又交织的问题亟待解决。在经典信号分子与细胞间通信模式基础上，新型信号分子与通信方式不断涌现。是否还存在尚未发现的信号分子与细胞－微环境交互、跨器官细胞间通信方式？细胞如何时空选择性地整合外界刺激，决定其命运与生物学功能，完成复杂生命活动？细胞间通信的异常如何导致人类疾病的发生？这些科学问题都尚待回答。随着研究体系与技术方法，特别是实时动态在体研究体系与技术方法的发展，我们将在新的层次上加深对细胞间通信及细胞与微环境的互作与功能的认识与理解。

一、新的信号分子与细胞间通信模式的发现

信号分子是细胞间通信的基础。得益于技术的发展，近年来研究人员发现了许多新型细胞间通信的信号分子。种类繁多、结构多样的代谢小分子如何作为信号分子，参与细胞间通信，调控细胞的结构与功能，还有待系统、深入的研究。生理和病理条件下，信号分子发生怎样特异性的修饰改变，信号分子的修饰又如何精细调控细胞间通信，尚未进行广泛的研究。解析信号分子的功能的精细调控与作用机理是亟须解决的科学问题。除了经典的水溶性、脂溶性信号分子、细胞外囊泡及细胞间的直接接触，是否还存在其他的细胞间通信方式，也是亟待回答的重要问题。细胞以自分泌、旁分泌、内分泌的形式与不同距离范围内其他细胞交换信息，细胞是如何实现不同距离信号的有效、精细递送，特别是在跨器官远距离运送信号分子过程中如何保护其生物活性，并在靶位点准确、有效地传递信息的，仍有待进一步研究。

二、信号分子分泌、接收的特异性与选择性

绝大部分信号分子的分泌和接收具有细胞、组织和器官特异性。细胞如何响应外界刺激，选择性地释放特定的信号分子，信号分子如何选择不同的传递方式，特定组织器官中的细胞如何选择性地接收信号；特定生理病理条件下，细胞如何选择单个信号分子的分泌和接收或以细胞外囊泡为代表的大量信号分子的批量分泌和接收方式，都是值得深入探究的科学问题。在大量新型信号分子和传递方式涌现的情况下，这些科学问题变得更为复杂，也更加亟须解答。受体是细胞感应外界环境变化而协调细胞功能响应的主要传感器，同时也是潜在的药物受体。系统地绘制发育、器官生理稳态和病理条件下，信号分子、受体单细胞精度时空特异表达图谱及其组织器官中的分布图谱，揭示其转运机制，系统筛选信号分子－受体配伍关系，解析信号分子、

受体结构和功能分类，从分子－细胞－组织等不同维度，绘制其作用网络，明晰其生物学功能与生理病理意义，具有重要的研究意义。

三、细胞间通信时空特异性调控网络

生命活动的复杂性决定细胞间通信必须由很多信号协同完成，同时信号分子之间必须存在广泛的相互调控关系，形成有机统一的整体。细胞间通信的信息流如何有效利用前馈、反馈机制精细调控信号？细胞作为细胞间通信信息流中的节点，如何整合复杂的上游信号，向下游分配和传递信号，通过信息流的传递－整合－分配，适应组织器官稳态维持或疾病发生发展的需要？这些都值得深入的探究。现有的技术研究手段仍有一定的局限性，无法做到对哺乳动物体内细胞间通信时空特异性调控网络的精细研究。建议在现有剖面式研究基础上，有效整合和发展生化、结构、分子影像与功能实验的研究体系及技术方法，发展实时动态的在体研究体系和技术方法，结合原位组织三维重建、单细胞空间组测序、高分辨率单细胞质谱技术等，从分子、细胞、组织器官、个体水平上全面立体、实时动态地研究发育、组织器官稳态和病理条件下细胞间信息流的关键节点与信息流传递－整合－分配的调控机制及其生理病理功能，剖析机体发育不同阶段、组织器官稳态和病理条件下细胞间通信的时空特异性调控网络。

四、靶向细胞间通信的复杂疾病诊断与防治

几乎所有的人类疾病都与细胞间通信异常相关。病理条件下，细胞－微环境间的交互、跨器官的细胞间通信有别于发育和组织稳态条件下的细胞间通信特征。研究癌症、代谢性疾病、神经退行性疾病、衰老等疾病发展过程中时空特异性的细胞间通信调控网络，发现疾病发展中关键的信号分子与受体、信号分子传递方式与细胞间通信模式，鉴定病理条件下细胞间通信调控网络中的关键节点，是发现疾病治疗靶标，发展高效、低毒的预防和诊治策略的关键前提。

第三节 重要研究方向的发展现状与趋势分析

一、细胞间通信方式

细胞间通信是细胞生物学的基本过程之一。细胞选择合适的通信方式进行信息的交流，以协调细胞的行为，共同完成复杂的生命活动。细胞间通信方式包含两个层面。第一个层面是信号分子的选择及信号的调控。细胞通过分泌水溶性、脂溶性的小分子化合物和多肽、蛋白质等大分子化合物，以自分泌、旁分泌、内分泌的方式调控其他细胞。细胞也可以通过细胞膜蛋白介导的瞬时或稳定的直接相互作用进行信息的传递。细胞间通信模式的选择是细胞间通信方式的第二个层面。除了单个小分子、大分子化合物信号分子与其受体一对一的作用模式，细胞还通过细胞间的亲水性孔道与细胞外囊泡，实现细胞之间信息的批量交流。

细胞通过信号分子与其受体之间的特异性识别，对不同的信号刺激做出特异性反应。细胞膜表面的受体蛋白与表达在其他细胞上的配体蛋白或可溶性信号分子发生相互作用，将细胞外化学信号、物料接触信号转变为细胞内部化学信号，进而调控多种细胞生理过程。除了经典的激素和多肽、蛋白质等生物大分子信号分子，近年来研究人员新发现了一系列信号分子。代谢物便是其中之一，代谢物除了参与经典代谢反应，满足细胞对能量和生物大分子的需求外，还作为胞外信号分子与受体结合，作为细胞间通信的中介体影响细胞行为。例如，乳酸与其受体 GPR81 结合，通过降低胞内环磷酸腺苷（cAMP）水平而减少脂肪分解（Ahmed et al.，2010）。乳酸激活 GPR81，协调肿瘤细胞中 MCT、CD147 和 PGC1α 的表达，确保乳酸的有效利用（Roland et al.，2014）。乳酸 -GPR81 通路同时也在免疫监视抑制中发挥作用。GPR81 的激活可降低肿瘤细胞内 cAMP 水平，抑制蛋白激酶 A 活性，进而通过转录因子 TAZ 激活 PD-L1/PD1 免疫检查点，导致 T 细胞功能缺陷（Feng et al.，

2017）。巨噬细胞特异表达的GPR132作为乳酸受体促进巨噬细胞M2表型转换，进而促进肿瘤转移（Chen et al.，2017）。肿瘤细胞中色氨酸双加氧酶（TDO）催化色氨酸分解生成犬尿酸原。犬尿酸原以自分泌或旁分泌的形式与芳香烃受体（AHR）结合抑制抗肿瘤免疫反应，促进肿瘤细胞存活和运动（Opitz et al.，2011）。细菌脂多糖、病毒DNA或RNA等病原体相关分子模式（pathogen associated molecular pattern，PAMP）作为信号分子，结合TLR、RIG-I等模式识别受体（pattern recognition receptor，PRR），激活天然免疫反应（Zindel and Kubes，2020）。组织或细胞受到损伤或应激刺激，释放尿酸和ATP等小分子化合物。这些内源性应激信号作为损伤相关分子模式（damage-associated molecular pattern，DAMP）被模式识别受体识别，诱导自身免疫或免疫耐受，在关节炎、动脉粥样硬化、系统性红斑狼疮等疾病的发生发展过程中发挥重要作用（Zindel and Kubes，2020）。细胞和组织通常表现出固体和液体具有的黏弹性性质，即在负载下会缓慢变形，在变形后移除负载呈新的形状。细胞受到张力和拉力、形变、流体剪切力、静态液压力等力学环境的刺激（Miller and Davidson，2013）。张力等机械力、组织刚性及细胞外基质的空间拓扑结构等物理信号作为另一类信号，通过整合素（integrin）、离子通道（ion channel）、连接子（connexon）及初级纤毛（primary cilia）等力学感受器影响细胞的基本行为（Chaudhuri et al.，2020）。生物力学在发育过程中为细胞命运的决定提供关键的位置信息，时空特异性地调控生理和病理过程中的基因表达和蛋白质活性。间充质干细胞具有分化成为包括成骨、软骨、脂肪细胞在内的多种细胞的潜能，细胞分化方向的决定受到细胞周围基质弹性模量的影响。机械拉伸刺激能够触发上皮细胞快速分裂（Gudipaty et al.，2017）。减弱细胞机械力转导能够抑制癌基因介导的细胞重编程和肿瘤的发生（Panciera et al.，2020）。

信号分子、受体会发生糖基化、脂质化等形式多样的翻译后修饰（post-translational modification，PTM），这些翻译后修饰极大地增加了信号分子、受体功能的多样性与复杂性，进而调控信号分子与其受体的识别。PD-1、PD-L1均发生复杂的糖基化，其糖基化水平与谱式不仅与其蛋白质稳定性和细胞表面定位相关，而且对于PD-1与PD-L1的相互识别至关重要（Benicky et al.，2021；Sun et al.，2020），靶向糖基化的PD-1/PD-L1开展免疫治疗，可以显著提高免疫治疗的效果。Notch 受体 O- 糖基化的谱式决定其结合 Delta/DLL 还是 Serrate/Jagged 配体（Pandey et al.，2019）。棕榈酰化修饰的 Wnt 蛋白与磷脂酰肌醇蛋白聚糖 Dlp 结合，增加 Wnt 蛋白的水溶性，使胞外的 Wnt 蛋白能够扩散并与受体结合，实现 Wnt 信号通路的高效传递，最终发挥形态发生素功能（McGough et al.，2020）。信号分子与受体结合，诱导受体发生构象变化与形式多样的翻译后修饰，并引起一系列下游信号转导事件、基因转录与蛋白质翻译，从而使细胞的功能发生改变，以适应微环境的变化和机体整体需要。除了磷酸化、泛素化、甲基化、乙酰化、蛋白水解等，近年来研究人员还发现了亚硝基化（*S*-nitrosylation）、ADP 核糖基化（ADP-ribosylation）等新的修饰类型。对于常见的磷酸化、泛素化等翻译后修饰，也发现了新的翻译后修饰形式，如发生于精氨酸上的磷酸化、半胱氨酸上的泛素化修饰。

信号分子的识别及其信号的传递也受到脂质分子、小分子化合物的非共价调控。例如，胆固醇一方面结合 T 细胞受体（T cell receptor，TCR）的 β 亚基，使 TCR 稳定在静息状态，从而调整 TCR 激活阈值（Swamy et al.，2016）；另一方面，胆固醇上调 T 细胞中的免疫检查点蛋白 PD-1、TIM-3 和 LAG-3 的表达，从而诱导 $CD8^+$ T 细胞的耗竭（Ma et al.，2019）。静息状态下，细胞质膜上的酸性磷脂屏蔽 TCR 和 CD28 的活性位点，TCR 激活后诱发的 Ca^{2+} 内流可以中和酸性磷脂的作用，释放 TCR 和 CD28 的活性位点，促进它们的信号转导。激活的 TCR 和 CD28 进一步加速 Ca^{2+} 内流，由此形成的正反馈环路迅速放大抗原刺激信号，从而为 T 细胞产生高抗原敏感性提供信号基础（Shi et al.，2013；Xu et al.，2008；Yang et al.，2017）。乳酸结合 N-Myc 下游调节蛋白（NDRG3）使其免受 PHD2 的羟化作用，进而导致 NDRG3 的积累，激活 Raf/ERK 介导的血管新生和细胞增殖，确保细胞适应持续低氧环境（Lee et al.，2015）。葡萄糖匮乏条件下，谷氨酸脱氢酶 GDH1 与 IKK 复合体相互作用，产生局部高浓度 α-KG。这些 α-KG 作为第二信使直接结合并激活 IKKβ 及下游 NF-κB 信号通路，上调葡萄糖转运体 GLUT1 的表达，并代偿性地增强肿瘤细胞的葡萄糖摄入，从而维持了肿瘤细胞在糖匮乏条件下的存活（Wang et al.，2019a）。乙酰辅酶 A 作为乙酰基供体调控组蛋白和非组蛋白的乙酰化（Lee et al.，2014）。细胞单碳代谢途径的产物 *S*- 腺苷甲硫氨酸

（SAM）作为甲基供体，调控组蛋白、DNA、mRNA的甲基化（Chiang et al.，2009；Shyh-Chang et al.，2013）。胞内α-酮戊二酸（α-KG）水平可以直接影响TET家族DNA去甲基化酶、Jumonji C家族组蛋白去甲基化酶、mRNA去甲基化酶FTO和ALKBH5以及脯氨酰羟化酶（PHD）的活性。野生型异柠檬酸脱氢酶（IDH1/2）负责将TCA循环代谢物异柠檬酸盐转化为α-KG，而IDH突变体则将α-KG催化为2-羟戊二酸（2-HG）。2-HG与α-KG结构类似，竞争性地抑制α-KG依赖的双加氧酶并调控细胞内整体的甲基化（Figueroa et al.，2010；Xu et al.，2011）。代谢小分子可以调控mRNA的稳定性从而调控靶基因的表达。尿苷二磷酸葡萄糖（uridine diphosphate glucose，UDPG）可以直接结合RNA结合蛋白HuR，从而抑制HuR和转录因子SNAI1 mRNA的结合，造成SNAI1 mRNA的不稳定及随后的降解，从而抑制肿瘤细胞迁移（Wang et al.，2019b）。

细胞也通过细胞膜蛋白介导的细胞间接触依赖的瞬时相互作用传递信号。淋巴细胞迁移过程中，细胞通过表面选择素、整合素等黏附分子与血管内皮表面地址素等配体结合，介导淋巴细胞黏附并穿越血管壁到达特定组织部位（Kechagia et al.，2019）。适应性免疫细胞如T细胞通过与抗原呈递细胞（antigen-presenting cell，APC）建立细胞间接触启动免疫反应，同时活化的T细胞通过接触包含病原体的目的细胞执行杀伤功能，调节性T细胞等可以通过细胞表面CTLA-4等蛋白与抗原呈递细胞直接作用，剥夺效应T细胞的共刺激信号，以达到免疫抑制的目的（Motz and Coukos，2011）。胚胎发育过程中细胞间接触对于神经系统发育，包括皮质前体细胞和成神经细胞的分裂、星形胶质细胞的形态变化及神经元突触的生成具有决定性作用（Tsai and McKay，2000），成骨细胞与间充质干细胞的相互作用可直接影响发育过程中的骨生成（Mao et al.，2016）。

除了配体-受体相互作用这一单个信号分子一对一的作用模式，细胞还通过细胞之间信息的批量交流实现细胞间通信与细胞命运、细胞功能的调控。相邻细胞通过细胞间隙连接（gap junction）等结构形成的亲水性孔道共享水溶性的小分子化合物。这一共享机制使得相邻细胞可以快速、可逆地实现对外界信号的协同反应。细胞可以通过分泌多种类型的细胞外囊泡与邻近细胞或远处的细胞进行信息沟通。根据它们的亚细胞起源，可将细胞外囊泡分成外泌体（exosome）和脱落微囊泡（shedding microvesicle）等多种不同的类型（Colombo et al.，2014）。细胞外囊泡具有脂质双分子层结构，其表面分子使得细胞外囊泡能够选择性地与特定靶细胞结合，经由受体－配体相互作用，或者通过内吞和 / 或吞噬作用与靶细胞膜内化或融合，将内容物递送到靶细胞的细胞质中。一个细胞分泌的细胞外囊泡可以被其他细胞（即旁分泌或内分泌）或该细胞本身（即自分泌）吞噬内化，将其中包含的大量蛋白质、mRNA、lncRNA、microRNA、脂质等多种物质同时导入靶细胞中，进行物质和信号的交流，由此改变受体细胞的生理状态，实现对靶细胞的批量、多维度协同调控。细胞外囊泡作为介导细胞间信息传递的重要方式，受到越来越多的关注。在生殖发育、机体免疫、代谢与心血管疾病、神经退行性疾病、癌症中，细胞外囊泡在其中均扮演重要角色（Kalluri and LeBleu，2020）。携带 MHC 多肽的外泌体可直接通过抗原递呈作用激活 T 细胞，在小鼠体内可以显著抑制肿瘤生长（Zitvogel et al.，1998）。肿瘤来源的外泌体携带 PD-L1，可直接抑制 $CD8^+$ 杀伤性 T 细胞的活化，抑制树突状细胞的成熟，并诱导 T 细胞耗竭，与肿瘤免疫耐受密切相关（Kalluri and LeBleu，2020）。肿瘤来源的外泌体还会诱导预转移微环境（premetastatic niche）的建立，促进胰腺癌的肝转移（Costa-Silva et al.，2015）。细胞在迁移过程中在其胞体后侧释放大于 1 μm 的小囊泡——迁移体（migrasome）（Ma et al.，2015），迁移体的形成与细胞迁移引发的 Tspan4 蛋白、胆固醇在收缩丝上的局部高度富集，以及该区域膜的弯曲刚度增加相关（Huang et al.，2019）。迁移体或是被释放到细胞外基质中或是被周围的细胞直接摄取。斑马鱼胚胎原肠运动时期会产生大量的迁移体，在胚胎发育过程中呈现特定的时空分布。迁移体中存在大量趋化因子、细胞因子和生长因子，这些信号因子随迁移体的分布形成局部区域信号中心，引导背部先驱者细胞（dorsal forerunner cell）的正确定位，从而参与调控斑马鱼胚胎器官形态发生（Jiang et al.，2019）。

信号分子是细胞间通信的基础。除了经典的小分子、大分子信号分子，近年来还发现了包括代谢物在内的许多新型信号分子。代谢物种类繁多，结构多样，并随着细胞功能状态的改变而动态变化。是否有其他的代谢物作为信号分子，参与细胞间通信，调控细胞的结构与功能，还有待系统、深入的研究。信号分子的修饰或精细或显著改变信号分子的构象。修饰引起的信号分子构象的改变调节其与受体的亲和力，甚至改变其与受体的结合特异性。生理病理条件下，信号分子的修饰发生怎样特异性的改变，是尚未广泛研究的重要问题。除了糖基化、羟基化等常见修饰，代谢物也可能作为多肽、蛋白质类信号分子的新的修饰方式，通过调节其受体选择性，参与细胞间通信。信号分子以自分泌、旁分泌、内分泌的方式与不同距离的细胞实现细胞间通信。信号分子的有效通信范围如何确定，特别是在信号分子远距离递送过程中，其生物活性如何受到保护，又如何准确、有效地递送到靶器官、靶细胞，并将信息特异、有效地传递给靶细胞；信号分子的修饰如何调节信号分子信息的准确、有效传递，这些问题都需要系统的阐明。细胞通过信号分子－受体一对一的作用模式传递信号，细胞也可以通过细胞外囊泡等方式实现细胞之间信息的批量交流，实现细胞间通信与细胞命运、细胞功能的调控。除了经典的水溶性、脂溶性信号分子、细胞间的直接接触及细胞外囊泡以外，新的细胞间通信方式的发现也将极大地帮助我们加深对细胞间通信机制的理解。

二、细胞－微环境时空特异性交互网络

胚胎发育及成体组织器官稳态的建立和维持过程中，细胞处于复杂、动态变化的微环境之中，受到所处微环境的调控。细胞接受并整合微环境信号，将胞外信号转导为胞内信号，并通过转录与表观遗传调控改变基因表达，最终决定增殖、分化、衰老、死亡的细胞命运和细胞类型特异性的生物学功能，维持多细胞生命体的生长发育和组织稳态。细胞感知微环境信号，做出适应性的改变，并反过来影响其微环境，细胞与其微环境交互调控，细胞与微环境构成时空特异性的互作网络。细胞－微环境相互作用的异常改变可导致发育异常、代谢性疾病、癌症、神经退行性疾病等疾病的发生。病理条件下，细胞与微环境互作网络有着不同于正常发育、稳态建立维持的特征。

细胞所处的微环境，根据其性质可以分成三大类，即生物微环境、化学微环境和物理微环境。组织器官由多种类型的细胞组成。成纤维细胞、炎症细胞、血管内皮细胞等间质细胞组成了上皮（功能）细胞的生物微环境（Tlsty and Coussens，2006）。这些间质细胞通过与上皮细胞直接接触的方

式，或是通过分泌大量的趋化因子、细胞因子、蛋白酶、胞外基质蛋白等间接方式，调控上皮细胞的活动。胞外基质（extracellular matrix，ECM），是由细胞合成、分泌的生物大分子在细胞表面或细胞之间构成的复杂网络结构（Sorokin，2010）。胞外基质不仅为上皮细胞提供了物理性支撑，同时还作为信号分子与细胞表面受体结合，并通过与之结合的细胞黏附分子、生长因子、细胞因子、趋化因子从多个方面调节细胞的存活、生长与死亡，细胞的分化以及细胞的形态决定，并参与对细胞迁移的调控。细胞的化学微环境，包括离子浓度、pH、O_2 浓度在极大程度上影响细胞的行为，细胞化学微环境的相对稳定对细胞正常功能的发挥至关重要，而化学微环境的微调精细调控了细胞对外界刺激的响应和细胞活动。Ca^{2+} 与细胞骨架蛋白结合，在细胞骨架动态调控中发挥重要作用，并调控整合素与其配体的亲和力（Becchetti and Arcangeli，2010）。低氧条件下细胞表现出更强的运动能力（Bosco et al.，2008）。血液和淋巴循环系统中快速的液体流动及细胞与细胞外基质间体液的慢速间隙渗流（interstitial flow）所产生的机械力是重要的细胞物理微环境（Rutkowski and Swartz，2007）。细胞感受流体静压、切变应力、压力和张力等微环境机械力，并与细胞内部细胞骨架的张力达到动态平衡。微环境机械力在胞外基质重塑、细胞迁移、细胞分化及组织形态发生与功能发挥中起到至关重要的作用（Butcher et al.，2009）。细胞外基质的刚性与空间拓扑结构作为细胞另一类物理微环境调控了细胞极性、细胞应力纤维的形成和细胞迁移（Gao et al.，2010；Petrie et al.，2009）。

细胞微环境呈现高度的异质性。例如，免疫微环境中的巨噬细胞高度异质，在组织稳态平衡、疾病进展和损伤修复等方面发挥多种功能。肝中的巨噬细胞既可来源于循环单核细胞的招募，也可来自胚胎来源的组织特异巨噬细胞（tissue-resident macrophage），即库普弗细胞（Kupffer cell）的自我更新。库普弗细胞感知组织损伤，引发炎症反应；而由浸润的 $Ly\text{-}6C^{+}$ 单核细胞衍生而来的巨噬细胞则与慢性炎症和肝纤维化密切相关，这些巨噬细胞在纤维化消退过程中进一步分化成 Ly-6C 低表达巨噬细胞，促进肝损伤修复过程（Tacke and Zimmermann，2014）。成纤维细胞在组织稳态维持、损伤修复过程中发挥着重要作用（Tomasek et al.，2002）。损伤修复过程中成纤维细胞短暂激活，而修复完成后成纤维细胞恢复到正常状态。与损伤修复不同，肿瘤中的成纤维细胞［即肿瘤相关成纤维细胞（cancer-associated fibroblast，CAF）］持续处于激活状态（Kalluri，2016）。成纤维细胞高度异质，依据分子标志物表达谱式的差异，CAF 可以分为 4 个亚群（Costa et al.，2018）；利用单细胞测序技术，其中的 S1 亚群又可被细分为至少 8 个亚群（Kieffer et al.，2020）。这些亚群不仅与乳腺癌不同的病理亚型相关，而且从不同的方面调控肿瘤免疫微环境与对免疫治疗的响应（Costa et al.，2018）。微环境中的细胞因子、趋化因子、生长因子是细胞 - 微环境信息交流的主要媒介。它们可由微环境中的上皮细胞、成纤维细胞、免疫细胞等直接产生，也可通过血管、淋巴管等从外周到达组织微环境，或由坏死或焦亡的细胞大量释放。肿瘤相关巨噬细胞（tumor-associated macrophage，TAM）、中性粒细胞等炎症细胞高表达肿瘤坏死因子 -α（tumor necrosis factor-α, TNF-α）等促炎因子，促进血管生成和肿瘤细胞侵袭（Moore et al.，1999），白细胞介素 -6（interleukin-6, IL-6）促进肿瘤细胞的增殖、抑制凋亡，加速肿瘤发展（Li et al.，2015）。髓源抑制性细胞（myeloid-derived suppressor cell, MDSC）和 T 细胞等产生大量的转化生长因子 -β（transforming growth factor-β, TGF-β）等抑炎因子抑制肿瘤细胞周期、促进凋亡（Morrison et al.，2013）。

细胞微环境同时呈现高度的可塑性。细胞感知微环境信号，并反作用于微环境，对微环境进行重塑。除了分泌细胞因子、生长因子等信号分子，细胞分泌的代谢产物也能通过自分泌和旁分泌途径作用于自身和免疫细胞，影响多种类型细胞之间的功能调控网络。肿瘤细胞高水平的糖酵解导致乳酸堆积，使得肿瘤微环境维持低 pH（Yang，2017）。乳酸作用于 T 细胞，诱导初始 T 细胞（naïve T cell）凋亡，抑制其活化及迁移，减少 T 细胞、自然杀伤细胞（natural killer cell，NK cell）产生 IFN-γ，抑制其效应功能（Brand et al.，2016）；乳酸作用于巨噬细胞也会促进 HIF-1α 稳定进而促进 M2 型巨噬细胞极化，产生更多抑炎型细胞因子（Vitale et al.，2019）。细胞的代谢物可以被其他细胞摄取再利用。肿瘤中靠近血管、营养供给充分的氧化型肿瘤细胞与远端低氧型肿瘤细胞建立了一种代谢共生关系。低氧型肿瘤细胞通过糖酵解产生大量乳酸，经由 MCT4 释放至微环境中，氧化型肿瘤细胞通过 MCT1 摄取乳酸，在乳酸脱氢酶的催化下产生丙酮酸，进而参与氧化代谢（Sonveaux et al.，2008）。肿瘤细胞和基质细胞也建立了以乳酸为基础的代谢耦联关系。CAF 激活葡萄糖转运体表达，增加葡萄糖摄取及 MCT4 介导的乳酸释放。与此同时，前列腺癌细胞的糖酵解活性被抑制，转而代谢 CAF 来源的乳酸（Fiaschi et al.，2012）。肺癌及胃癌 CAF 均可以利用肿瘤细胞来源的乳酸（Apicella et al.，2018）。大脑中也存在上述代谢共生关系。星形胶质细胞分泌的乳酸被神经元摄取用作营养物质（Machler et al.，2016）。肿瘤细胞与其微环境中的免疫细胞、基质细胞竞争氧气及营养物质。肿瘤细胞代谢倾向于糖酵解。T 细胞的活化及 M1 型巨噬细胞的促炎功能依赖于糖酵解过程及葡萄糖的摄取。肿瘤细胞与免疫细胞竞争葡萄糖使后者功能受限。肿瘤细胞和免疫细胞对不同种类氨基酸有着各自的偏好。T 细胞响应 TCR 信号依赖于谷氨酰胺、半胱氨酸、色氨酸、精氨酸等多种氨基酸的摄取及分解（Geiger et al.，2016）。M2 型肿瘤相关巨噬细胞更倾向于摄取谷氨酰胺，限制 M2 型巨噬细胞的谷氨酰胺摄入会使其重新向 M1 型极化（Palmieri et al.，2017）；M2 型巨噬细胞高表达精氨酸酶 1（arginase 1），分解精氨酸。肿瘤细胞代谢偏好谷氨酰胺、精氨酸。肿瘤细胞与 T 细胞竞争摄取谷氨酰胺、精氨酸，从而限制 T 细胞效应功能的发挥与巨噬细胞的极化。

除了经典的细胞凋亡（apoptosis）之外，微环境信号还会诱导细胞发生受到严格调控的坏死样的程序性死亡，细胞的这种死亡方式被称作程序性坏死（programmed necrosis）。例如，在中枢系统炎症和神经退行性疾病中，小神经胶质细胞通过释放 TNF、IL-1α/β 和 IFN-γ 等细胞因子诱导邻近的少突胶质细胞发生细胞裂亡（Ito et al.，2016）；组织中的病毒 RNA 可以通过 ZBP1 诱导细胞裂亡（Zhang et al.，2020b）；免疫微环境中的 T 细胞和其他免疫细胞则可以通过释放 Granzyme A/E 等诱导癌细胞发生焦亡（Zhang et al.，2020c；Zhou et al.，2020b）。细胞死亡并不是一个被动的过程。死亡的细胞以可控的方式主动释放信号分子，重塑其微环境，改变细胞与微环境的交互，发挥重要调控功能。无论细胞凋亡还是细胞坏死都可以促进对 T 淋巴细胞的抗原提呈（Galluzzi et al.，2017）。死亡的细胞通过释放 HMGB1、ATP、HSP 家族蛋白、组蛋白、线粒体蛋白和核苷酸等损伤相关分子模式引发炎症反应。化疗药物致 DAMP 的释放增强免疫原性，提高肿瘤微环境中杀伤性与调节性 T 细胞的比例，一定程度上纠正机体免疫抑制状态（He et al.，2018; Tang et al.，2019）。细胞死亡可以代偿性地促进存活细胞的增殖。肌纤维细胞的程序性坏死通过调节骨骼肌损伤后的微环境，促进肌肉干细胞的增殖和肌肉组织的损伤修复过程（Zhou et al.，2020a）。Caspase 介导的质膜PANX1 通道开放，以可控的方式在细胞凋亡过程中持续释放代谢物。凋亡代谢物分泌组（secretome）诱导邻近健康细胞的特定基因表达，进一步调节机体的炎症反应、细胞增殖、胚胎发生与组织再生（Medina et al.，2020）。

细胞选择性接受并整合高度复杂异质的微环境信号，并主动重塑其所处的微环境。细胞与其微环境交互调控，以时空特异性的复杂互作网络模式调控胚胎发育、成体组织器官稳态建立、维持，及疾病发生发展。干细胞干性的维持与分化的命运决定受到微环境中间质细胞及其分泌的 Wnt、BMP 等信号蛋白的调控。位于小肠隐窝区底部的多能性 LGR5+ 隐窝基底柱状细胞（crypt-base columnar cell，CBC）分化产生吸收性细胞系（肠上皮细胞）、潘氏细胞（Paneth cell）、簇细胞（tuft cell）、肠内分泌细胞（enteroendocrine cell）、杯状细胞（goblet cell）等分泌细胞系与静息态细胞，驱动肠上皮的更替（McCarthy et al.，2020a）。间质细胞分泌配体蛋白质 Wnt，通过激活 Wnt 信号通路促进小肠干细胞自我更新，抑制其分化（de Lau et al.，2014）。与 Wnt 信号通路相反，BMP 抑制了小肠干细胞的自我更新，促进其向功能细胞分化（Davis et al.，2015；Haramis et al.，2004）。间质细胞沿隐窝–绒毛轴产生相反的 Wnt、BMP 梯度，将小肠干细胞限制在隐窝基部，促进过渡扩增细胞（transit amplifying cell，TA 细胞）分裂，驱动小肠上皮细胞的命运决定（Kosinski et al.，2007；McCarthy et al.，2020b; Powell et al.，2011）。肺泡由肺泡Ⅰ型（AT1）上皮细胞和肺泡Ⅱ型（AT2）细胞组成。AT2 上皮细胞同时可作为祖细胞，进行自我更新和分化为 AT1 上皮细胞维持肺泡上皮稳态，这一过程受到 Wnt、BMP 信号的拮抗调控（Zepp et al.，2019）。$Axin2^+$AT2 上皮细胞在损伤后，受到 Wnt 信号调控，自我更新并分化成熟的 AT1 和 AT2 上皮细胞，在急性损伤后的肺泡修复中起到主要作用（Nabhan et al.，2018；Zacharias et al.，2018）。与 Wnt 信号通路相反，肺泡发育过程中 BMP 信号限制 AT2 上皮细胞自我更新，促进其向 AT1 上皮细胞分化（Chung et al.，2018）。肺泡间质由表达 PDGFRa 的成纤维细胞和间质肺泡龛细胞（mesenchymal alveolar niche cell，MANC）等间质细胞组成。MANC 细胞位于 AT2 上皮细胞和 $Axin2^+$ 肌成纤维前体细胞周围，是肺泡再生的关键。

MANC 响应损伤刺激，产生 FGF7、IL-6 及 NBL1、GREM2、FSTL1 等 BMP 信号的拮抗剂促进 AT2 上皮细胞的增殖和自我更新（Zepp et al.，2017）。损伤修复时，肺血管内皮细胞上调基质金属蛋白酶 MMP14 的表达。MMP14 通过降解细胞外基质释放的表皮生长因子样配体，刺激上皮细胞生长（Rafii et al.，2015）。除了间质细胞，上皮细胞（如 AT2 细胞）也会产生 CCL2、IL-33 等趋化因子和细胞因子，从而募集单核细胞和先天淋巴样细胞，抑制急性炎症反应，调控巨噬细胞到 M2 细胞系的极化，进而通过重塑胞外基质促进 AT2 细胞增殖（Lechner et al.，2017）。

肿瘤的发生发展与其所处的微环境息息相关（Gilkes et al.，2014；Hanahan and Weinberg，2011；Kalluri，2016；Lu et al.，2012；Tlsty and Coussens，2006）。肿瘤发展过程中，肿瘤微环境发生适应性的改变，并促进了肿瘤的发展和转移（Gilkes et al.，2014；Hanahan and Weinberg，2011；Kalluri，2016；Lu et al.，2012；Polyak and Kalluri et al.，2010；Tlsty and Coussens，2006）。肿瘤细胞分泌的 TGF-β、PDGF 和 FGF2 是 CAF 激活的主要诱导因子（Kalluri，2016；Kalluri and Zeisberg，2006）。CAF 的招募、增殖与侵袭均依赖于肿瘤微环境中的 TGF-β。CAF 处于低氧、低 pH 等微环境中。HIF-1α 及 XBP-1 是 CAF 激活及维持其活化状态的重要转录因子（Hernandez-Gea et al.，2013；Madsen et al.，2015）。成纤维细胞的活化同时受到微环境的机械信号调控。刚性微环境信号导致成纤维细胞的活化及 CAF 状态的维持，并进一步促进了肿瘤微环境细胞外基质的重塑及刚性微环境，形成一个正反馈的调节环路（Calvo et al.，2013；Machado et al.，2015）。细胞微环境协同发挥功能，形成的复杂网络在多个层面上调控了肿瘤发生、发展、转移。与胚胎干细胞或成体干细胞相似，微环境对肿瘤干细胞性状的维持及肿瘤的发生至关重要。CAF 分泌的骨膜蛋白（periostin）增强了乳腺癌干细胞中的 Wnt 信号，从而促进乳腺癌的转移（Malanchi et al.，2011）。在结直肠癌中，CAF 分泌的 HGF，激活了临近肿瘤干细胞中的 β- 联蛋白（β-catenin）信号通路，并维持了其干性特征（Vermeulen et al.，2010）。CD44 等细胞外基质蛋白受体被认为是肿瘤干细胞的分子标志物，并在肿瘤的发生中发挥重要功能（Todaro et al.，2014）。CAF 不仅通过分泌 VEGF（Kim et al.，2012）、CXCL12（Orimo et al.，2005）、CTGF（Yang et al.，2005）等细胞因子诱导血管新生，而且通过细胞外基质重塑调控肿瘤血管新生过程。血管生成依赖于血管内皮细胞的基底膜。低氧诱导 LOXL2 的表达，进而通过交联基底膜中的Ⅳ型胶原蛋白促进血管生成（Bignon et al.，2011）。抑制 LOX 家族氧化酶活力抑制了肿瘤血管新生（Ingber and Folkman，1988）。肿瘤处于免疫抑制性的微环境中。M2 型肿瘤相关巨噬细胞通过调控肿瘤细胞存活、血管新生、肿瘤免疫等多个方面促进肿瘤发展（Joyce and Pollard，2009；Ruffell and Coussens，2015）。CAF 分泌的 CXCL12、IL-6 等细胞因子调控巨噬细胞的招募及其向促肿瘤的 M2 类型的分化（Joyce and Pollard，2009；Ruffell and Coussens，2015）。除了影响固有免疫外，CAF 还调控适应性免疫。CAF 表达的 CXCL12，导致肿瘤中 T 细胞排斥（Feig et al.，2013）。TGF-β 在 CAF 中高表达，除了通过自分泌方式进一步调控 CAF 的活化状态外，TGF-β 诱导杀伤性 T 细胞的凋亡（Sanjabi et al.，2009），促进调节性 T 细胞的存活（Ouyang et al.，2010），抑制 Th1 细胞的分化（Laouar et al.，2005），从而调控肿瘤炎性微环境与肿瘤免疫。

细胞处于复杂、动态变化的微环境之中。不同的组织器官，发育、疾病发展的不同阶段，细胞所处微环境的组成、组织结构与信号发生均处于时空特异性的动态变化中。微环境信号影响细胞的命运决定与功能状态，同时细胞向其微环境传递信号。细胞感知微环境信号，做出适应性的改变，并反过来动态重塑其微环境。细胞与其微环境交互，形成反馈环路，共同演化。疾病的发生发展既可以认为是细胞 - 微环境异常的结果，也可以认为是细胞及其微环境共演化达到新的细胞 - 微环境平衡的结果。恶性肿瘤等疾病的发生在多大程度上是由于微环境信号的异常改变导致正常细胞的基因突变的？肿瘤复发是因为肿瘤细胞没有被完全清除还是因为微环境持续处于异常状态？这些都是值得研究的问题。疾病的治疗也可以认为是一个细胞 - 微环境交互正常化的过程。细胞处于复杂的微生态系统之中，细胞同时接受复杂多样的微环境信号。生命活动的复杂性决定了细胞间通信由很多信号协同完成，同时信号分子之间存在广泛的相互调控关系，形成一个复杂的交互式调控网络。细胞 - 微环境互作网络的有效整合，是细胞命运决定与细胞复杂行为的基础。迄今为止，发育、组织稳态建立与维持，疾病发生发展过程中细胞命运决定与功能，细胞微环境的动态变化及细胞 - 微环境时空特异性的交互规律还没有得到很好的阐明。对发育、组织稳态建立与维持中细胞－微环境时空特异性交互共性和个性规律的全景式理解，将有助于阐明生命的基本规律。鉴定细胞－微环境交互网络的关键节点及其作用规律，可以作为剖析细胞－微环境时空特异性的交互规律的有效抓手与切入点，也有望成为疾病预防、早期诊断和治疗有效的干预靶标。

三、跨器官细胞间通信及细胞与微环境互作

生命体是一个有机统一的整体。细胞命运的决定与功能的发挥不仅取决于其所处的局部微环境，还受到来源于机体其他组织器官信号的调控。无论是生理条件下的发育及衰老，还是病理条件下的疾病发生发展，跨器官细胞间通信都在其中发挥着重要作用。

从 20 世纪发现胰岛素开始，人们陆续发现了数十种激素通过体液跨组织传递信息。这些激素及相关分泌器官构成了经典意义上的内分泌系统，精确地调控代谢、发育等许多生理过程。类固醇和多肽激素的跨组织传递主要依赖于循环系统。靶细胞的激素受体负责接收这些激素信号。多肽类激素的受体主要分布在细胞膜上，激素与之结合后通过一系列的信号转导过程使靶细胞激活或抑制某些特定的信号通路；类固醇类的激素是脂溶性的，可以通过细胞膜进入细胞质，与胞浆受体结合形成复合物后进入细胞核，与核内受体结合，参与基因表达的调控。胚胎发育过程中，形态发生素（morphogen）在特定组织中有着特定的时空分布，控制不同器官的发育（Steinhauer and Treisman，2009）。形态发生素在成体阶段也扮演着重要的角色。成体干细胞的维持与分化、伤口愈合，乃至癌症的发生发展都受到形态发生素的紧密调控。与激素不同，形态发生素并非来自于某个特定的器官或腺体，而是由不同器官中的某些特定细胞分泌。

炎症反应不仅在消除病原体和组织修复的过程中起着关键作用，越来越多的研究表明适当的炎症反应对于维持机体的正常生理稳态也至关重要，而这一切都依赖于炎症因子的精确调控。炎症因子主要来源于各组织内的免疫细胞。在病理条件下，炎症因子功能失调与许多疾病的发展密切相关。在肿

瘤组织中，炎症因子可以帮助肿瘤细胞重塑微环境，使得肿瘤细胞逃过免疫系统的监督并更加容易转移到其他组织。炎症因子的失衡也会导致神经退行性疾病、心血管疾病等疾病的发生。

在很多不同类型的组织细胞中，研究人员还发现 microRNA 和 lncRNA 等非编码 RNA 也可以被跨组织传递，并在靶细胞中调控基因表达。这些非编码 RNA 的来源更为复杂，很有可能是一种原始但更普遍存在的细胞间通信方式。

除了上述这些有机的生物活性分子以外，一些无机的化学分子也可以跨器官进行细胞间通信。血液中的一氧化氮（NO）在心血管疾病以及神经系统中都发挥着信号分子的作用（Snyder，1992）。在进一步研究上述这几类信号分子在靶器官作用的分子机制时，人们发现许多经典的信号通路（如胰岛素 / 胰岛素样生长因子通路、Wnt 通路等），在细胞间与细胞内构成了复杂交错的通信网络调控各种生命活动。干预通信网络是治疗诸多疾病的有效手段，如使用胰岛素治疗糖尿病的经典方法和新近热门的基于 PD-1 抗体的癌症治疗。

生命体作为一个有机的整体，其代谢过程除了在胞内受到严格的调控之外，各个组织器官之间也存在多种形式的相互作用。胰腺 α 胰岛细胞分泌胰高血糖素，加速肝糖原分解、促进糖异生、间接抑制周围组织摄取葡萄糖而使血糖升高，并促进脂肪细胞分解脂肪。在葡萄糖、乳糖、核糖、精氨酸等刺激下，胰腺 β 胰岛细胞分泌胰岛素，促进肝细胞、肌肉细胞将葡萄糖转化为糖原，增强脂肪酸的合成和脂肪酸的酯化作用，降低蛋白质和脂肪降解，降低葡萄糖的合成。脂肪组织、肝组织和肌肉组织之间，存在着一个器官因子调节网络，通过自分泌、旁分泌与内分泌形式，参与体内代谢过程。脂肪组织通过释放瘦素（leptin）、脂联素、抵抗素、血清内脂素、视黄醇结合蛋白 -4 等调节能量平衡。瘦素由肥胖基因（*ob*）编码，是脂肪细胞分泌的脂源性蛋白类激素，通过与下丘脑弓状核神经元上的瘦素受体结合，抑制促食欲的神经肽 AgRP 和 NPY 的表达，或升高抑食类神经肽 POMC 和 CART 的表达，以降低食物摄入、增加能量消耗（Friedman，2019）。瘦素作用的丧失是造成肥胖的重要原因之一。肝脏因子（hepatokine）是一类主要或者全部由肝脏分泌的细胞因子或活性多肽，通过旁分泌或者自分泌的形式作用于肝脏或肌肉、脂肪等组织，调节糖、脂代谢，引发系统性亚临床炎症反应，在非酒精性脂肪性肝病、糖尿病、动脉粥样硬化等病理过程中发挥重要调控作用（Watt et al.，2019）。目前已知的肝脏因子主要有成纤维细胞生长因子 21（fibroblast growth factor 21，FGF21）、胰岛素样生长因子、性激素结合球蛋白、胎球蛋白 A、硒蛋白 P、血管生成素相关生长因子和 GPNMB 等。FGF21 主要在肝脏和胰腺中表达（Geng et al.，2020）。饥饿状态下，脂肪组织分解释放游离脂肪酸，游离脂肪酸通过激活肝脏中的 PPARα 诱导 FGF21 表达。FGF21 参与调控血糖、脂质代谢和能量稳态。FGF21 通过增加脂肪细胞的葡萄糖摄取，促进脂联素的分泌，减少循环系统中的游离脂肪酸和组织脂毒性，促进肝糖原生成，调控血糖。FGF21 抑制肝脏细胞中转录因子 SREBP2 的活性，降低胆固醇的生物合成，具有抗动脉粥样硬化的保护功能。肝脏产生的 FGF21 还以内分泌方式进入循环系统，通过 FGFRl / β-klotho-P13K-Akt-BAD 信号通路对心脏发挥保护作用。骨骼肌是人体最大的器官，在全身广泛分布。骨骼肌分泌超过 600 种肌细胞因子（myokine）。通过这些肌细胞因子，骨骼肌与中枢神经、肝脏、脂肪等多种组织器官进行信息交流，相互作用，协调各器官的功能，维持机体稳态，共同参与各种生理、病理活动（Lee et al.，2019）。鸢尾素（irisin）是由含Ⅲ型纤连蛋白域蛋白 5（fibronectin type Ⅲ domain containing protein 5 ，FNDC5）经过切割形成的小肽（Teufel et al.，2002）。运动后，骨骼肌产生鸢尾素。鸢尾素促进白色脂肪向棕色脂肪转变，进而加强糖、脂代谢，减少脂肪的积累（Bostrom et al.，2012）。鸢尾素还可以通过血脑屏障，降低神经细胞中的活性氧浓度，保护神经细胞，提高脑梗后神经细胞的存活率，促进脑功能的修复（Uysal et al.，2018）。骨骼肌在收缩时向血液中释放肌联素（myonectin）。肌联素的功能与胰岛素相似，能够有效降低血糖浓度（Seldin et al.，2012）。肌联素可以抑制肝细胞的自噬，提高肝的血糖调节能力（Seldin et al.，2013）。肌因子在衰老的“传染”过程中传递衰老信号，是重要的抗衰老靶点。由衰老骨骼肌分泌至血液中的 Dkk3 能够作用于其他尚未衰老的骨骼肌，导致骨骼肌功能衰退，造成整体衰老（Yin et al.，2018）。

肿瘤的发展与转移受其微环境的影响，同时也受到远程跨器官细胞信号的调控。神经系统活动调节组织发育、组织稳态与组织再生。神经系统与肿瘤细胞、肿瘤微环境的相互作用在肿瘤发生发展中的关键调节器受到越来越多的重视。生活在丰富环境（enriched environment）中的小鼠，肿瘤生长明显减缓。研究人员发现，丰富环境可选择性地调节下丘脑脑源性神经营养因子（brain derived neurotrophic factor，BDNF）的表达，BDNF通过交感神经－肾上腺素信号通路调节脂肪细胞分泌瘦素，抑制肿瘤生长（Cao et al.，2010）。谷氨酸能神经元通过分泌神经配蛋白-3（neuroligin-3）以旁分泌形式激活肿瘤细胞中的PI-3K-mTOR信号通路，并通过神经元和胶质瘤相互作用形成AMPA受体依赖的神经元－胶质瘤突触与神经胶质瘤细胞膜电位的去极化，驱动恶性神经胶质瘤的生长（Venkatesh et al.，2015；Venkatesh et al.，2019）。肿瘤微环境中的交感神经、副交感神经通过神经递质调节肿瘤的发生、发展与转移。胰腺癌肿瘤微环境中神经递质5-羟色胺（5-hydroxytryptamine，5-HT）失调，5-HT通过调控有氧糖酵解促进胰腺癌细胞在代谢应激条件下的生长，进而促进胰腺癌的恶性进展（Jiang et al.，2017）。特定神经类型的功能在不同类型肿瘤中的功能不尽相同。胆碱能神经信号抑制胰腺癌的发展和转移（Renz et al.，2018），但促进胃腺癌的发展（Hayakawa et al.，2017）。肿瘤细胞与其微环境信号同时反作用于神经系统，引起神经系统重塑和功能的障碍。肿瘤细胞在神经纤维周围浸润导致周围神经的重塑和慢性疼痛综合征的发生。

肿瘤的转移是一个多因素、多步骤协同作用的演变过程。肿瘤细胞－微环境的交互作用在肿瘤转移的各个步骤均发挥重要调控作用。原位肿瘤转移前在转移靶器官诱导形成有利于肿瘤细胞转移的微环境，即预转移微环境。预转移微环境是循环肿瘤细胞定植的先决条件。原位肿瘤的肿瘤细胞、CAF分泌细胞因子和细胞外基质重塑相关蛋白，在肿瘤将要转移的位点建立了以细胞外基质、炎症细胞作为主要组分的预转移微环境。原发肿瘤分泌的TNF-a、TGFb、CXCL12、VEGFA等细胞因子或外泌体上调预转移微环境中的纤连蛋白、S100A8、S100A9、MMP2、MMP9和LOX，进而招募骨髓$CD34^{+}VEGFR1^{+}$造血祖细胞和$CD11b^{+}$髓样细胞。LOX介导肺部Ⅳ型胶原蛋白的共价交联，并招募骨髓来源的$CD11b^{+}$髓系细胞，VEGFA诱导肺部的血管新生，从而形成预转移微环境（Erler et al.，2009；O'Connell et al.，2011）。肝纤维化和活化的肝星状细胞表达骨膜蛋白是胰腺癌肝转移的先决条件（Nielsen et al.，2016）。细胞外基质蛋白空间结构的改变在肿瘤细胞休眠中也发挥着调控作用。LOX参与招募巨噬细胞，并促进肺部Ⅰ型胶原蛋白的共价交联和纤维化，继而促进了乳腺癌的肺转移（Barkan et al.，2010；Cheng et al.，2014）。肿瘤细胞在转移靶器官定植后，可长期维持在休眠状态（Aguirre-Ghiso，2007）。微环境的是否平衡决定了肿瘤细胞是否脱离休眠，起始快速增殖过程（Aguirre-Ghiso，2007）。在休眠微环境和转移微环境中，细胞外基质蛋白的表达有着很大的差异（Ghajar et al.，2013）。肿瘤转移过程中，肿瘤细胞定植在组织的微血管系统。血管内皮细胞表达血小板应答蛋白-1（thrombospondin-1）形成一个维持肿瘤休眠的微环境，抑制乳腺癌细胞生长和转移灶形成。当这些血管内皮细胞的尖端萌出时，新生血管中的TGF-β1和骨膜蛋白将休眠微环境转变为转移微环境，促进乳腺癌细胞的生长和转移灶的形成（Ghajar et al.，2013）。

组织器官的局部细胞－微环境交互网络作为机体的组成部分，相互耦合，交互调控，形成有机统一的整体。组织器官之间信息的交换形成了无形的信息流。在信息流传递—整合—分配的基础上，组织器官的细胞－微环境交互“局域网络”信号互联互通构成了机体整体的“互联网络”。细胞如何实现信号的跨器官长距离准确、有效递送，特别是运送信号分子过程中其生物活性的保护，并仅在靶位点有效传递信息？细胞接受跨器官信号后，如何适应组织器官稳态维持或疾病发生发展的需要，利用前馈、反馈机制适时放大或衰减信号？经典的内分泌、神经系统信号如何与微环境信号协调？细胞作为信息流中的节点，如何整合复杂的上游信号，选择性地释放特定的信号分子，向下游分配和传递信号？这些都是值得深入探究的科学问题。

四、细胞与微环境互作研究技术方法

从20世纪初托马斯·亨特·摩尔根（Thomas Hunt Morgan）选择黑腹果蝇作为研究对象，奠定了经典遗传学的基础开始，科学家利用线虫、果蝇、斑马鱼、小鼠等模式生物开展了广泛的细胞研究。特别是近30年来，研究人员借助现代遗传学的手段在特定细胞类型、特定组织的组成型或诱导型的基因敲低/敲除或转基因过表达，结合遗传示踪手段，研究信号分子、受体、重要信号转导通路、转录因子、表观遗传调控因子等在生物体发育、损伤修复及疾病发生发展过程中的功能及机制；利用 Axin2-LacZ、Gli1-LacZ 等报告基因小鼠模型，在组织层面研究了效应细胞对 Wnt、Hedgehog 等信号分子时空特异性的响应。这些研究体系和技术手段的建立极大地推动了对细胞的基本框架及其如何调控细胞命运决定与生物学功能发挥的理解。细胞与细胞之间的相互作用是多细胞生物的基本特征。近年来，利用合成生物学的手段，通过人工合成信号通路实现了相邻细胞的标记和转录调控，并通过遗传操控元件实现对特定细胞的示踪及基因表达调控，大大提升了体内细胞及其与微环境互作的研究水平和精度，为解决细胞间通信及调控生理功能等基本的生物学问题提供了重要的技术平台及研究手段。人工合成 Notch 信号通路（Synthetic Notch，SynNotch）技术利用人工合成的遗传元件结合经典的 Notch 信号通路跨膜段，将合成生物学技术应用到了真核细胞的示踪和功能调控中（Morsut et al.，2016）。当分别表达配体和受体的两个细胞邻近时，配体和受体发生特异性结合，引起 Notch 跨膜段的构象改变与 γ 分泌酶（γ-secretase）酶介导的剪切，释放胞内区的人工合成遗传元件，如转录调控因子 Gal4、tTA 等进入细胞核，调控下游报告基因（如红色荧光蛋白基因）的表达，从而标记表达合成受体的细胞。利用合成受体及其配体，研究人员在果蝇体内实现了通过细胞与细胞接触标记邻近细胞（He et al.，2017；Huang et al.，2016）。在哺乳动物小鼠体内，研究人员利用细菌中的转肽酶 Sortase A 将 G5 标签标记到相互接触的免疫细胞膜上，实现了活体内短时程的细胞标记及其命运追踪，这种 LIPSTIC 示踪技术有助于解析细胞－细胞之间的交流及其生物学功能，为体内细胞的命运可塑性及其调控机制研究提供了新的重要技术手段（Pasqual et al.，2018）。

近年来，利用基于标签（barcode）的单细胞识别策略，可以通过单细胞测序同时获得一个组织中成千上万个单细胞的信息，揭示每个细胞特有的个性化特征。单细胞测序技术不仅使得不同细胞类型得以更加精细的区分，揭示了全新的细胞类型，而且可以预测细胞的分化轨迹。同时，单细胞测序技术使得在单细胞水平开展细胞间通信研究成为可能。利用 SingleCellSignalR、iTALK、NicheNet 等分析方法，通过分析单细胞水平的信号分子－受体表达水平与配伍情况，研究人员深入分析了多种组织器官生理、病理情况下，如正常肝、非酒精性脂肪性肝炎、胆管细胞癌中肝细胞/肿瘤细胞与微环境之间信号交互（Xiong et al., 2019；Zhang et al., 2020a）。进一步加深基于单细胞测序数据的细胞间通信研究。还有待于新的计算生物学算法、生物信息学分析平台的开发。一方面，信号分子与受体的相互作用数据库尚需要进一步的更新与修正；另一方面，现有的生物信息学方法往往基于一个配体/一个受体的配伍关系，忽略了许多受体作为多亚基复合物才能发挥功能，对细胞膜上的共受体以及激动剂、拮抗剂在信号转导与细胞间通信中的贡献也考虑不足。复杂生命活动的细胞间通信由很多信号协同完成，同时信号分子之间存在广泛的调控关系，形成有机统一的网络。定量化地研究细胞间通信的网络调控也是生物信息学分析的重点和难点。单细胞测序虽然提供了海量的数据，但受限于技术本身，这些数据缺少重要的空间位置信息。通过结合成像和单细胞测序技术的单细胞空间转录组将为基因表达提供丰富的空间位置信息，促进细胞间通信的研究。需要注意的是，转录水平与蛋白质水平并不完全是线性对应的关系。信号分子、受体等发生时空特异性的翻译后修饰，这些翻译后修饰精细调控配体-受体选择性与下游信号事件。在单细胞测序基础上，发展高分辨率单细胞质谱技术，将有助于更加全面地研究发育、组织器官稳态和病理条件下细胞间通信的关键节点与时空特异性调控网络。

我国科学家对细胞间通信的早期研究主要集中在胞内信号转导通路，在Wnt、TGF-b、EGFR等G蛋白耦联受体、受体酪氨酸激酶信号通路等领域做出了杰出的工作，大大拓展了我们对信号转导通路的认识。近年来，科学技术部、国家自然科学基金委员会启动了一批细胞间通信的基础研究项目，有效地推动了国内细胞间通信领域的发展。我国科学家在新的细胞间通信信号分子、细胞间通信模式、细胞-微环境互作及其生理病理功能方面做出了卓越的科研成果，揭示小分子代谢物作为胞外信号分子，以类似于神经递质和激素的方式行使功能，作为信号通路第二信使、关键酶的别构调节剂或辅因子调控关键信号通路、Tet双加氧酶、甲基转移酶等表观遗传修饰调控酶与基因表达；揭示机械力门控离子通道PIEZO感知微环境机械信号的结构基础及其在骨发育、组织纤维化中的生理病理功能。在新的细胞间通信模式方面，我国研究人员揭示了ESCRT依赖与非依赖的外泌体生成途径，及其在恶性肿瘤、组织纤维化、心肌梗死恢复、免疫稳态等生理病理过程中的作用机制，探讨了外泌体用于液体活检与治疗的可行性；发现了细胞迁移过程迁移体小囊泡这一全新的细胞间通信细胞器，阐明了迁移体的特征与生成机制，及其在胚胎器官形态发生中的功能；开展了广泛深入的细胞–微环境互作研究，在微环境调控干细胞干性维持与定向分化、微环境调控胚胎发育与器官形成、肿瘤微环境、肿瘤免疫、造血微环境等方向都取得了重要成果；发展了双同源重组的谱系示踪技术、分子影像、单细胞测序、高分辨率质谱技术与从线虫到小鼠的完备的研究体系和技术方法。基于这些研究成果，我国科学家在疾病的生物治疗、免疫治疗、靶向治疗的靶点确证、药物研发方面也取得可喜进展。这些相关研究成果及优秀的科研团队的涌现，极大地提高了我国细胞间通信及细胞与微环境互作研究领域的研究水平，使得我国在该领域的国际地位逐渐上升，为进一步发展打下了良好的基础。

第四节　我国发展战略与重点方向

一、发展总体思路和发展目标

1. 总体发展思路

虽然近年来我国在细胞间通信及细胞与微环境的互作的功能基础与应用领域取得了一些成绩，但是整体研究水平距与欧美等国家仍存在相当大的差距。我国要在瞄准生命科学重大问题和世界前沿领域的基础上，密切结合我国实际，以新型信号分子与细胞间通信方式为抓手，以细胞–微环境互作与跨器官细胞间通信为核心，发展实时动态在体的新研究体系与新技术方法，通过多学科交叉融合，整合关键技术，阐明细胞间通信时空特异性调控网络的调控机制及其生理病理功能，为靶向细胞间通信的复杂疾病诊断与防治研究和应用打下坚实基础。

2. 发展目标

系统解析生理条件下细胞间通信时空特异性调控网络的动态变化特征、调控机制及其生理功能，阐明复杂疾病发生发展过程中细胞间通信的功能及时空特异性调控网络，为疾病的精确诊断和治疗提供新的策略。在此基础上，加大细胞间通信研究前沿新技术，特别是实时动态在体技术方法的开发力度，加强细胞间通信相关药物的研发，加快细胞间通信即细胞与微环境互作基础研究成果的临床转化应用，提升我国生物医药领域的持续创新能力。

二、优先发展领域或重要研究方向

希望能够整合国内的优势研究力量组建创新团队和跨学科创新平台，通过学科深度交叉融合与高水平联合攻关，优先开展下述研究方向。

1. 鉴定新的信号分子与细胞间通信模式

鉴定新的信号分子，系统厘清新型信号分子与信号分子修饰参与细胞间通信的功能与作用机理，鉴定新的细胞间通信模式，阐明生理病理过程中细胞如何实现不同距离，特别是跨器官信号传递过程中信号的有效、精细运送，及信号传递的特异性与选择性。

2. 发展新的实时动态的在体研究体系与技术方法

结合体内成像技术与谱系示踪技术，建立新型的信号分子－受体特异性相互作用信息指导下的遗传操控技术，实现信息交换的细胞标记示踪及细胞命运操纵，结合原位组织三维重建、单细胞空间组测序、高分辨率质谱技术等多学科技术体系、方法，实现细胞间通信实时动态的在体观察与分析。

3. 阐明细胞间通信间时空特异性调控网络的调控机制及其功能

在现有剖面式研究基础上，从分子、细胞、组织器官、个体水平上全面立体、实时动态地研究发育、器官生理稳态和病理条件下的信号分子－受体相互作用、细胞接触动态变化特征、细胞－微环境互作、跨器官细胞间通信，阐明细胞间信息流的传递－整合－分配规律，剖析细胞间通信的时空特异性调控网络。

4. 发展靶向细胞间通信的复杂疾病诊断与防治的新策略、新技术、新药物

研究癌症、代谢性疾病、神经退行性疾病、衰老等疾病发展过程中时空特异性的细胞间通信调控网络有别于发育、组织稳态维持的基本特征，剖析疾病发展中细胞间通信调控网络的关键节点，发现疾病治疗靶标，发展靶向细胞间通信的复杂疾病诊断与防治的新策略、新技术、新药物。

本章参考文献

Aguirre-Ghiso J A. 2007. Models, mechanisms and clinical evidence for cancer dormancy. Nat Rev Cancer, 7（11）: 834-846.

Ahmed K, Tunaru S, Tang C, et al. 2010. An autocrine lactate loop mediates insulin-dependent inhibition of lipolysis through GPR81. Cell Metab, 11（4）: 311-319.

Apicella M, Giannoni E, Fiore S, et al. 2018. Increased lactate secretion by cancer cells sustains non-cell-autonomous adaptive resistance to MET and EGFR targeted therapies. Cell Metab, 28（6）: 848-865 e846.

Barkan D, El Touny L H, Michalowski A M, et al. 2010. Metastatic growth from dormant cells induced by a col-I-enriched fibrotic environment. Cancer Res, 70（14）: 5706-5716.

Becchetti A, Arcangeli A. 2010. Integrins and ion channels in cell migration: implications for neuronal development, wound healing and metastatic spread. Adv Exp Med Biol, 674: 107-123.

Benicky J, Sanda M, Kennedy Z B, et al. 2021. PD-L1 glycosylation and its impact on binding to clinical antibodies. J Proteome Res, 20（1）: 485-497.

Bignon M, Pichol-Thievend C, Hardouin J, et al. 2011. Lysyl oxidase-like protein-2 regulates sprouting angiogenesis and type Ⅳ collagen assembly in the endothelial basement membrane. Blood, 118（14）: 3979-3989.

Bosco M C, Puppo M, Blengio F, et al. 2008. Monocytes and dendritic cells in a hypoxic environment: Spotlights on chemotaxis and migration. Immunobiology, 213（9-10）: 733-749.

Bostrom P, Wu J, Jedrychowski M P, et al. 2012. A PGC1-alpha-dependent myokine that drives brown-fat-like development of white fat and thermogenesis. Nature, 481（7382）: 463-468.

Brand A, Singer K, Koehl G E, et al. 2016. LDHA-associated lactic acid production blunts tumor immunosurveillance by T and NK cells. Cell Metab, 24（5）: 657-671.

Butcher D T, Alliston T, Weaver V M. 2009. A tense situation: forcing tumour progression. Nat Rev Cancer, 9（2）: 108-122.

Calvo F, Ege N, Grande-Garcia A, et al. 2013. Mechanotransduction and YAP-dependent matrix remodelling is required for the generation and maintenance of cancer-associated fibroblasts. Nat Cell Biol, 15（6）: 637-646.

Cao L, Liu X, Lin E J, et al. 2010. Environmental and genetic activation of a brain-adipocyte BDNF/leptin axis causes cancer remission and inhibition. Cell, 142（1）: 52-64.

Chaudhuri O, Cooper-White J, Janmey P A, et al. 2020. Effects of extracellular matrix viscoelasticity on cellular behaviour. Nature, 584（7822）: 535-546.

Chen P, Zuo H, Xiong H, et al. 2017. Gpr132 sensing of lactate mediates tumor-macrophage interplay to promote breast cancer metastasis. Proc Natl Acad Sci U S A, 114（3）: 580-585.

Cheng T, Liu Q, Zhang R, et al. 2014. Lysyl oxidase promotes bleomycin-induced lung fibrosis through modulating inflammation. J Mol Cell Biol, 6（6）: 506-515.

Chiang E P, Wang Y C, Chen W W, et al. 2009. Effects of insulin and glucose on cellular metabolic fluxes in homocysteine transsulfuration, remethylation, *S*-adenosylmethionine synthesis, and global deoxyribonucleic acid methylation. J Clin Endocrinol Metab, 94（3）: 1017-1025.

Chung M I, Bujnis M, Barkauskas C E, et al. 2018. Niche-mediated BMP/SMAD signaling regulates lung alveolar stem cell proliferation and differentiation. Development, 145（9）: dev163014.

Colombo M, Raposo G, Thery C. 2014. Biogenesis, secretion, and intercellular interactions of exosomes and other extracellular vesicles. Annu Rev Cell Dev Biol, 30（1）: 255-289.

Costa A, Kieffer Y, Scholer-Dahirel A, et al. 2018. Fibroblast heterogeneity and immunosuppressive environment in human breast cancer. Cancer Cell, 33（3）: 463-479 e410.

Costa-Silva B, Aiello N M, Ocean A J, et al. 2015. Pancreatic cancer exosomes initiate pre-metastatic niche formation in the liver. Nat Cell Biol, 17（6）: 816-826.

Davis H, Irshad S, Bansal M, et al. 2015. Aberrant epithelial GREM1 expression initiates colonic tumorigenesis from cells outside the stem cell niche. Nat Med, 21（1）: 62-70.

de Lau W, Peng W C, Gros P, et al. 2014. The R-spondin/Lgr5/Rnf43 module: regulator of Wnt signal strength. Genes Dev, 28（4）: 305-316.

Erler J T, Bennewith K L, Cox T R, et al. 2009. Hypoxia-induced lysyl oxidase is a critical mediator of bone marrow cell recruitment to form the premetastatic niche. Cancer Cell, 15（1）: 35-44.

Feig C, Jones J O, Kraman M, et al. 2013. Targeting CXCL12 from FAP-expressing carcinoma-associated fibroblasts synergizes with anti-PD-L1 immunotherapy in pancreatic cancer. Proc Natl Acad Sci U S A, 110（50）: 20212-20217.

Feng J, Yang H, Zhang Y, et al. 2017. Tumor cell-derived lactate induces TAZ-dependent upregulation of PD-L1 through GPR81 in human lung cancer cells. Oncogene, 36（42）: 5829-5839.

Fiaschi T, Marini A, Giannoni E, et al. 2012. Reciprocal metabolic reprogramming through lactate shuttle coordinately influences tumor-stroma interplay. Cancer Res, 72（19）: 5130-5140.

Figueroa M E, Abdel-Wahab O, Lu C, et al. 2010. Leukemic IDH1 and IDH2 mutations result in a hypermethylation phenotype, disrupt TET2 function, and impair hematopoietic differentiation. Cancer Cell, 18（6）: 553-567.

Friedman J M. 2019. Leptin and the endocrine control of energy balance. Nat Metab, 1（8）: 754-764.

Galluzzi L, Buque A, Kepp O, et al. 2017. Immunogenic cell death in cancer and infectious disease. Nature Reviews Immunology, 17（2）: 97-111.

Gao Y, Xiao Q, Ma H, et al. 2010. LKB1 inhibits lung cancer progression through lysyl oxidase and extracellular matrix remodeling. Proc Natl Acad Sci U S A, 107（44）: 18892-18897.

Geiger R, Rieckmann J C, Wolf T, et al. 2016. *L*-arginine modulates T cell metabolism and enhances survival and anti-tumor activity. Cell, 167（3）: 829-842 e813.

Geng L, Lam K S L, Xu A. 2020. The therapeutic potential of FGF21 in metabolic diseases: from bench to clinic. Nat Rev Endocrinol, 16（11）: 654-667.

Ghajar C M, Peinado H, Mori H, et al. 2013. The perivascular niche regulates breast tumour dormancy. Nat Cell Biol, 15（7）: 807-817.

Gilkes D M, Semenza G L, Wirtz D. 2014. Hypoxia and the extracellular matrix: drivers of tumour metastasis. Nat Rev Cancer, 14（6）: 430-439.

Gudipaty S A, Lindblom J, Loftus P D, et al. 2017. Mechanical stretch triggers rapid epithelial cell division through Piezo1. Nature, 543（7643）: 118-121.

Hanahan D, Weinberg R A. 2011. Hallmarks of cancer: the next generation. Cell, 144（5）: 646-674.

Haramis A P, Begthel H, van den Born M, et al. 2004. De novo crypt formation and juvenile polyposis on BMP inhibition in mouse intestine. Science, 303（5664）: 1684-1686.

Hayakawa Y, Sakitani K, Konishi M, et al. 2017. Nerve growth factor promotes gastric tumorigenesis through aberrant cholinergic signaling. Cancer Cell, 31（1）: 21-34.

He L, Huang J, Perrimon N. 2017. Development of an optimized synthetic Notch receptor as an *in vivo* cell-cell contact sensor. Proc Natl Acad Sci U S A, 114（21）: 5467-5472.

He S, Wang X. 2018. RIP kinases as modulators of inflammation and immunity. Nat Immunol, 19（9）: 912-922.

Hernandez-Gea V, Hilscher M, Rozenfeld R, et al. 2013. Endoplasmic reticulum stress induces fibrogenic activity in hepatic stellate cells through autophagy. J Hepatol, 59（1）: 98-104.

Huang T H, Velho T, Lois C. 2016. Monitoring cell-cell contacts *in vivo* in transgenic animals. Development, 143（21）: 4073-4084.

Huang Y, Zucker B, Zhang S, et al. 2019. Migrasome formation is mediated by assembly of micron-scale tetraspanin macrodomains. Nat Cell Biol, 21（8）: 991-1002.

Ingber D, Folkman J. 1988. Inhibition of angiogenesis through modulation of collagen metabolism. Lab Invest, 59（1）: 44-51.

Ito Y, Ofengeim D, Najafov A, et al. 2016. RIPK1 mediates axonal degeneration by promoting inflammation and necroptosis in ALS. Science, 353（6299）: 603-608.

Jiang D, Jiang Z, Lu D, et al. 2019. Migrasomes provide regional cues for organ morphogenesis during zebrafish gastrulation. Nat Cell Biol, 21（8）: 966-977.

Jiang S H, Li J, Dong F Y, et al. 2017. Increased serotonin signaling contributes to the Warburg effect in pancreatic tumor cells under metabolic stress and promotes growth of pancreatic tumors in mice. Gastroenterology, 153（1）: 277-291 e219.

Joyce J A, Pollard J W. 2009. Microenvironmental regulation of metastasis. Nat Rev Cancer, 9（4）: 239-252.

Kalluri R. 2016. The biology and function of fibroblasts in cancer. Nat Rev Cancer, 16（9）: 582-598.

Kalluri R, LeBleu V S. 2020. The biology, function, and biomedical applications of exosomes. Science, 367（6478）: eaau6977.

Kalluri R, Zeisberg M. 2006. Fibroblasts in cancer. Nat Rev Cancer, 6（5）: 392-401.

Kechagia J Z, Ivaska J, Roca-Cusachs P. 2019. Integrins as biomechanical sensors of the microenvironment. Nature Reviews Molecular Cell Biology, 20（8）: 457-473.

Kieffer Y, Hocine H R, Gentric G, et al. 2020. Single-cell analysis reveals fibroblast clusters linked to immunotherapy resistance in cancer. Cancer Discov, 10（9）: 1330-1351.

Kim J W, Evans C, Weidemann A, et al. 2012. Loss of fibroblast HIF-1alpha accelerates tumorigenesis. Cancer Res, 72（13）: 3187-3195.

Kosinski C, Li V S, Chan A S, et al. 2007. Gene expression patterns of human colon tops and basal crypts and BMP antagonists as intestinal stem cell niche factors. Proc Natl Acad Sci U S A, 104（39）: 15418-15423.

Laouar Y, Sutterwala F S, Gorelik L, et al. 2005. Transforming growth factor-beta controls T helper type 1 cell development through regulation of natural killer cell interferon-gamma. Nat Immunol, 6（6）: 600-607.

Lechner A J, Driver I H, Lee J, et al. 2017. Recruited monocytes and type 2 immunity promote lung regeneration following pneumonectomy. Cell Stem Cell, 21（1）: 120-134 e127.

Lee D C, Sohn H A, Park Z Y, et al. 2015. A lactate-induced response to hypoxia. Cell, 161（3）: 595-609.

Lee J H, Jun H S. 2019. Role of myokines in regulating skeletal muscle mass and function. Front Physiol, 10: 42.

Lee J V, Carrer A, Shah S, et al. 2014. Akt-dependent metabolic reprogramming regulates tumor cell histone acetylation. Cell Metab, 20（2）: 306-319.

Li W, Xiao J, Zhou X, et al. 2015. STK4 regulates TLR pathways and protects against chronic inflammation-related hepatocellular carcinoma. J Clin Invest, 125（11）: 4239-4254.

Lu P, Weaver V M, Werb Z. 2012. The extracellular matrix: a dynamic niche in cancer progression. J Cell Biol, 196（4）: 395-406.

Ma L, Li Y, Peng J, et al. 2015. Discovery of the migrasome, an organelle mediating release of cytoplasmic contents during cell migration. Cell Res, 25（1）: 24-38.

Ma X, Bi E, Lu Y, et al. 2019. Cholesterol induces CD8（+）T cell exhaustion in the tumor microenvironment. Cell Metab, 30（1）: 143-156 e145.

Machado M V, Michelotti G A, Pereira T A, et al. 2015. Accumulation of duct cells with activated YAP parallels fibrosis progression in non-alcoholic fatty liver disease. J Hepatol, 63（4）: 962-970.

Machler P, Wyss M T, Elsayed M, et al. 2016. *In vivo* evidence for a lactate gradient from astrocytes to neurons. Cell Metab, 23（1）: 94-102.

Madsen C D, Pedersen J T, Venning F A, et al. 2015. Hypoxia and loss of PHD2 inactivate stromal fibroblasts to decrease tumour stiffness and metastasis. EMBO Rep, 16（10）: 1394-1408.

Malanchi I, Santamaria-Martinez A, Susanto E, et al. 2011. Interactions between cancer stem cells and their niche govern metastatic colonization. Nature, 481（7379）: 85-89.

Mao A S, Shin J W, Mooney D J. 2016. Effects of substrate stiffness and cell-cell contact on mesenchymal stem cell differentiation. Biomaterials, 98: 184-191.

McCarthy N, Kraiczy J, Shivdasani R A. 2020a. Cellular and molecular architecture of the intestinal stem cell niche. Nat Cell Biol, 22（9）: 1033-1041.

McCarthy N, Manieri E, Storm E E, et al. 2020b. Distinct mesenchymal cell populations generate the essential intestinal bmp signaling gradient. Cell Stem Cell, 26（3）: 391-402 e395.

McGough I J, Vecchia L, Bishop B, et al. 2020. Glypicans shield the Wnt lipid moiety to enable signalling at a distance. Nature, 585（7823）: 85-90.

Medina C B, Mehrotra P, Arandjelovic S, et al. 2020. Metabolites released from apoptotic cells act as tissue messengers. Nature, 580（7801）: 130-135.

Miller C J, Davidson L A. 2013. The interplay between cell signalling and mechanics in developmental processes. Nat Rev Genet, 14（10）: 733-744.

Moore R J, Owens D M, Stamp G, et al. 1999. Mice deficient in tumor necrosis factor-alpha are resistant to skin carcinogenesis. Nat Med, 5（7）: 828-831.

Morrison C D, Parvani J G, Schiemann W P. 2013. The relevance of the TGF-beta paradox to EMT-MET programs. Cancer Lett, 341（1）: 30-40.

Morsut L, Roybal K T, Xiong X, et al. 2016. Engineering customized cell sensing and response behaviors using synthetic Notch receptors. Cell, 164（4）: 780-791.

Motz G T, Coukos G. 2011. The parallel lives of angiogenesis and immunosuppression: cancer and other tales. Nat Rev Immunol, 11（10）: 702-711.

Nabhan A N, Brownfield D G, Harbury P B, et al. 2018. Single-cell Wnt signaling niches maintain stemness of alveolar type 2 cells. Science, 359（6380）: 1118-1123.

Nielsen S R, Quaranta V, Linford A, et al. 2016. Macrophage-secreted granulin supports pancreatic cancer metastasis by inducing liver fibrosis. Nat Cell Biol, 18（5）: 549-560.

O'Connell J T, Sugimoto H, Cooke V G, et al. 2011. VEGF-A and Tenascin-C produced by $S100A4^+$ stromal cells are important for metastatic colonization. Proc Natl Acad Sci U S A, 108（38）: 16002-16007.

Opitz C A, Litzenburger U M, Sahm F, et al. 2011. An endogenous tumour-promoting ligand of the human aryl hydrocarbon receptor. Nature, 478（7368）: 197-203.

Orimo A, Gupta P B, Sgroi D C, et al. 2005. Stromal fibroblasts present in invasive human breast carcinomas promote tumor growth and angiogenesis through elevated SDF-1/CXCL12 secretion. Cell, 121（3）: 335-348.

Ouyang W, Beckett O, Ma Q, et al. 2010. Transforming growth factor-beta signaling curbs thymic negative selection promoting regulatory T cell development. Immunity, 32（5）: 642-653.

Palmieri E M, Menga A, Martin-Perez R, et al. 2017. Pharmacologic or genetic targeting of glutamine synthetase skews macrophages toward an M1-like phenotype and inhibits tumor metastasis. Cell Rep, 20（7）: 1654-1666.

Panciera T, Citron A, Di Biagio D, et al. 2020. Reprogramming normal cells into tumour precursors requires ECM stiffness and oncogene-mediated changes of cell mechanical properties. Nat Mater, 19（7）: 797-806.

Pandey A, Harvey B M, Lopez M F, et al. 2019. Glycosylation of specific Notch EGF repeats by O-Fut1 and fringe regulates Notch signaling in drosophila. Cell Rep, 29（7）: 2054-2066 e2056.

Pasqual G, Chudnovskiy A, Tas J M J, et al. 2018. Monitoring T cell-dendritic cell interactions in vivo by intercellular enzymatic labelling. Nature, 553（7689）: 496-500.

Petrie R J, Doyle A D, Yamada K M. 2009. Random versus directionally persistent cell migration. Nat Rev Mol Cell Biol, 10（8）: 538-549.

Polyak K, Kalluri R. 2010. The role of the microenvironment in mammary gland development and cancer. Cold Spring Harb Perspect Biol, 2（11）: a003244.

Powell D W, Pinchuk I V, Saada J I, et al. 2011. Mesenchymal cells of the intestinal lamina propria. Annu Rev Physiol, 73: 213-237.

Rafii S, Cao Z, Lis R, et al. 2015. Platelet-derived SDF-1 primes the pulmonary capillary vascular niche to drive lung alveolar regeneration. Nat Cell Biol, 17（2）: 123-136.

Renz B W, Tanaka T, Sunagawa M, et al. 2018. Cholinergic signaling via muscarinic receptors directly and indirectly suppresses pancreatic tumorigenesis and cancer stemness. Cancer Discov, 8（11）: 1458-1473.

Roland C L, Arumugam T, Deng D, et al. 2014. Cell surface lactate receptor GPR81 is crucial for cancer cell survival. Cancer Res, 74（18）: 5301-5310.

Ruffell B, Coussens L M. 2015. Macrophages and therapeutic resistance in cancer. Cancer Cell,

27（4）: 462-472.

Rutkowski J M, Swartz M A. 2007. A driving force for change: interstitial flow as a morphoregulator. Trends Cell Biol, 17（1）: 44-50.

Sanjabi S, Mosaheb M M, Flavell R A. 2009. Opposing effects of TGF-beta and IL-15 cytokines control the number of short-lived effector $CD8^+$ T cells. Immunity, 31（1）: 131-144.

Seldin M M, Lei X, Tan S Y, et al. 2013. Skeletal muscle-derived myonectin activates the mammalian target of rapamycin（mTOR）pathway to suppress autophagy in liver. J Biol Chem, 288（50）: 36073-36082.

Seldin M M, Peterson J M, Byerly M S, et al. 2012. Myonectin（CTRP15）, a novel myokine that links skeletal muscle to systemic lipid homeostasis. J Biol Chem, 287（15）: 11968-11980.

Shi X, Bi Y, Yang W, et al. 2013. $Ca2^+$ regulates T-cell receptor activation by modulating the charge property of lipids. Nature, 493（7430）: 111-115.

Snyder S H. 1992. Nitric oxide: first in a new class of neurotransmitters. Science, 257（5069）: 494-496.

Shyh-Chang N, Locasale J W, Lyssiotis C A, et al. 2013. Influence of threonine metabolism on S-adenosylmethionine and histone methylation. Science, 339（6116）: 222-226.

Sonveaux P, Vegran F, Schroeder T, et al. 2008. Targeting lactate-fueled respiration selectively kills hypoxic tumor cells in mice. J Clin Invest, 118（12）: 3930-3942.

Sorokin L. 2010. The impact of the extracellular matrix on inflammation. Nat Rev Immunol, 10（10）: 712-723.

Steinhauer J, Treisman J E. 2009. Lipid-modified morphogens: functions of fats. Curr Opin Genet Dev, 19（4）: 308-314.

Sun L, Li C W, Chung E M, et al. 2020. Targeting glycosylated PD-1 induces potent antitumor immunity. Cancer Res, 80（11）: 2298-2310.

Swamy M, Beck-Garcia K, Beck-Garcia E, et al. 2016. A cholesterol-based allostery model of T cell receptor phosphorylation. Immunity, 44（5）: 1091-1101.

Tacke F, Zimmermann H W. 2014. Macrophage heterogeneity in liver injury and fibrosis. J Hepatol, 60（5）: 1090-1096.

Tang D, Kang R, Berghe T V, et al. 2019. The molecular machinery of regulated cell death. Cell Res, 29（5）: 347-364.

Teufel A, Malik N, Mukhopadhyay M, et al. 2002. Frcp1 and Frcp2, two novel fibronectin type Ⅲ repeat containing genes. Gene, 297（1-2）: 79-83.

Tlsty T D, Coussens L M. 2006. Tumor stroma and regulation of cancer development. Annu Rev Pathol, 1: 119-150.

Todaro M, Gaggianesi M, Catalano V, et al. 2014. CD44v6 is a marker of constitutive and reprogrammed cancer stem cells driving colon cancer metastasis. Cell Stem Cell, 14（3）: 342-356.

Tomasek J J, Gabbiani G, Hinz B, et al. 2002. Myofibroblasts and mechano-regulation of connective tissue remodelling. Nat Rev Mol Cell Biol, 3（5）: 349-363.

Tsai R Y, McKay R D. 2000. Cell contact regulates fate choice by cortical stem cells. J Neurosci, 20（10）: 3725-3735.

Uysal N, Yuksel O, Kizildag S, et al. 2018. Regular aerobic exercise correlates with reduced anxiety and incresed levels of irisin in brain and white adipose tissue. Neurosci Lett, 676: 92-97.

Venkatesh H S, Johung T B, Caretti V, et al. 2015. Neuronal activity promotes glioma growth through neuroligin-3 secretion. Cell, 161（4）: 803-816.

Venkatesh H S, Morishita W, Geraghty A C, et al. 2019. Electrical and synaptic integration of glioma into neural circuits. Nature, 573（7775）: 539-545.

Vermeulen L, de Sousa E M F, van der Heijden M, et al. 2010. Wnt activity defines colon cancer stem cells and is regulated by the microenvironment. Nat Cell Biol, 12（5）: 468-476.

Vitale I, Manic G, Coussens L M, et al. 2019. Macrophages and metabolism in the tumor microenvironment. Cell Metab, 30（1）: 36-50.

Wang X, Liu R, Qu X, et al. 2019a. alpha-Ketoglutarate-activated NF-kappaB signaling promotes compensatory glucose uptake and brain tumor development. Mol Cell, 76（1）: 148-162 e147.

Wang X, Liu R, Zhu W, et al. 2019b. UDP-glucose accelerates SNAI1 mRNA decay and impairs lung cancer metastasis. Nature, 571（7763）: 127-131.

Watt M J, Miotto P M, de Nardo W, et al. 2019. The liver as an endocrine organ-linking NAFLD and insulin resistance. Endocr Rev, 40（5）: 1367-1393.

Xiong X, Kuang H, Ansari S, et al. 2019. Landscape of Intercellular crosstalk in healthy and NASH liver revealed by single-cell secretome gene analysis. Mol Cell, 75（3）: 644-660 e645.

Xu C, Gagnon E, Call M E, et al. 2008. Regulation of T cell receptor activation by dynamic membrane binding of the CD3epsilon cytoplasmic tyrosine-based motif. Cell, 135（4）: 702-713.

Xu W, Yang H, Liu Y, et al. 2011. Oncometabolite 2-hydroxyglutarate is a competitive inhibitor of alpha-ketoglutarate-dependent dioxygenases. Cancer Cell, 19（1）: 17-30.

Yang F, Tuxhorn J A, Ressler S J, et al. 2005. Stromal expression of connective tissue growth factor promotes angiogenesis and prostate cancer tumorigenesis. Cancer Res, 65（19）: 8887-8895.

Yang L V. 2017. Tumor microenvironment and metabolism. Int J Mol Sci, 18（12）: 2729.

Yang W, Pan W L, Chen S K, et al. 2017. Dynamic regulation of CD28 conformation and signaling by charged lipids and ions. Nature Structural & Molecular Biology, 24（12）: 1081-1092.

Yin J, Yang L, Xie Y, et al. 2018. Dkk3 dependent transcriptional regulation controls age related skeletal muscle atrophy. Nat Commun, 9（1）: 1752.

Zacharias W J, Frank D B, Zepp J A, et al. 2018. Regeneration of the lung alveolus by an evolutionarily conserved epithelial progenitor. Nature, 555（7695）: 251-255.

Zepp J A, Morrisey E E. 2019. Cellular crosstalk in the development and regeneration of the respiratory system. Nat Rev Mol Cell Biol, 20（9）: 551-566.

Zepp J A, Zacharias W J, Frank D B, et al. 2017. Distinct mesenchymal lineages and niches promote epithelial self-renewal and myofibrogenesis in the lung. Cell, 170（6）: 1134-1148 e1110.

Zhang M, Yang H, Wan L, et al. 2020a. Single-cell transcriptomic architecture and intercellular crosstalk of human intrahepatic cholangiocarcinoma. J Hepatol, 73（5）: 1118-1130.

Zhang T, Yin C, Boyd D F, et al. 2020b. Influenza virus Z-RNAs induce ZBP1-mediated necroptosis. Cell, 180（6）: 1115-1129 e1113.

Zhang Z, Zhang Y, Xia S, et al. 2020c. Gasdermin E suppresses tumour growth by activating anti-tumour immunity. Nature, 579（7799）: 415-420.

Zhou S, Zhang W, Cai G, et al. 2020a. Myofiber necroptosis promotes muscle stem cell proliferation via releasing tenascin-C during regeneration. Cell Res, 30（12）: 1063-1077.

Zhou Z, He H, Wang K, et al. 2020b. Granzyme A from cytotoxic lymphocytes cleaves GSDMB to trigger pyroptosis in target cells. Science, 368（6494）: eaaz7548.

Zindel J, Kubes P. 2020. DAMPs, PAMPs, and LAMPs in immunity and sterile inflammation. Annu Rev Pathol, 15: 493-518.

Zitvogel L, Regnault A, Lozier A, et al. 1998. Eradication of established murine tumors using a novel cell-free vaccine: dendritic cell-derived exosomes. Nat Med, 4（5）: 594-600.

第十章

分子生物网络与“数字化”细胞

第一节 概　　述

生命体是由细胞构成的，而细胞又是由一系列分子构成的，包括水和离子等小分子，以及蛋白质、核酸（RNA和DNA）、糖类等大分子。这些分子之间会存在多种多样的相互作用。例如，根据分子生物学的中心法则，DNA转录得到信使RNA（mRNA），mRNA编码蛋白质，蛋白质又反过来调控DNA的表达。如果把生物分子当作点，这些分子之间的相互作用当成线，那么就可以构建出一个生物分子网络（Chen，2010）。研究生物分子网络或网络生物学（network biology）对于描述细胞内分子之间的相互作用，进而在系统水平上阐释细胞的功能至关重要（Chen et al.，2009）。

在基因水平上，转录因子（transcription factor）通过与DNA进行结合与解离来控制基因的表达与否，即是否合成mRNA。由于这些转录因子本身也是基因表达的产物，因此这种调控实际上是基因和基因之间相互调控，这样形成的网络也称作基因调控网络（gene regulatory network）。在蛋白质水平，蛋白质和蛋白质之间的相互作用是许多细胞功能的基础，如和染色体转录激活功能相关的组蛋白甲基化就是由赖氨酸甲基转移酶和组蛋白相互作用形成的，由蛋白质－蛋白质相互作用构成的网络称作蛋白质相互作用网络（protein interaction network）。此外，细胞是在不断代谢的，代谢产物在酶的催化下发生生化反应形成新的代谢产物，而代谢网络（metabolic network）则是描述这类反应的网络。不仅如此，细胞为了生存，需要对外界环境做出响应，当外界环境发生变化时，这些变化信号会从细胞外传到细胞内，然后诱发一系列不同层次上的相互作用，包括前面讲到的蛋白－蛋白相互作用和 DNA- 蛋白相互作用等，来调控基因的表达，以此来适应外界变化。这个过程中所涉及的相互作用所构成的网络称为信号转导网络（signaling transduction network）。除了上述 4 种经典的分子生物网络，近年来分子生物网络的研究领域还涌现出了一批非常具有前景的研究方向，比如表观遗传调控网络（epigenetic regulation network），所谓的表观遗传指的是 DNA 序列在不发生改变的情况下，基因功能却发生了可遗传的改变，如组蛋白修饰等。

总而言之，细胞是由一系列生物分子和小分子构成的，细胞的各种功能是通过这些分子之间的相互作用，也就是生物分子网络实现的。因此，研究分子生物网络的结构和动力学是认识细胞功能的基础。特别是，一些重大疾病（如癌症和糖尿病）的发病机制只有在生物分子网络的水平上才可能被解释，这是因为这类疾病往往并不是个别基因突变所导致的，而是由整个网络水平上的变化引起的（Chen et al.，2009）。

借助于计算机仿真计算和可视化能力的提高，以及数学、信号与信息处理科学领域的发展，20 世纪 90 年代中期以后多种“数字化”细胞研究技术相继诞生（Tomita et al.，1999）。相关的技术，包括在计算机上模拟真实细胞的结构、物质组成、生命活动的动力学行为和生命现象的电子细胞（也称虚拟细胞或细胞模型）；也包括 DNA 折纸术（DNA origami）（Nangreave et al.，2010），即通过将一条长的 DNA 单链与一系列经过设计的短的 DNA 片段进行碱基互补，从而组装成纳米级的不同结构或者框架；同时还包括人工合成细胞（synthetic cell）（Kendall and Powell，2018），即利用蛋白质、脂质、核酸等细胞所需要的组分重新构建出一个活细胞；甚至包括 DNA 存储技术（Panda et al.，2018），即用人工合成的脱氧核糖核酸（DNA）存储文本文档、图片和声音文件等数据，随后完整读取的技术。这些相应的新兴技术具有划时代的意义，可以在生命的运行机制、DNA 纳米材料的生成、人体病变细胞的修复和数据生物存储等方面得到重大应用。

第二节　关键科学问题

分子生物网络或网络生物学的研究问题主要可以分为三类。第一类是如何从已有的知识和数据中构建出网络，也就是网络的重构问题；第二类是如何分析构建出的网络，也就是网络的分析问题；第三类是如何应用网络来解决生物学问题，也就是网络的应用问题。下面对这三类问题分别展开讨论。

一、网络的重构

2001 年，人类基因组计划（Human Genome Project，HGP）的完成标志着我们进入了后基因组时代（Venter et al.，2001），自此以后，科学家们发展了各种各样的高通量方法获得了大量的生物学数据，这些数据使得分子生物网络的构建成为可能。从生物分子水平来看，这类方法可以从整个基因组水平上对特定的生物分子进行测量，从而产生大量的组学数据，包括基因组、转录组、蛋白质组、代谢组等，这类数据是构建分子生物网络的基础。从生物分子间相互作用的水平上看，高通量方法也可以获得整个蛋白质 -DNA、蛋白质 - 蛋白质、蛋白质 - 配体之间的相互作用数据，这类数据可以构建网络的骨架结构。最后，从功能的水平上看，因为生物分子间存在多种相互作用类型，如基因调控和信号传导等，仅仅只知道是否存在相互作用并不意味着知道了相互作用的功能，这类数据则是分子生物网络的“血肉”。正是得益于这些大数据，我们才有机会从更宏观的角度，也就是网络的角度来认识细胞及其功能。但这也带来了许多新的挑战，因为当前我们所面对的生物数据是十分巨大的，而且这些数据不仅噪声大，假阳性率高，而且不同类别的数据差异非常大，因此如何对不同类别的数据进行整合以及如何高效地处理和分析这些数据是我们分子生物网络中的一个重要问题。

二、网络的分析

当分子网络重构好以后，需要对网络的结构、功能和动力学进行分析，也就是网络分析的问题。我们都知道分子生物网络非常复杂，那么在这种复杂性之中是否存在共性呢？答案是存在的，目前人们已经发现不同的网络中存在一些相同的结构单元，也就是模体（motif），这些模体是实现网络调控功能的关键因素。例如，前馈模体或反馈模体可以加速或延迟目标基因的表达。不仅如此，模体可能对网络整体的稳定性或自适应性都有重要影响，如带缓冲节点的负反馈环路。除了静态分析，还可以对网络进行动态分析，例如何控制网络的动态变化，目前已经发展出一些基于节点或者模体的网络控制方法，这对我们认识生物系统的工作机制有重要意义。然而目前分析分子生物网络的方法还主要是线性和静态体系，存在不少缺陷，这也需要我们去发展更有效的基于非线性和动力学数学模型及计算工具来进行分析（Chen C et al.，2020；Chen P et al.，2020; Huang et al.，2020；Leng et al.，2020；Liu et al.，2019；Ma et al.，2018；Yang et al.，2018；Zhang et al.，2015; Zhao et al.，2016）。

三、网络的应用

最后一个也是最具有实际意义的问题是，如何应用分子生物网络来解决实际的生物学及医药学问题。目前已经有一系列的代表性工作。在寻找疾病相关基因方面，传统的方法是去找差异表达基因，但这种方法不仅鲁棒性差，而且因为找出的差异表达基因往往较多，可解释性也欠佳，而基于分子生物网络的方法找出的网络标志物能够更准确地找到疾病相关基因。从诊断和量化的角度，与传统的分子生物标志物（molecular biomarker）不同，网络生物标志物（network biomarker）可稳定地诊断或表征疾病状态或生物系统状态，真正实现由网络诊断疾病状态（Venter et al.，2001）。从预测的角度，动态网络标志物（dynamic network biomarker）可表征生物系统的临界状态，从而可实现疾病状态的预警，即由动态网络标志物来预测疾病（Venter et al.，2001）。在药物设计方面，传统的药物设计方法通常是一个药物针对一个基因，也就是所谓的“一药一靶”，但实际中一种疾病往往和多个基因相关，所以这种方法所设计出的药物治疗效果并不理想；而基于分子网络的方法能够同时考虑到多个靶点，因此设计出的多靶点药物相较于传统的药物具有更好的疗效，这也是网络药理学（network pharmacology）的基本理念。不过即使找到了多个靶点，要设计出有效的药物还是非常困难的，目前通常采用药物组合的方法来加速药物研发。考虑到已有药物的组合是个天文数字，实验上不可能去对所有的组合进行筛选，所以如何利用分子网络来开发计算方法并用于筛选有效的组合药物也是分子生物网络研究的重要课题。

第三节　重要研究方向的发展现状与趋势分析

一、基因调控网络

基因虽然是遗传信息载体，但基因本身并不直接执行各种生物学功能，它需要通过转录过程将遗传信息传递到mRNA，再通过翻译过程进一步用mRNA上的遗传信息编码蛋白质，最后由蛋白质来执行各种功能，这个过程也称为基因的表达。由于细胞功能是动态变化的，基因的表达也必须随之改变，因此细胞有一套机制来调控基因的表达，而参与这套机制的重要一员就是转录因子。转录因子本身就是蛋白质，也就是基因表达的产物。因此可以说，基因和基因之间是可以互相调控的，这样就形成了一个基因调控网络（Hasty et al.，2001）。需要说明的是，这里所指的基因调控网络是包含基因产物的，虽然这样的网络通常应该称为转录调控网络（transcription regulatory network），但考虑到这两种网络是紧密关联的，所以后文中对这两种网络不加区分。

基因调控网络研究的核心问题就是如何基于生物数据对调控网络进行重

构。目前的重构所使用的输入数据有两类，分别是表达数据和原始数据。因为 mRNA 是基因的直接表达产物，它的数目能够直观反映基因的表达水平，而当一个转录因子同时调控多个基因时，这些基因通常是共表达的，所以我们可以用表达数据来重构基因调控网络。基因芯片（Duggan et al.，1999）和 RNA-seq（Marioni et al.，2008）是两种比较成熟的获取基因表达数据的技术，它们都能够在转录水平上对整个基因组的表达量进行测量，测得的结果通常称作表达谱。然而，这类技术所测得的表达谱是样品中所有细胞的平均表达水平，这样就丢失掉了细胞间的异质性信息，为了解决这一问题，Tang 等（2009）开发了单细胞测序技术 scRNA-seq，该技术可以获得单个细胞基因的表达水平，为我们研究细胞之间的差异乃至网络水平的差异提供了可能。目前，用基于 scRNA-seq 得到的表达数据来重构基因调控网络是分子生物网络研究的热点（Chen et al.，2019）。另一类可用作重构输入的数据为原始数据，它描述的是转录因子和目标基因之间的关系，之所以可以用该数据作为重构网络的输入数据是因为基因在转录水平上主要是由基因和转录因子的相互作用调控的。目前获取这类数据最常用的技术为 ChIP-chip（Zheng et al.，2007）和 ChIP-seq（Johnson et al.，2007），这两项技术分别是染色质免疫沉淀（chromatin immunoprecipitation，ChIP）技术与微阵列技术和第二代高通量测序技术的整合，它们的本质都是识别结合了特定蛋白质的 DNA 序列，不过由于当前测序技术成本低且灵活性强，ChIP-seq 技术已经逐步取代了 ChIP-chip 技术。

在获取输入数据之后，我们便可以开始重构基因调控网络，重构有以下基本步骤：对输入数据预处理、计算推理和优化及验证。在第一步中，输入数据的误差通常来源于两种，即系统误差和随机误差，系统误差可以通过数据归一化近乎消除，而随机误差则可以通过重复测量来降低。第二步是重构中最核心的一步，即用一个特定的模型将处理后的数据映射成网络结构，这也是所谓的网络推理。目前主流的推理模型有基于信息论的模型、布林网络模型、微分方程模型、贝叶斯预测模型和神经网络模型，这些模型都各有优缺点。例如，Zheng 等（2016）基于条件互信息理论开发了 CMIP 软件用来重构调控网络，Orlando 等（2008）构建了一个布林网络模型来描述酵母的细胞周期，这两类模型比较简单，所以可以用来重构大规模的调控网络，不过基于信息论的模型只适用于静态数据，而布林网络模型只能处理离散值。微分方程模型可以处理连续且动态的数据，如 Matsumoto 等（2017）用基于微分方程模型的算法 SCODE 实现了对细胞分裂过程中的调控网络的重构。不过微分方程模型对噪声比较敏感，而贝叶斯预测模型则能很好地处理噪声，如 Chekouo 等（2015）基于贝叶斯预测模型成功从噪声数据中识别了肾癌的标志基因。但是，贝叶斯预测模型中不允许出现反馈环，而基于神经网络模型不仅可以处理噪声数据，还能重构出任何类型的网络，如 Rubiolo 等（2018）设计了名为 ELM 的有监督网络模型，重构效果要显著优于其他常用模型。不过神经网络模型也有自己的局限性，比如需要进行大量的调参，只适用于小网络的重构等。在第三步中，我们还需要对重构的网络进行优化和验证。根据已有的知识，分子生物网络应该是稀疏且无尺度的（sacle-free），同时还应该满足高层次性，这些性质都可以作为约束来对重构的网络进行优化。完成优化后的网络还需要进行验证，也就是与参考网络进行比对（Marbach et al., 2012）。

总之，目前已经有了许多重构基因调控网络的方法，其中基于 scRNA-seq 数据的网络重构是一个大的趋势。但是由于 scRNA-seq 数据存在较多的缺失信息以及噪声大的缺陷，因此如何对 scRNA-seq 数据进行预处理和分析是当前基因网络重构研究中的热点问题之一。我国已经有部分课题组对该问题进行了系统深入的研究，并取得了一系列突出的成果。例如，Chen 课题组（Dai et al., 2019）基于 scRNA-seq 数据发展了一套构建细胞特异性网络的方法，也就是为一个细胞构建一个网络，首次从单个细胞的水平上实现了基因调控网络的研究，而且这种方法得到的网络同样可以用现有的方法进行分析，如寻找差异基因及其关联。此外，考虑到许多疾病是由不同基因表达关联的变化引起的，也就是基因之间相互作用的改变，而不是基因本身表达量的变化，因此，Chen 课题组还提出了边网络的概念（Yu et al., 2014），相较于传统的关注节点的网络，边网络能够更有效地识别疾病相关的基因。不仅如此，考虑到实际的生物分子网络是一个动态系统，仅仅从静态的角度是无法实现生物学功能的。因此，Chen 课题组进一步提出了动态网络标志物（dynamic network biomarker, DNB）的概念（Chen et al., 2012），所谓的 DNB 指的是与系统状态变化紧密相关的一些基因，这些基因在临界处具有高涨落、高关联的动力学特性。与传统的静态分子标志物和网络标志物相比，DNB 能够更有效地对疾病进行早期预警，为预防疾病提供了有力的理论工具。

二、蛋白质相互作用网络

蛋白质相互作用的类型多种多样。按相互作用的特异性来分，可以分为特异性相互作用和非特异性相互作用。按相互作用的时间长短来分，可以分为持续相互作用和短暂相互作用；而在短暂相互作用里又可细分为强相互作用和弱相互作用。甚至，还可以根据相互作用的蛋白质是否一样将相互作用分为同源多聚体和异质多聚体相互作用。之所以要介绍相互作用的类型是因为不同类型的相互作用往往对应不同的功能。例如，分子伴侣 GroEL 是由多个蛋白质构成（Xu et al.，1997），这些蛋白质的相互作用就是持续相互作用，这样可以让 GroEL 有稳定的结构；而传递信号的相互作用往往是短暂且动态的，如 Ras 和 Raf 两个蛋白质的相互作用（Plowman and Hancock，2005）。下面介绍蛋白质相互作用网络研究的几个热点问题。

蛋白质相互作用网络研究的首要问题就是如何构建网络，更具体地说就是判断蛋白质与蛋白质之间是否有相互作用，目前主要有两类方法来处理这一问题。第一类方法是直接采用实验手段测量，经典的方法包括免疫共沉淀方法和酵母双杂交实验（yeast two-hybrid，Y2H），这两种方法都可以在不阻断蛋白质与其他蛋白质相互作用的前提下对其相互作用进行高通量的检测。除了这两种经典的实验方法，近年来发展的亲和纯化质谱方法（AP-MS）和蛋白质芯片（protein chip）技术也已经被广泛使用。考虑到当前的实验方法众多，Braun 等（2009）提出了一种置信度分数对这些方法进行了评估，发现这些方法各有优劣，可以对这些互补的方法进行整合。Thakur 和 Movileanu（2019）基于生物纳米孔技术实现了单分子精度的蛋白质相互作用实时测量，进一步丰富了测量的实验手段。但是目前的实验方法还存在不足，不仅需要消耗大量的人力物力，得到的结果也可能有较高的假阳性率，因此科学家开发了大量基于理论和计算的方法，也就是我们将要讨论的第二类方法。根据所使用的信息不同，这类方法又可细分为若干个类别。对于仅使用序列信息的方法，通常的做法（Shen et al.，2007）是对氨基酸进行分类，然后进一步将分类后的序列信息进行编码，这样就可以用传统的机器学习方法，如支持向量机（support vector machine，SVM），来对蛋白质间的相互作用进行分类。不过基于序列信息的算法能够达到的精度比较有限，因此序列的共进化信息也被用来预测两个蛋白质间的相互作用（Weigt et al.，2009），这类方法的基本思想是如果两个蛋白质通过相互作用来执行某项功能，那么在进化的过程中，为了保留该功能，参与这个相互作用的那些残基应该是一起突变的，也就是所谓的共进化。事实上，在加入共进化信息后，蛋白质相互作用预测的精度获得了显著提升。当然，三维结构信息也非常重要，例如，研究人员提出了基于蛋白质三维结构信息的蛋白质互作预测（predicting protein-protein interaction，PrePPI）法，使用该方法在酵母和人中分别预测了超过 30 000 个和 300 000 个相互作用，其预测精度远超其他预测方法，达到了与高通量实验相当的精度，研究者将精度的大幅提高归功于同源建模方法和蛋白质间的空间关系（Zhang et al., 2012）。除了前面一些基于序列和结构信息来进行预测的方法，还有一类同源映射的方法，它直接假定不同物种的蛋白质相互作用是保守的，所以一个物种中已知的相互作用可以迁移到另一个物种中来进行预测，其中的代表方法为 BIPS（Garcia-Garcia et al.，2012），不过这类方法对短暂的相互作用的预测精度要远低于持续的相互作用。需要特别说明的是，随着近年来蛋白质相互作用数据的指数级增加和计算能力的大幅提高，深度学习方法也常被用来预测蛋白质相互作用并取得了非常好的结果，如 Zhang 等（2019）采用 EDNN 训练的模型在酿酒酵母的蛋白质相互作用预测中达到了 95% 以上的精度。

在构建好蛋白质－蛋白质相互作用网络后，我们便可以讨论网络的功能及应用，特别是和疾病及药物相关的方面。例如，在致病基因预测问题上，Wu 等（2008）通过整合蛋白质相互作用网络和表型网络提出了预测疾病基因的新方法 CIPHER，该方法在分子相互作用机制不明确的情况下依旧有较好的预测效果。Vanunu 等（2010）提出的 PRINCE 算法在此基础上进一步提高了致病基因的预测精度。但是由于蛋白质相互作用的假阳性率较高，导致预测精度难以更进一步提高。为了解决这一问题，Wang 等（2012）在传统的蛋白质相互作用网络中加入了蛋白质结构信息，即在蛋白质相互作用的界面，搭建出了三维的蛋白质相互作用网络，基于该网络，作者对大量疾病的致病基因进行了预测，发现和疾病相关的基因大都集中在蛋白质相互作用界面，不仅如此，该方法还找出了182种疾病中292个新的致病基因。在药物设计领域，蛋白质–蛋白质相互作用的研究也发挥着重要的指导作用，如Baker小组（Tinberg et al.，2013）基于蛋白质相互作用网络设计了17个地高辛（digoxigenin）结合蛋白，其中一个名为DIG10的蛋白与样地黄毒苷的实际结合位点与计算结果吻合得较好。Xie等（2009）在全基因组的水平上研究了蛋白质–配体间的结合特性，并据此发展了一套确定蛋白质–药物相互作用的方法，最后成功将该方法应用到了胆固醇酯转运蛋白（cholesteryl ester transfer protein，CETP）抑制剂上，解释了其药物不良反应的分子机制。

综上所述，蛋白质–蛋白质相互作用网络的研究目前有两个比较热门的方向。一个是蛋白质–蛋白质相互作用的预测，特别是在数据量和算力都有了大幅提高的今天，我们如何基于深度学习方法进一步提高相互作用的预测精度；另一个问题是如何基于蛋白质–蛋白质相互作用网络去解决生物学问题，特别是在药物设计方向，如何基于蛋白质–蛋白质相互作用网络去设计有效的多靶点药物是涉及人类健康的重大问题。目前我国学者在蛋白质–蛋白质相互作用的预测上取得了一些不错的成果。例如，Du等（2017）基于深度神经网络提出了DeepPPI算法，让网络去有效地学习蛋白质的表示，并最终取得了92.5%的预测精度。不过在药物设计的方向，国内虽然也有一些科研成果发表，但整体水平与国际相比还是存在着较大的差距。

三、代谢网络

代谢网络指的是由生物体内所有的代谢反应或途径，以及相关的调节过程所构成的复杂网络系统（Caetano-Anolles et al.，2007；Goldford and Segrè，2018）。生物体通过代谢途径交叉点上的关键中间产物，来实现各代谢途径的联系与转化。

代谢网络作为一个复杂的系统，具有如下五个方面的特性（Jeong et al.，2000）。第一，整体性。组成和结构具有整体性，网络中的节点为代谢物，各节点间的连线为酶促反应、调节反应和物质转运反应，它们作为一个整体共同构成代谢网络。代谢活动和调节也具有整体性，食物中的混合营养成分经消化吸收到体内后的代谢是同时进行的，且各种物质代谢之间和各条代谢途径之间彼此互相联系，或相互转变，或相互依存，或相互制约，从而构成统一的整体。第二，复杂性。代谢网络由若干不同的代谢途径所构成，包括直线途径、分支途径、循环途径、反馈途径、交叉途径等形式，从而使代谢网络的结构具有复杂性。物质代谢与能量代谢、合成代谢与分解代谢，代谢反应与调节反应相耦联也使代谢网络表现出不同于单一代谢途径的新的功能属性。第三，可调性。由神经、体液、细胞内酶、细胞间隔等组成的一套调控系统，对代谢网络进行精细、完善而又复杂的调控。第四，特异性。组织细胞和发育阶段存在特异性。第五，无尺度性。在代谢网络中，个别代谢物比其他代谢物参与了更多的化学反应，它们在网络中处于更为重要的地位，如ATP、NAD+、NADPH、乙酰辅酶A、6-磷酸葡萄糖等。

代谢网络可以在不同的水平下进行讨论（Guimera and Amaral，2005；Ravasz et al.，2002），包括细胞水平、器官水平和整体水平。细胞水平的代谢网络由细胞内所有的代谢物及其相关的代谢反应和代谢途径、调节反应和物质转运反应所构成。通过细胞水平的代谢网络，糖、脂肪、氨基酸和核苷酸四大类物质之间可以在一定程度上实现相互转变。器官水平的代谢网络指利用体液（主要是血液）的物质转运过程相互耦联，并通过体液因素（主要是激素）进行调节，在不同组织器官之间形成复杂的代谢网络。它的作用主要是维持生物体内葡萄糖代谢的平衡，以便为脑和红细胞提供稳定的葡萄糖供应，同时也使能源物质在充分时能够被储存，而在短缺或需求过多时能够被有效地动员或补充。整体水平的代谢网络指由神经系统、内分泌系统及全身各器官、组织和细胞参与构成的物质代谢网络，对于肥胖或不同的应激状态，整体水平的代谢网络会做出一系列相应的反应。

分析和理解代谢网络的一个重要途径是代谢组学的研究（Dettmer et al.，2007；Fiehn，2002；Singh，2020）。代谢组学是一门通过定量检测生物体在病理生理刺激或遗传变异情况下，动态的、多参数的代谢应答改变，获得生物体特定的代谢模式，从而实现对基因表达、物质代谢及代谢调节过程进行整体分析和理解的学科。它着重研究生物整体、器官或者组织的内源性代谢物质的代谢途径及其所受内、外在因素的影响及随时间变化的规律，并揭示代谢物与生理病理变化之间的关系。通过代谢组学分析获得的代谢物谱，更能迅速并准确地反映机体生命活动的变化，是判断细胞、组织器官或生物体的基因表达调控、物质代谢调节及病理生理效应的有效指标之一。

代谢组学的分析流程包括样品制备、代谢物成分分析鉴定和数据分析与解释（Chao et al.，2008；Field and Sansone，2006）。因为代谢物的种类繁多，而目前可用的成分检测和数据分析方法又多样，所以根据研究对象的不同，采用的样品制备、分离鉴定手段及数据分析方法也各不相同。色谱 / 质谱联用与核磁共振为当前代谢组学研究中分析鉴定代谢物成分的两大核心技术（Dettmer et al.，2007；Dunn and Ellis，2005）。色谱 / 质谱联用包括气相色谱 / 质谱联用（GC/MS）和液相色谱 / 质谱联用（LC/MS）。首先，利用色谱仪对样品进行分离；接着，利用质谱仪对气相色谱中的每一个峰以 1s 的间隔进行扫描，获得每个峰的质谱图。但是，质谱并不能检测出所有的代谢产物。这并不是因为质谱的灵敏度不够，而是由于质谱只能检测离子化的物质。对于那些在质谱仪中不能被离子化的代谢产物，采用核磁共振的方法，可以弥补质谱分析的不足（Blaise et al.，2009）。如今，这两种方法在代谢组学研究中已经普遍使用。研究人员通常将这两种方法结合使用以获得大量数据用于分析。

在代谢组学应用中，数据处理一般包括预处理、模式识别、模型验证及变量筛选等步骤（Lazar et al.，2015）。在得到原始图谱后，首先需要对数据进行预处理，保留与分类有关的大部分信息，消除多余的干扰因素的影响，使得数据结构标准化（Zhu et al.，2010）。数据预处理过程包括归一法、数据转化、中心化、标准化等步骤，是后续单维和多维统计分析、生物标志物筛选的重要前提。单维统计用于评价单个变量和变化，主要有参数检验和非参数检验两种。多维分析能够同时考虑所有变量，并寻找变量内部的关联，主要的方法分为非监督性学习方法和监督性学习方法。非监督性学习方法用来探索完全未知的数据特征，对原始数据信息依据样本特性进行归类，并采用相应的可视化技术直观地表达出来。常用的方法有：主成分分析（PCA）（Moore et al.，1981）、非线性映射（NLM）（Sammon，1969）、聚类分析（CA）（Kaufman and Rousseeuw，1990）等。监督性学习方法是在已有的可供学习的训练样本的基础上建立信息组，并利用所建立的组对未知数据进行辨识、归类和预测。常用的方法有：线性判别分析（Guo，2007）、偏最小二乘法－显著性分析（Geladi and Kowalski，1985）、人工神经元网络（Jain et al.，2015）等。

代谢网络在整个医药研究领域具有重要的应用和发展前景（Mardinoglu and Nielsen，2012）。第一，在代谢途径描述中的应用。用代谢组学手段可以方便地找到调节位点（Kümmel et al.，2006），因为在该处，底物和关键酶的浓度将会发生相反的变化。第二，在疾病机制研究中的应用。代谢组学在疾病机制中的研究主要应用于肿瘤、遗传病和一些常见病。例如，有研究者应用核磁共振（nuclear magnetic resonance，NMR）研究高血压大鼠与正常大鼠尿液的代谢谱差异，结果发现牛磺酸、肌酸及一些未鉴定的代谢物存在明显的差异。第三，在疾病诊断中的应用。机体的病理生理变化引起代谢产物发生相应的改变，通过对某些代谢产物进行分析，并与正常人的代谢产物相互比较，有助于发现疾病的生物标志物，达到辅助临床诊断的目的。第四，在药理学研究中的应用。代谢组在药理学方面的研究集中于毒理学研究。第五，在基因功能研究中的应用。有时，利用代谢组学方法检测代谢物的变化可以判断基因表达水平的变化，从而推断基因的功能及其对代谢物的影响。

四、信号转导网络

在生物学中，信号转导是指一个细胞将信号或刺激转送到另一细胞的过程（Kim et al.，2010b; Nakanishi et al.，2009）。细胞信号转导是指细胞通过细胞膜（或胞内）感受外界刺激或信号，并将细胞信号进行转换，然后通过一系列的生化反应将信号传递到细胞核内，诱导特定基因的表达，从而引起细胞生理上的变化。当外界环境发生变化，单细胞生物直接对外界环境变化做出反应，而多细胞生物体则通过细胞间和细胞内的更为复杂的信号转导网络（STN）来传递信息，作出反应，从而调控生物体活动。高等生物由以亿计的细胞组成的有机体，细胞不与外界直接接触，因此，多细胞生物对外界的刺激（如物理、化学因素），需要细胞间复杂的信号传递系统来传递（Kim et al.，2010a）。

实现信号转导的一系列特定反应系统称为信号转导通路（Gupta et al.，

1995）。信号转导通路是多细胞生物细胞间进行通信的关键过程，细胞内的各种生物过程都与各种各样的信号通路有关。大多数信号转导过程都涉及细胞内系列生化反应的指令序列，这些生化反应是由酶来催化的，被第二信使激活，从而形成一条信号转导路径。若干条信号转导路径就可组成信号转导网络（Kim et al.，2010b）。信号转导过程十分复杂，一个信号往往不是单一传导，而是存在许多其他蛋白质或信号去增强它或抑制它，构成了一个复杂的信号转导网络。

现有研究表明，许多疾病的发生是由信号通路紊乱所导致的。例如，胰岛素信号通路的失调容易引起糖尿病和肥胖症（Wang et al.，2016）等。Wnt/Notch 信号通路失调与胰腺癌有关（Nakhai et al.，2008；Shaw et al.，2006）。因此，很多科学家一直致力于寻找与疾病相关的信号通路，并将其作为可能的生物标记。例如，Ideker 等和 Tan 等利用信号通路信息和基因表达数据，找到了与疾病相关的信号通路（Kuo et al.，2010）。Levin 等和李霞等通过综合考虑信号通路和基因关联数据，找到了与类风湿关节炎发生相关的信号通路（Eleftherohorinou et al.，2011）。由此可见，信号转导通路可以帮助人们理解疾病的产生过程，并可以帮助开发有效的治疗药物。

目前，虽然公共数据库，如 KEGG（Gerlich and Neumann，2000）和 BioCarta 等，保存有一些已知的信号通路，但是这些信号通路还只是实际信号通路的一部分，而且即便是这一小部分通路，其中的很多信号分子也是未知的（Koenigsberg and Ondrey，2020）。因此，在有关信号通路的研究中，发现信号通路中的新组分和确定信号通路的网络结构是非常重要也是必要的一步。目前，生物学家利用各种各样的实验手段，像 RNAi、基因敲除或抑制、高通量单分子成像技术等，可以确定信号通路中的组分（Bouché and Bouchez，2001；Jackson et al.，2003），但是通过实验手段来确定信号通路是一个费时费力而又代价高昂的过程。随着信号转导数据的不断积累，有必要建立大规模的数据库收集已知的信号通路，为解释生物过程和疾病机理提供帮助。针对不同需求，人们已经建立了多种生物通路和分子相互作用的数据库，数据存储和展现形式各不相同。例如，BioCarta 是目前覆盖范围最广的信号通路数据库，包含了大量的通路细节知识，方便进行单个分子的查询，但是单个通路规模较小，不提供批量下载。KEGG 和 Reactome（Jassal et al.，2020；

Sidiropoulos et al.，2017）作为经典的信号通路数据库建立时间较早，图示清楚，下载方便，但与 BioCarta 相比包含的通路数据不够全面。

由于信号转导通路在生命中的重要性，在过去半个世纪中，科学家一直致力于信号通路的研究（Honigberg et al.，2003；Li et al.，2006；Vabulas et al.，2002）。与信号通路相关的学术论文在过去的 20 年中每年也在呈指数级增长。与此同时，多位科学家由于在信号通路研究方面的突出贡献荣获诺贝尔奖。

近十几年来，快速发展的生物信息学越来越引起人们的重视（Likic et al.，2010）。与生物实验相比，利用生物信息学方法识别信号通路大大降低了实验的成本和周期，并且可以为实验提供指导。在信号转导过程中，信号的传递基本上是由一系列的蛋白质相互作用来实现的。因此，蛋白质相互作用数据能有助于理解和揭示信号通路的形成和作用机制（Bila et al.，2019）。近年来，随着高通量生物实验技术的进步，越来越多模式生物的蛋白质相互作用数据开始涌现。这使得生物信息学家利用蛋白质相互作用网络开发精确的计算模型来识别信号通路及其所涉及的组件成为可能。

信号转导网络中广泛存在的交叉调控关系是生物复杂性的体现（Ivashkiv，2009），与其承担的重要生物学功能有关，同时也给信号转导网络的生物信息学分析带来了许多困难。未来几年该领域可能出现以下几个研究热点：第一，随着现有通路不断的补充和扩展，可能发现通路之间更多的交叉联系，组成大规模的信号转导网络。这使得对于大规模信号转导网络的结构属性分析成为可能，从而揭示信号转导的一般作用机制和规律。第二，为了满足实验人员的需要，可能出现更多的专门针对信号转导网络设计的生物信息学分析工具，寻找信号转导通路中的结构、功能模块和薄弱环节，发现可能的疾病标志物以及药物靶标。第三，随着多种信号通路的作用机制被揭示，信号转导通路的定量数据不断积累，为动态模拟信号转导过程提供了基础。动态信号转导数据库的规模将扩大，同时信号转导的动态模拟和分析将成为主要的研究方向。

此外，信号转导过程控制细胞的生存和凋亡，与多种生物学功能及疾病密切相关，因此有必要对于信号转导动态过程进行深入研究（Li et al.，2006）。系统生物学方法期望通过建立细胞信号转导过程的模型，找到参与此过程的各种分子之间相互作用的网络，阐明其在基因调控、疾病发生中的作用，为疾病治疗和药物发现提供依据。一些可能用到的定量模型包括布尔网络、逻辑模型、概率模型、佩特里网（Petri net）、常微分方程和偏微分方程模型等（Chen et al.，2009）。

五、表观调控网络

表观遗传学是指在DNA序列不改变的情况下可遗传的基因表达的改变，包括DNA甲基化、组蛋白修饰、非编码RNA等。DNA甲基化、组蛋白修饰参与转录调控，而非编码RNA主要影响转录后调控。表观遗传学是生物分子网络的重要组成部分，是完整刻画生物系统动态变化所必需的。表观基因组学数据的积累是构建表观遗传学网络的基础。目前国际上的三大表观遗传学计划包括FANTOM 5、DNA元件百科全书计划（Encyclopedia of DNA Elements，ENCODE）和BLUEPRINT，分别由日本、美国和欧洲主导。FANTOM 5计划是由日本理化学研究所（RIKEN）主持的大型多国协作计划，截至2014年已经积累了400多种细胞类型的启动子、增强子、miRNA和lncRNA数据（Consortium et al.，2014）。Marbach等（2016）通过整合FANTOM 5计划中的CAGE（cap analysis of gene expression）数据，构建了394种细胞/组织特异性的转录调控网络，是FANTOM 5在转录调控网络领域标志性的贡献。CAGE技术能够鉴定全长mRNA的5′端序列并进行定量，从而实现高通量的基因表达分析以及转录起点的图谱分析，非常适合用来构建转录调控网络。目前，科学界积累更多的数据是常规的基因表达谱数据，如表达谱芯片数据和RNA-seq数据等。基于基因表达谱，转录调控网络也可以通过逆向工程来实现。哥伦比亚大学的系统生物学家Andrea Califano带领研究组在2006年开发了一个针对高等生物的转录调控网络构建工具ARACNE（Margolin et al.，2006）。该工具使用互信息（mutual information，MI）来衡量基因和转录因子之间的依赖关系，并用信息科学中的数据处理不等式（data processing inequality，DPI）来去除间接的调控关系。ARACNE是影响力最大的转录调控网络构建工具之一，到目前为止已经被引用2000次以上，它的广泛应用表明，即使单纯使用基因表达谱数据也能构建精准的转录调控网络。

2015年，Olga Troyanskaya研究组采用贝叶斯法分析了14 000套表达谱数据，构建了144种组织特异性基因网络（Greene et al.，2015）。ENCODE是表观遗传学领域的国际重大项目，鉴定了100多种组织中的转录因子结合位点、DNA甲基化、组蛋白修饰，为最终解析组织特异性转录调控网络提供了坚实的基础（Consortium et al.，2020）。利用这些海量数据，Manolis Kellis研究组开发了ChromHMM软件，采用隐马尔可夫模型（Hidden Markov Model）的方法将基因组区间进行模块化聚类，每一类代表一种染色质状态（Ernst and Kellis，2012），这种模块化的表观修饰方式也提示着表观遗传学网络是高度结构化的，受到严格的调控。有研究者通过整合表达谱和ChIP-seq数据，在乳腺癌中鉴定了受表观遗传学调控的转录调控网络（Zhu et al.，2020）。长非编码RNA（lncRNA）可以介导增强子和启动子之间的结合（Orom et al.，2010），有研究者利用ChIA-PET（chromatin interaction analysis by paired-end tag sequencing）数据构建了lncRNA介导的调控网络（Thiel et al.，2019）。也有研究者整合DNA甲基化和表达谱数据来构建转录调控网络，并鉴定关键调控因子（Silva et al.，2019）。表观遗传学领域另一个重大的国际项目是欧洲的BLUEPRINT（https://www.blueprint-epigenome.eu/），该计划系统描述了造血系统的表观基因组学。利用该计划中的ATAC-Seq数据，研究者构建了慢性粒细胞白血病亚型特异性转录调控网络及关键调控因子（Rendeiro et al.，2016）。转录调控网络的一个重要用途是用来预测转录因子在不同生物条件下的活性变化。转录因子的活性在很大程度上受到表观遗传学的调控，如DNA甲基化直接决定CTCF能否结合在DNA上（Ong and Corces，2014）。Andrea Califano团队开发了一个转录因子活性预测算法VIPER，该算法读入ARACNE构建的转录调控网络，通过比较不同生物条件下转录因子靶基因的表达变化，来预测转录因子活性的变化（Alvarez et al.，2016）。然而由于转录调控的复杂性，不同转录因子的靶基因可能存在相当程度的重合，因而对每个转录因子单独分析会产生大量的假阳性。为了解决这个问题，Amin等（2019）开发了EPEE（effector and perturbation estimation engine）方法，能够同时刻画和比较所有转录因子的活性，从而大幅减低假阳性率，并提高敏感度。

上文主要概述了表观遗传学数据库以及它们在转录调控网络构建、转录因子活性预测方面的基础性作用。由于表观遗传修饰的泛在性和高动态性，使得构建包含 DNA 甲基化、组蛋白修饰的表观遗传学网络非常困难，目前的努力主要集中在构建非编码 RNA 参与的转录调控网络。非编码 RNA 特别是 microRNA 能够对基因进行转录后调控，是基因表达网络动态变化的重要调节因子。中山大学屈良鹄研究组开发了 StarBase 数据库，收集整理了海量的非编码 RNA 相互作用数据（http://starbase.sysu.edu.cn），目前已被引用 4000 次以上（Li et al.，2014）。不同的 microRNA 可以调控相同的基因，从而形成转录后调控网络（Anastasiadou et al.，2018；Bracken et al.，2016）。如同共表达网络一样，非编码 RNA 网络也可以作为疾病的分子标志物。广西医科大学的研究者利用 StarBase 构建了一个由环状 RNA、小 RNA 和 mRNA 组成的子网络，是冠状动脉疾病潜在的标志物（Miao et al.，2019）。DNA 甲基化网络构建描述 DNA 甲基化和疾病之间的关系（Yuan et al.，2019）。

六、大、小分子相互作用网络

在生命活动所处的自然环境中，生物体不可避免地会接触到一些无机和有机小分子化合物。同时在有机体中，除了需要生物大分子的相互作用来维持生命的有机结构以及有序的生命活动过程，也少不了小分子的协同作用。并且，生物大分子也是由小分子有序组成的。所以纵观整个生命过程，无非是这些生物小分子与大分子的相互作用网络，构成高度有机的生物体。

分子间的相互作用形式很多，就化学水平来看，主要有共价的相互作用以及非共价的相互作用。共价的相互作用在生物体内的表现形式首先是直接的修饰，包括甲基化，乙酰化，磷酸化修饰等；其次是各种酶促反应，如各种磷酸化酶激酶（phosphorylase kinase）调节磷酸化修饰，脱氢脱羧酶的催化反应。非共价的相互作用主要有离子键（如金属离子和蛋白质的相互作用）和分子间相互作用（配体与受体相互作用）等。正是这些复杂而有序的相互作用，构成生物体的小分子与生物大分子的相互作用网络。想要具体细致地描述这个网络显然很难，可以从以下几个方面去简单了解其相互作用，以及如何研究这些相互作用网络。

首先，是小分子对生物大分子的修饰。生物大分子高度有序，且功能多样。就生物体所处的环境来看，要应对高度复杂的外部环境就要求生物体能够有足够的响应分子。而生物大分子从组成上看，高度保守，作为响应分子显然有些不够。因此，生物大分子不仅仅只是靠改变一级结构去改变其功能，同时还有丰富的修饰，如核酸的甲基化修饰，蛋白质翻译后修饰等。这些丰富的修饰在调控生物大分子的功能上起着不可忽视的作用。

核酸作为遗传物质，在生命系统中起着至关重要的作用（Dahm，2008），核酸的多态性造就了世界的五彩斑斓，生命的多姿多态。在过去的几十年中，已经有过大量研究，这些研究表明，核酸中异常表观遗传修饰的出现总是与许多疾病的发生，包括癌症、糖尿病、阿尔茨海默病和许多其他与年龄有关的疾病息息相关（Dolinoy et al.，2007；Hatchwell and Greally，2007；Petronis，2010；Portela and Esteller，2010）。

蛋白质具有多种生化功能，并且在所有生命过程中都至关重要，包括代谢、信号转导、转录、翻译、细胞结构完整性和细胞运动。所有生命体的一个共同特征是能够适应不断变化的环境并在其中生存。为了应对环境变化，必须改变生物体的蛋白质或蛋白质组。蛋白质翻译后修饰是在翻译过程中或翻译后几乎每种蛋白质都发生共价变化的过程。蛋白质翻译后修饰遵循各种信号通路，通过诱导蛋白质与新的功能化学基团（如磷酸、乙酰基、甲基、糖基和泛素）的共价键，引起酶的激活，这些酶在调节蛋白质的活性、稳定性、定位、相互作用或折叠方面起着至关重要的作用（Chiang and Gack，2017；Deribe et al.，2010；Lin and Begley，2011）。蛋白质翻译后修饰是现代生物学和医学领域的前沿，对于许多生物学和病理学过程至关重要（Krueger and Srivastava，2006），如细胞分化和增殖、蛋白质降解、信号转导、转录调控、基因表达调控和蛋白质-蛋白质相互作用。例如，临床上以痴呆为特征的阿尔茨海默病呈现出形态丰富的神经原纤维病变。微管相关蛋白 tau 的功能主要是稳定中枢神经系统神经元中的微管结构。研究证据表明，tau 蛋白的异常过度磷酸化与阿尔茨海默病中神经原纤维的成对螺旋丝和直丝结构有关（Flores-Rodríguez et al.，2015；Šimić et al.，2016）。除阿尔茨海默病外，癌症与蛋白质翻译后修饰高度相关（Stowell et al.，2015）。例如，在某些人类癌症中已观察到抑癌蛋白质 p53 的过度磷酸化或乙酰化（Bode and Dong，2004）。蛋白质糖基化直接影响细胞的生长和存活，并促进肿瘤诱导的免疫调节和最终转移。癌变和肿瘤进展总是伴随着异常的糖基化，这为我们提供了至关重要的诊断和预后信息。迄今，在临床医学中应用的大多数癌症生物标志物都是糖蛋白（Reis et al.，2010）。

其次，除了小分子的修饰外还有一类作用机制是非共价的相互作用，包括一些金属离子与蛋白质或者核酸的相互作用。这类金属离子对维持蛋白质的特殊功能起着非常重要的作用。此外，还有胞外小分子配体与膜受体的非共价相互作用，这种相互作用正是细胞响应外界刺激并做出相应反应的基础。

谈到金属离子，我们自然会想到血红蛋白中铁离子对血红蛋白载氧功能的调节。转录因子中，锌指结构域中氨基酸残基与锌离子的相互作用形成稳定的结构。正是这种稳定结构赋予蛋白质能够结合 DNA 或者 RNA 调控基因表达。还有钙调蛋白通过钙离子与蛋白质的相互作用来调节蛋白质的结构（Ardura et al.，2020）。这种相互作用的最终结果是维持蛋白质呈一个稳定的二级或者三级结构去行使特定的功能。以钙离子在生物体内的功能为例，我们简单地描述一下钙离子与生物大分子的相互作用网络在生命活动过程中的重要作用。在漫长的进化过程中，哺乳动物细胞需要维持细胞质（约 100 nmol/L）中极低浓度的钙离子 [相对胞外（1～1.5 mmol/L）有 10～15 000 倍差异]，而这个浓度的维持主要由两个 ATP 依赖的钙泵实现。一个是细胞膜钙 ATP 酶（PMCAs），该酶负责将细胞内的钙排出细胞外；另一个是内质网钙 ATP 酶（SERCAs），该酶主要负责将细胞内的钙吸收到内质网中。在细胞中 Ca^{2+} 本身也是第二信使，能够与 Ca^{2+} 相互作用的蛋白质大多含有相关的钙调蛋白结构域。而这些蛋白质的功能一方面受 Ca^{2+} 浓度的影响，另一方面反过来影响 Ca^{2+} 的浓度，来维持细胞的钙稳态（Marchi et al.，2020）。

此外还有一些胞外配体与相关受体的相互作用网络。膜受体作为最直接的外界信号接触单元，可以感受各种各样的外界刺激，如力、光、电、热等各种我们可以想象到的外界刺激条件，这些自然也包括各种自然界的小分子。如钙敏感受体，谷氨酸受体等 G 蛋白耦联受体（Leach et al.，2020）。这类受体的作用机制是胞外小分子配体与受体的可逆相互作用，改变受体蛋白构象，并传递信号到胞内。

小分子与生物大分子的相互作用十分复杂。生命活动的主要功能分子是生物大分子，理论上作用于分子间的相互作用网络会调控相应的生理过程。在此基础上延伸出另外一个研究领域即药物小分子研究。大多数药物靶标是细胞蛋白质，而小分子药物也正是通过与这些靶标蛋白的选择性地相互作用来诊断或治疗疾病（Imming et al.，2006）。研究表明，经典的治疗药物靶标包含约 130 个蛋白质家族（Hopkins and Groom，2002；Overington et al.，2006），如酶、G 蛋白耦联受体、离子通道和转运蛋白、核激素受体（Drews，2000；Hopkins and Groom，2002）。据估计，人类基因组中有 6000～8000 个具有药理学意义的靶标，但是只有一小部分靶标开发出了相应的可以临床应用的药物（Drews，2000；Landry and Gies，2008；Overington et al.，2006）。大量潜在的药物靶点仍有待验证。

基于小分子与大分子相互作用的重要性，研究哪些小分子会参与生物大分子的功能调节，以及小分子如何与生物大分子作用并调节其功能就显得尤为重要。前文中总结的一些相关的研究显示，小分子参与生命活动的每一个过程，我们目前观察到的只是冰山一角；生物大分子的修饰种类多样，大量未知的参与修饰生物大分子的小分子需要研究人员去探索；越来越多的新修饰被发现，这些修饰到底是随机的分子碰撞还是在生理过程中起十分重要的功能还需要研究人员继续跟进。随之而来的问题还有怎样探索新的修饰分子，如何研究小分子与大分子的相互作用？此外，小分子与生物大分子的相互作用网络还缺少比较完整的研究体系，从复杂性来看，这种作用网络天然就是需要多组学的参与的。生物大分子维度，如蛋白质，虽然还有待进一步解析但总是有限的，而小分子的维度几乎无穷。好在我们依然能从有限的大分子维度去探索相互作用的小分子。期待相关的研究能一方面揭示生命的奥秘，另一方面解决人类面临的生老病死问题。

七、“数字化”细胞

一个多世纪以来，人们一直十分重视对细胞的研究，近几十年，随着计算机仿真计算和可视化能力的提高，以及在数学、信号与信息处理科学领域的发展，于 20 世纪 90 年代中期以后诞生了一种研究细胞的新的技术手段——电子细胞（electronic cell，E-Cell）（Tomita et al.，1999）。

电子细胞也称细胞模型（cellular model），是在计算机上模拟真实细胞的结构、物质组成、生命活动的动力学行为和生命现象，用虚拟现实的方式实现了人机交互，以便研究人员构造细胞结构和其内部环境物质组成，考察、记录细胞实验现象和功能，再现细胞生命活动和发现新的生物学现象规律。因此，电子细胞也称人工细胞，是人工生命的重要基础部分（Tomita et al.，2000）。

建立电子细胞一直是系统生物学和数学生物学中一个特别具有挑战性的任务，需要开发高效的算法、数据结构、可视化和通信等工具，将大量生物数据与计算机建模相结合。它还与生物信息学、计算生物学和人工生命直接相关（Schaff and Loew，1999；Tomita et al.，1999）。建立电子细胞涉及使用计算机模拟许多细胞中的子系统。例如，代谢产物和酶的网络包括代谢、信号转导途径和基因调控网络，以分析和可视化这些细胞过程的复杂联系。生物化学反应、转运过程及它们之间的空间组织构成了一个复杂网络，使得开发活细胞预测模型成为 21 世纪的一项重大挑战。

电子细胞的基本问题是细胞的生长、发育、繁殖和凋亡过程的建模模拟（Tomita et al.，2000；Tomita et al.，1999），涉及细胞内外环境和细胞相关物质对细胞的生命活动的作用表现，这包括离子通道、信号转导、细胞的组成结构及相应功能，细胞内外物质活动和生化反应、基因表达、蛋白质、酶、能量物质等相互作用的动力学行为的真实再现。另外，涉及细胞的医学、病理学、药学、生物学上的功能，以及对环保、能源等领域的作用，也是电子细胞的重要研究范畴。

电子细胞主要有如下几个研究方向（Schaff and Loew，1999；Tomita et al.，2000）：①细胞组成、结构和功能建模。主要涉及细胞的各组成成分物理结构、生化功能、物质流动、信号生成等的建模，是对构成整个细胞的模拟有作用的各个有机成分及其相互结构功能作用的研究。②细胞生命活动相关数据库及相应管理软件。主要涉及搜索、发现、整理和存储细胞生命活动的相关试验测试数据，建立数据库，在此基础上创造出知识库，涉及基因、蛋白质、生化反应及其他反应链、代谢过程、细胞各物质组成及其相互作用等。这是逼真地再现和预测细胞生命活动的基础。③细胞间通信和信号转导。主要涉及对细胞生命活动信号的产生机理、转导机理和细胞间相互作用的建模、模拟和有关现象规律的研究。④染色体和基因表达。主要涉及细胞染色体活动的模拟、基因表达的产生条件、产生过程、动力学模型、基因的功能、遗传、变异规律等。⑤蛋白质合成、结构分析预测。主要涉及从细胞基因出发，应用分子力学和分子动力学的原理建模，并观察发现蛋白质的结构规律、合成规则，分析已探明蛋白质的生成机理，预测和模拟产生新的蛋白质。⑥细胞代谢过程模拟和分析。主要涉及细胞物质代谢和能量代谢过程的机理、生化反应、动力学建模、模拟分析等。⑦分子进化和比较基因组学。主要涉及利用不同物种中基因序列的异同研究细胞中氨基酸序列甚至相关蛋白质的功能结构，模拟这一过程，通过对比来研究分子进化和细胞进化。如今各种高通量及单细胞测序技术的出现，也为从整个基因组及单细胞角度来研究分子和细胞进化提供了条件。⑧疾病诊疗。生物的病变在很大程度上反映在细胞的病变，通过研究细胞的物质成分组成和表型变化等来诊断病变，并采用多种手段在电子细胞上再现治疗控制过程，将对发现新的治疗和诊断方法有十分重要的作用。⑨药物设计。电子细胞的目的之一是可视化再现和阐明细胞生命活动的规律，刻画和预测细胞中各种成分的结构功能、相互作用以及与各种人类疾病之间的关系，以寻求各种防治药物和方法。在电子细胞上进行药物设计和药理实验，既经济又快速安全。

电子细胞的发展始于20世纪90年代。日本和美国自开始提出和重视计算机模拟细胞的计算研究，到现在为止，已经开发出了多个版本的软件。两个重要的计划分别是日本的E-cell（Tomita et al.，2000）和美国的Virtual Cell（Schaff and Loew，1999）。E-cell计划发起于1996年，主要的发起单位是日本的庆应义塾大学（Keio University），研究人员率先提出了世界上第一个电子细胞模型，并开发出整个电子细胞仿真的软件环境E-Cell system，目前的最新版本是version 4（https://ecell4.e-cell.org）。美国于1999年发布了开源软件Virtual Cell，用于对细胞进行建模和仿真，供从实验细胞生物学到理论生物物理学等广大科学家的使用，目前的最新的稳定版本是2018年发布的版本7.1（https://vcell.org）。

目前国外的不同团队已开发出了不同的电子细胞模型。日本庆应义塾大学的Masaru Tomita教授是电子细胞的先驱者，他领导的研究组以原核微生物支原体为对象，在1997年建立了世界上第一个虚拟电子细胞模型（Tomita et al.，1999）。该电子细胞以支原体生殖糖代谢过程为主，选取了127个与代谢过程相关的基因，对细胞内与代谢过程密切相关的糖、脂肪、甘油和能量代谢过程及机理建立了数据模型。1999年，美国学者James Schaff和Leslie Loew建立了真核细胞钙转运的模型Virtual Cell（Schaff and Loew，1999）。利用该虚拟细胞可以完成神经母细胞瘤的钙动力学、受精卵细胞的钙离子流、RNA的转运、线粒体的作用和细胞核膜的作用等一些关于真核细胞的生物学活动和功能的研究。2012年，来自斯坦福大学的研究人员首次模拟了来自人类的病原菌——支原体的整个生命周期，提出了一个全细胞计算机模型，这一模型囊括了这个病原菌的所有分子组，以及相互作用，这将有助于促进细胞生物学的发展。随后，哥伦比亚大学Tavazoie教授等评价道：这项研究提出了首个整合动态计算机模式，为系统生物学研究提供了全细胞的定量预测模型（Freddolino and Tavazoie，2012）。癌细胞转移是癌症患者死亡的主要原因之一，该过程中所涉及的大量信号远远超出了任何研究人员预测细胞行为的能力。为了解决该问题，Selvaggio等建立了虚拟细胞，对癌细胞进行了预测。在这些预测中，Selvaggio研究小组证实，表达癌基因*Src*的混合癌细胞如果位于更坚硬的细胞外基质上，就会变成间充质细胞（Selvaggio et al.，2020）。

此外，随着科技的发展还涌现出了许多其他的“数字化”细胞技术。比如，DNA折纸术（DNA origami），该词是2006年由Rothumend教授课题组提出的，该课题成果发表在2006年的*Nature*上（Nangreave et al.，2010；Rothemund，2006）。DNA折纸术是DNA组装（由DNA组装成不同的结构或者框架）的一个分支，在2006年之前大部分课题组主要集中在短DNA分子砖（DNA tile）组装上，也因此面临许多的问题：短DNA之间的比例影响对最终结果影响太大，需要多步实验，需要多次提纯。而DNA折纸术恰恰能解决这些问题。简而言之，由一根DNA长链作为主链，再加入许多短链引导着这根长链组装成预定的结构。该技术在新兴的纳米领域中具有广泛的潜在应用。科学家清楚地知道，很多材料和器件在纳米尺度范围内具有全新的物理性能，当前生命科学的前沿亦必定是纳米或分子水平上的研究，DNA折纸术的发展，无疑具有至关重要的意义（Bila et al.，2019）。2018年，中国科学院国家纳米科学中心和美国亚利桑那州立大学（ASU）的科学家团队利用

DNA 折纸术制造出了世界上第一种智能抗癌的纳米机器人（Li et al., 2018），这种纳米机器人可以在人体内自行找到给肿瘤供血的血管，随后释放药物制造血栓阻塞血管，从而“饿死”肿瘤，在动物实验中体现了良好的疗效和安全性。

另一个相关的技术是合成细胞，或者说人造细胞（Kendall et al., 2018; Loose et al., 2008）。二十多年来，科研人员一直努力地创造人造细胞，他们的思路就是将各种生物分子按照某种方式组合起来。虽然方法不尽相同，但是大致可以分为以下三大类：①想办法将生物分子包裹在某一个空间内；②研究足以维持生命的生物化学系统；③研究细胞行为的信息控制、储存和管理问题。得益于近年来微流体技术的进步，科研人员可以精细调控细胞组分之间的微小活动，所以人工合成细胞这方面的工作也在加速开展之中。目前科研人员已经找到了一些方法，可以将细胞样的小泡塑造成我们需要的样子，也创造出了一些最基本的细胞代谢能力，而且还成功地将人造基因组转移到活体细胞里。该技术同样具有重大意义，可以培育器官组织并移植到患者体内，对致命病毒实现免疫化控制，如埃博拉病毒和寨卡病毒等，甚至可以在实验室内培育抗癌细胞。

2017 年 9 月，荷兰 17 家实验室的科研人员一起组成了“构建人造细胞组织”（Venter and Gibson, 2010）。据该组织的负责人兼荷兰代尔夫特理工大学的生物物理学家 Dogterom 介绍，他们的目标是在 10 年之内，人工合成出一个细胞样的、可生长和可分裂的新东西。目前，他们已经得到了 1880 万欧元（约合 2130 万美元）的荷兰引力基金的支持。2018 年 9 月，美国国家科学基金会宣布了他们的第一个人工合成细胞项目，并表示将为此投入大约 1000 万美元。包括 Schwille 在内的多个欧洲的研究人员也都建议拥有 10 亿欧元资金的欧盟委员会未来新兴技术项目对人工合成细胞给予经费支持。从事从头合成（de novo synthesis）的生物学家都预计第一批具有真正意义的人工合成细胞将在大约 10 年之后诞生。

DNA 存储也是一种与“数字化”细胞有关的新技术（Panda et al., 2018），即用人工合成的脱氧核糖核酸（DNA）存储文本文档、图片和声音文件等数据，随后完整读取的技术（Organick et al., 2018; Rutten et al., 2018）。DNA 存储技术作为数字存储媒介的显著优点之一是容量大。DNA 分子是一种令人难以置信的密集存储介质，1g DNA 能够存储大约 2 千万亿字节（petabyte，PB），相当于大约 300 万张 CD。

用 DNA 存储数据保存时间可能长达数千年。与硬盘、磁带等存储介质不同的是，DNA 不需要经常维护。就读取方式而言，DNA 存储不涉及兼容问题。这项技术有重要的实际应用。首先，可以把各种材料的生产方式放到材料里面，比如可以在植入物中放入患者的诊疗信息。若干年后，可能患者的病历已经难以找到，但只需从这名患者体内拿出这个植入物就可以了解其过去的健康情况。我们还可以用同样的方式来生产汽车部件。其次，可应用于隐藏信息。我们可以把一些常见的东西或物件当作隐藏信息的载体，如鞋带、镜片、衬衣、扣子等，所有这些物件都可以帮助我们储存和隐藏信息。甚至我们可以像喝饮料一样，把硅珠喝到肚子里，需要的时候再把它排泄出来。再次，可应用于自我复制的机器人。通常机器人是无法自我复制的，但是自我复制机器人有能力把自己的生产方式传递给下一代机器人。

DNA 用于数据存储可以上溯到 1988 年（Extance，2016；Wang et al.，2014）。Davis 等用 DNA 编码了一个古日耳曼符文。此后，利用 DNA 进行信息存储取得了系列突破性进展。第一个比较成功的案例是，2012 年 Church 等利用 DNA 合成技术存储了一本书，包括 53 426 个单词、11 幅图片和 1 个 JavaScript 程序，共 5.27 MB；随后，他们又将一张动态 gif 信息存储入 DNA 中并导入大肠杆菌中进行自我复制（Church et al.，2012）。2016 年，华盛顿大学和微软研究院合作把莎士比亚的《十四行诗》、马丁 • 路德 • 金（Martin Luther King）的演讲原声、医学论文等资料共计 739 kB 的数据编码成了 DNA 序列并存储起来，并通过精心设计特定的引物，标记每一个文件在 DNA 序列上的地址进行数据的随机快速读取。2017 年，Erlich 等也开展了一次 DNA 存储实验，不仅存储密度更高，并且编码了一个操作系统（movie 1），其能够运行起来玩扫雷游戏（Shipman et al.，2017）。2020 年，哥伦比亚大学 Yaniv Erlich 联合苏黎世联邦理工学院专家等通过 3D 打印制作了一只兔子模型，并且将这只兔子模型的三维结构数据以双链 DNA 结构形式内置在打印材料中。通过编码和解码，这只 3D 打印的兔子模型实现了其自身数据的 DNA 存储和传递（Koch et al.，2020）。

DNA 有可能是我们终极的存储设备。它的存储密度要比其他存储技术高，存储时间也更久，而且提取简单。人类只要使用一台普通的DNA测序仪，就可以把信息解析出来。同时，由于没有材料或形状的限制，我们可以把DNA注入日常物件中，使存储信息成为我们生活的一部分。

目前，全面和真实地模拟一个细胞尚有不少的困难。首先，最主要的问题之一是缺少定量的数据。大多数生物学的知识是以定性的方式提供的，而实现模拟需要定量化。其次，我们对于细胞的了解甚少，许多生理、化学模型还无法用数学模型去描述。但随着高通量测序技术的出现、相关数据库的不断增长以及计算机运行速度的显著提高，相信未来“数字化”细胞仍然具有很大的发展前景。

第四节　我国的发展战略与重点方向

一、我国发展对该领域的需求、我国该领域研究面临的问题与挑战

我国发展对生物分子网络与“数字化”细胞等生命与健康科学领域的需求主要是信息技术与生物技术的融合。《“十三五”卫生与健康科技创新专项规划》中也指出：第一，推进健康中国建设迫切需要科技支撑；第二，引领健康产业发展迫切需要加强科技创新；第三，推进科技强国建设需要卫生与健康科技的创新突破。生物分子网络与相关的基础研究领域、疾病防控领域、新药创制领域、预防康复领域和中医药现代化领域具有密切关系。

目前，在生物分子网络与“数字化”细胞领域所面临的主要问题是生物分子通路、模块信息整合难度大。分子网络描述了生物分子间的相互作用关系，包括通路、模块、整体三个层次。通路和模块研究已经产生了海量的生物分子及其相互作用数据，同时也建立了大量的数据库。高通量测序技术的出现和不断改进，以及科研工作者在该领域进行的广泛研究，使得相关的数据和信息仍在不断快速增加中。但是，这些信息多具有生物体局部系统性差

异。例如，代谢通路仅研究细胞中代谢物在酶的作用下转化为新的代谢物过程中所发生的一系列化学反应；网络模块表述了许多分子相互结合，形成的具有稳定结构和功能的复合体，其对生物体系统和疾病的理解远远不够，现有信息整合对生物分子网络中更多重要的内在规律揭示不足。

基于上述的主要问题，生物分子网络或网络生物学面临的最大挑战就是模型的建立和分析，这也是基础研究转化应用的瓶颈。一方面，生物分子数据的整合和对生物学意义的结果的解释需要设计新的计算和试验平台。生物分子网络研究是一个数据主导的领域，是在实验数据的基础上研究复杂的系统。生物体，尤其是人体疾病是一种动态过程，在环境中不同时间、不同尺度的刺激下时刻发生着变化，所以需要改进现有计算模型，设计能提供空间和时间动态数据的实验模型。另一方面，现有的计算模型缺少功能性生物学终点，如细胞繁殖或凋亡，对人类生物学研究来讲，这些功能性生物学终点非常重要，有助于临床上对药物效果的预测。类似的功能性生物学终点与计算机模型相结合，将增加大规模计算机模拟生物应答的实用性，并为临床实践提供更准确的预测工具。

二、发展总体思路和发展目标

1. 发展总体思路

分子生物网络与系统生物学密切相关，它是基于网络的视角来研究生物学，其主要思想是整合。生物分子网络是在生物系统中包含很多不多层次和不同组织形式的整合型大网络，包括基因转录调控网络、生物代谢网络、信号转导网络、蛋白质相互作用网络。并且各个网络之间并未相互独立，而是相互关联、相互影响的整体。因此，分子生物网络是多信息的整合，其发展也应用整合的研究思维，将基因组学、蛋白质组学和其他各种组学的水平型研究和垂直型研究整合起来，成为一种三维的网络生物学研究。此外，分子生物网络的出现也是各个学科交叉整合的结果，是生物学、物理学、化学、数学、统计学、计算机科学等学科的共同发展导致了它的诞生，而它的研究成果出现之后又进一步促进了上述学科的发展。

2. 发展目标

分子生物网络或网络生物学是生命科学的新研究领域，其目的是在系统水平上以网络视角理解生物体。分子生物网络是以整体性研究为特征的整体性生物学，其目标就是要得到一个表征生物系统状态的网络模型，使其理论预测能够反映出生物系统的状态和真实性。分子生物网络不同于以往仅关心个别的基因和蛋白质的实验生物学，它要研究一个生物学系统中所有组成成分（包括基因、mRNA、蛋白质、生物小分子等）的构成，以及在指定条件下这些组分间，特别是所有的基因、所有的蛋白质组分间的相互关系的学科。显然，分子生物网络是以整体性研究为特征的，它是在基因组学、蛋白质组学发展的基础上，立足生命科学研究领域的一门覆盖“组学”、计算技术和转基因方法等的系统生物技术，因此，发展网络生物学是生物学今后发展的重要目标。

三、优先发展领域或重要研究方向

生命活动本身的复杂性和迅速增加的海量数据资源要求生命现象必须要在成千上万个生物分子组成的复杂系统层面上予以认识。不同层次、不同尺度上的分子生物网络为揭示数量巨大的生物分子及其间的相互作用在复杂生存环境中行使生物学功能上，都提供了不同于传统生物学研究手段的新技术，但最重要的两个研究方向还是基因调控网络和蛋白质互作网络。这两个网络最为重要的原因在于，网络由许多不同的参与生物过程的分子元件组成，其中最重要的便是在不同组织和细胞型中的基因与蛋白质网络谱。

1. 单细胞层面的基因调控网络谱

所有生物在生长发育和分化过程中，以及在对外部环境的反应中，各种相关基因有条不紊的表达起着至关重要的作用。与原核生物相比，真核生物基因表达的调控更为复杂，真核生物基因表达的调控主要是指编码蛋白质的 mRNA 产生和行使生物功能过程中的调节与控制。从理论上讲，基因表达调控可以发生在遗传信息传递过程的各个水平上，其中转录调控是基因表达调控中最重要、最复杂的一个环节，也是当前研究的重点。现在国际上已有单细胞谱的计划，但在组织和细胞层面的分子网络更为重要，建立不同组织和细胞型中的基因网络谱可揭示复杂生物现象的调控机制。

2. 单细胞层面的蛋白质互作网络谱

蛋白质是构成生物体的重要物质，也是形成生物功能的重要生物大分子。多个蛋白质通过彼此之间的相互作用构成蛋白质相互作用网络来参与生物信号传递、基因表达调节、能量和物质代谢及细胞周期调控等生命过程的各个环节。系统分析大量蛋白质在生物系统中的相互作用关系，对于了解生物系统中蛋白质的工作原理，了解疾病等特殊生理状态下生物信号和能量物质代谢的反应机制，以及了解蛋白质之间的功能联系都有重要意义（Li et al.，2018）。因此，建立不同组织和细胞型中的蛋白质互作网络谱可在网络层面上揭示复杂生物现象的分子机制。

3. 刻画生物系统状态的网络标志物理论

生物标志物是生物系统状态表征及医学定量检测的最基本工具，但传统的分子标志物一般来说随着时间和条件变化，其稳定性和准确性不能满足当前生物状态表征和医疗检测的需求。而从系统的观点来看，网络是表征生物系统状态的稳定标志。由观测的分子数据，建立单样本网络构建方法，可由单个样本构建网络标志物（Liu et al.，2016；Liu et al.，2019；Yu et al. 2014；Dai et al.，2019），因此可以实现由网络或网络熵定量表征生物系统状态或诊断疾病（Liu et al.，2016；Liu et al.，2019；Huang et al.，2020）。不同于分子的表达量等，网络标志物是由一群分子的关联性来表征生物系统状态的，所以网络标志物相对于单分子标记具有维度更高等特点，也使得网络标志物具有比传统的单分子标志物稳定性和准确性更高、误差率更低等优点。

4. 刻画生物系统临界状态的动态网络标志物理论

复杂的生物现象或过程是典型的非线性动态过程，由一个状态逐渐转化为另一状态，如细胞分化过程、慢性疾病过程、复杂性状演化等。这些过程都具有一个共同节点，即从一个状态到另一状态间存在一个“临界状态”（critical point or tipping point）。该临界状态是生物动态过程的关键节点，因此理解临界状态不仅可揭示复杂动态生物过程的分子机制而且可实现疾病预警，有重要的生物医学意义。基于动力学的分叉理论建立的动态网络标志物（dynamic network biomarker，DNB）理论和方法（Chen et al.，2012；Yang et al.，2018；Liu et al.，2019），可由网络波动特征预测生物动态过程或疾病的临界状态，由此可定量地检测临界状态或预警疾病，从而为分析复杂生物过程和实现“未病”预测提供理论。临界理论还可为健康人群的健康状态定量评估提供理论与方法，实现预防性医学，不仅可以显著延迟健康状态恶化的临界状态的到来，从而极大改善人类健康及提高生存质量，还可应用于流行性疾病的预警和生物非线性进化的研究。

四、我国未来研究与布局建议

进入后基因组时代，生命科学发生了重大转变。一方面，生命科学的研究可以转化为巨大的生产力，形成了一个新的产业，即生命科学工业。另一方面，人类基因组计划已开始进入由结构基因组学向功能基因组学，特别是网络生物学（如ENCODE计划）过渡、转化的过程。因此，我国应该在网络生物学领域的基础研究和生产力转化两方面加强布局。在加大资金投入和政策支持力度吸引多学科、多实验室协作研究的同时，也要意识到网络生物学研究将成为我国生命科学知识创新和技术创新，以及产业发展新的生成点。

本章参考文献

Alvarez M J, Shen Y, Giorgi F M, et al. 2016. Functional characterization of somatic mutations in cancer using network-based inference of protein activity. Nat Genet, 48（8）: 838-847.

Amin V, Agac D, Barnes S D, et al. 2019. Accurate differential analysis of transcription factor activity from gene expression. Bioinformatics, 35（23）: 5018-5029.

Anastasiadou E, Jacob L S, Slack F J. 2018. Non-coding RNA networks in cancer. Nat Rev Cancer, 18（1）: 5-18.

Ardura J A, Álvarez-Carrión L, Gutiérrez-Rojas I, et al. 2020. Role of calcium signaling in prostate cancer progression: effects on cancer hallmarks and bone metastatic mechanisms.

Cancers（Basel），12（5）：1071.

Bila H, Kurisinkal E E, Bastings M M C. 2019. Engineering a stable future for DNA-origami as a biomaterial. Biomater Sci, 7（2）：532-541.

Blaise B J, Shintu L, Elena B N D, et al. 2009. Statistical recoupling prior to significance testing in nuclear magnetic resonance based metabonomics. Analytical Chemistry, 81（15）：6242-6251.

Bode A M, Dong Z. 2004. Post-translational modification of p53 in tumorigenesis. Nature Reviews Cancer, 4（10）：793-805.

Bouché N, Bouchez D. 2001. Arabidopsis gene knockout: phenotypes wanted. Current Opinion in Plant Biology, 4（2）：111-117.

Bracken C P, Scott H S, Goodall G J. 2016. A network-biology perspective of microRNA function and dysfunction in cancer. Nat Rev Genet, 17（12）：719-732.

Braun P, Tasan M, Dreze M, et al. 2009. An experimentally derived confidence score for binary protein-protein interactions. Nat Methods, 6（1）：91-97.

Caetano-Anolles G, Kim H S, Mittenthal J E. 2007. The origin of modern metabolic networks inferred from phylogenomic analysis of protein architecture. Proc Natl Acad Sci U S A, 104（22）：9358-9363.

Chao Z, Ping H, Liang Q L, et al. 2008. Integration of metabonomics technology and its application in modernization of traditional Chinese medicine. Acta Pharmaceutica Sinica, 43（7）：683-689.

Chekouo T, Stingo F C, Doecke J D, et al. 2015. miRNA–target gene regulatory networks: A Bayesian integrative approach to biomarker selection with application to kidney cancer. Biometrics, 71（2）：428-438.

Chen C, Li R, Shu L, et al. 2020. Predicting future dynamics from short-term time series using an Anticipated Learning Machine. National Science Review, 7（6）：1079-1091.

Chen G, Ning B, Shi T. 2019. Single-cell RNA-seq technologies and related computational data analysis. Frontiers in Genetics, 10: 317.

Chen L. 2010. Modeling Biomolecular Networks in Cells: Structures and Dynamics London, New York: Springer.

Chen L, Liu R, Liu Z-P, et al. 2012. Detecting early-warning signals for sudden deterioration of complex diseases by dynamical network biomarkers. Sci Rep, 2（1）：1-8.

Chen L, Wang R S, Zhang X S. 2009. Biomolecular networks: methods and applications in systems biology. New York: John Wiley & Sons.

Chen P, Liu R, Aihara K, et al. 2020. Autoreservoir computing for multistep ahead prediction based on the spatiotemporal information transformation. Nat Commun, 11（1）: 4568.

Chiang C, Gack M U. 2017. Post-translational control of intracellular pathogen sensing pathways. Trends Immunol, 38（1）: 39-52.

Church G M, Gao Y, Kosuri S. 2012. Next-generation digital information storage in DNA. Science, 337（6102）: 1628.

Consortium E P, Moore J E, Purcaro M J, et al. 2020. Expanded encyclopaedias of DNA elements in the human and mouse genomes. Nature, 583（7818）: 699-710.

Consortium F, Pmi P, Clst, et al. 2014. A promoter-level mammalian expression atlas. Nature, 507（7493）: 462-470.

Dahm R. 2008. Discovering DNA: Friedrich Miescher and the early years of nucleic acid research. Hum Genet, 122（6）: 565-581.

Dai H, Li L, Zeng T, et al. 2019. Cell-specific network constructed by single-cell RNA sequencing data. Nucleic Acids Res, 47（11）: e62.

Deribe Y L, Pawson T, Dikic I. 2010. Post-translational modifications in signal integration. Nat Struct Mol Biol, 17（6）: 666-672.

Dettmer K, Aronov P A, Hammock B D. 2007. Mass spectrometry-based metabolomics. Mass Spectrometry Reviews, 26（1）:51-78

Dolinoy D C, Weidman J R, Jirtle R L. 2007. Epigenetic gene regulation: linking early developmental environment to adult disease. Reprod Toxicol, 23（3）: 297-307.

Drews J. 2000. Drug discovery: a historical perspective. Science, 287（5460）: 1960-1964.

Du X, Sun S, Hu C, et al. 2017. DeepPPI: boosting prediction of protein-protein interactions with deep neural networks. J Chem Inf Model, 57（6）: 1499-1510.

Duggan D J, Bittner M, Chen Y, et al. 1999. Expression profiling using cDNA microarrays. Nat Genet, 21（1）: 10-14.

Dunn W B, Ellis D I. 2005. Metabolomics: Current analytical platforms and methodologies. Trends in Analytical Chemistry, 24（4）: 285-294.

Eleftherohorinou H, Hoggart C J, Wright V J, et al. 2011. Pathway-driven gene stability selection of two rheumatoid arthritis GWAS identifies and validates new susceptibility genes in receptor mediated signalling pathways. Hum Mol Genet, 20（17）: 3494-3506.

Ernst J, Kellis M. 2012. ChromHMM: automating chromatin-state discovery and characterization.

Nat Methods, 9（3）: 215-216.

Extance A. 2016. Digital DNA. Nature, 537（7619）: 150.

Fiehn O. 2002. Metabolomics - the link between genotypes and phenotypes. Plant Molecular Biology, 48（1-2）: 155-171.

Field D, Sansone S A. 2006. A Special Issue on Data Standards. Omics-a Journal of Integrative Biology, 10（2）: 84-93.

Florcs-Rodríguez P, Ontiveros-Torres M A, Cárdenas-Aguayo M C, et al. 2015. The relationship between truncation and phosphorylation at the C-terminus of tau protein in the paired helical filaments of Alzheimer's disease. Front Neurosci, 9: 33.

Freddolino P L, Tavazoie S. 2012. The dawn of virtual cell biology. Cell, 150（2）: 248-250.

Garcia-Garcia J, Schleker S, Klein-Seetharaman J, et al. 2012. BIPS: BIANA Interolog Prediction Server. A tool for protein–protein interaction inference. Nucleic Acids Res, 40（W1）: W147-W151.

Geladi P, Kowalski B R. 1985. Partial least-squares regression: a tutorial. Analytica Chimica Acta, 185（1）: 1-17.

Gerlich M, Neumann S. 2000. KEGG: kyoto encyclopedia of genes and genomes. Nuclc Acids Research, 28（1）: 27-30.

Goldford J E, Segrè D. 2018. Modern views of ancient metabolic networks. Current Opinion in Systems Biology: 117-124.

Greene C S, Krishnan A, Wong A K, et al. 2015. Understanding multicellular function and disease with human tissue-specific networks. Nat Genet, 47（6）: 569-576.

Guimera R, Amaral L A N. 2005. Functional cartography of complex metabolic networks. Nature, 433（7028）: 895-900.

Guo Y. 2007. Regularized linear discriminant analysis and its application in microarrays. Biostatistics, 8（1）: 86-100.

Gupta S, Campbell D, Derijard B, et al. 1995. Transcription factor ATF2 regulation by the JNK signal transduction pathway. Science, 267（5196）: 389-393.

Hasty J, McMillen D, Isaacs F, et al. 2001. Computational studies of gene regulatory networks: In numero molecular biology. Nature Reviews Genetics, 2（4）: 268-279.

Hatchwell E, Greally J M. 2007. The potential role of epigenomic dysregulation in complex human disease. Trends Genet, 23（11）: 588-595.

Honigberg, M. S. 2003. Signal pathway integration in the switch from the mitotic cell cycle to meiosis in yeast. Journal of Cell Science, 116（11）: 2137-2147.

Hopkins A L, Groom C R. 2002. The druggable genome. Nature Reviews Drug Discovery, 1（9）: 727-730.

Huang Y, Chang X, Zhang Y, et al. 2020. Disease characterization using a partial correlation-based sample-specific network. Briefings in Bioinformatics, 22（6）: bbaa062

Imming P, Sinning C, Meyer A. 2006. Drugs, their targets and the nature and number of drug targets. Nature reviews Drug discovery, 5（10）: 821-834.

Ivashkiv L B. 2009. Cross-regulation of signaling by ITAM-associated receptors. Nat Immunol, 10（4）: 340-347.

Jackson A L, Bartz S R, Schelter J, et al. 2003. Expression profiling reveals off-target gene regulation by RNAi. Nature Biotechnology, 21（6）: 635-637.

Jain A K, Mao J, Mohiuddin K M. 2015. Artificial neural networks: a tutorial. Computer, 29（3）: 31-44.

Jassal B, Matthews L, Viteri G, et al. 2020. The reactome pathway knowledgebase. Nucleic Acids Res, 48（D1）: D498-D503.

Jeong H, Tombor B, Albert R, et al. 2000. The large-scale organization of metabolic networks. Nature, 407（6804）: 651-654.

Johnson D S, Mortazavi A, Myers R M, et al. 2007. Genome-wide mapping of *in vivo* protein-DNA interactions. Science, 316（5830）: 1497-1502.

Kaufman L, Rousseeuw P J. 1990. Finding Groups in Data: An Introduction to Cluster Analysis. New York: John Wiley.

Kendall, Powell. 2018. How biologists are creating life-like cells from scratch. Nature, 563（8）: 172-175.

Kim T H, Böhmer M, Hu H, et al. 2010b. Guard Cell Signal Transduction Network: Advances in Understanding Abscisic Acid, CO_2, and Ca^{2+} Signaling. Annual Review of Plant Biology, 61（1）: 561-591.

Koch J, Gantenbein S, Masania K, et al. 2020. A DNA-of-things storage architecture to create materials with embedded memory. Nat Biotechnol, 38（1）: 39-43.

Koenigsberg C, Ondrey F G. 2020. Genomic database analysis for head and neck cancer prevention targets: MTOR signal transduction pathway. Anticancer Res, 40（10）: 5417-5421.

Krueger K E, Srivastava S. 2006. Posttranslational protein modifications: current implications for cancer detection, prevention, and therapeutics. Mol Cell Proteomics, 5（10）: 1799-1810.

Kümmel A, Panke S, Heinemann M. 2006. Putative regulatory sites unraveled by network-embedded thermodynamic analysis of metabolome data. Molecular Systems Biology, 2（1）: 2006.0034

Kuo D, Licon K, Bandyopadhyay S, et al. 2010. Coevolution within a transcriptional network by compensatory trans and cis mutations. Genome Res, 20（12）: 1672-1678.

Landry Y, Gies J P. 2008. Drugs and their molecular targets: an updated overview. Fundam Clin Pharmacol, 22（1）: 1-18.

Lazar A G, Romanciuc F, Socaciu M A, et al. 2015. Bioinformatics tools for metabolomic data processing and analysis using untargeted liquid chromatography coupled with mass spectrometry. Bulletin of University of Agricultural Sciences & Veterinary Medicine Cluj Napoca Animal Science & Biotechnologies, 72（2）: 103-115.

Leach K, Hannan F M, Josephs T M, et al. 2020. International Union of Basic and Clinical Pharmacology. CVIII. calcium-sensing receptor nomenclature, pharmacology, and function. Pharmacol Rev, 72（3）: 558-604.

Leng S, Ma H, Kurths J, et al. 2020. Partial cross mapping eliminates indirect causal influences. Nat Commun, 11（1）: 2632.

Li J H, Liu S, Zhou H, et al. 2014. starBase v2.0: decoding miRNA-ceRNA, miRNA-ncRNA and protein-RNA interaction networks from large-scale CLIP-Seq data. Nucleic Acids Res, 42（Database issue）: D92-97.

Li S, Assmann S M, Albert R. 2006. Predicting essential components of signal transduction networks: a dynamic model of guard cell abscisic acid signaling. PLoS Biol, 4（10）: e312.

Li S P, Jiang Q, Liu S L, et al. 2018. A DNA nanorobot functions as a cancer therapeutic in response to a molecular trigger *in vivo*. Nature Biotechnology, 36（3）: 258.

Likic V A, McConville M J, Lithgow T, et al. 2010. Systems biology: the next frontier for bioinformatics. Adv Bioinformatics: 268925.

Lin H, Begley T. 2011. Protein posttranslational modifications: chemistry, biology, and applications. Molecular BioSystems, 7（1）: 14-15.

Liu X, Chang X, Leng S, et al. 2019. Detection for disease tipping points by landscape dynamic network biomarkers. National Science Review, 6（4）: 775-785.

Liu X, Wang Y, Ji H, et al. 2016. Personalized characterization of diseases using sample-specific networks. Nucleic Acids Research, 44（22）: e164.

Loose M, Fischer-Friedrich E, Ries J, et al. 2008. Spatial regulators for bacterial cell division self-organize into surface waves *in vitro*. Science, 320（5877）: 789-792.

Ma H, Leng S, Aihara K, et al. 2018. Randomly distributed embedding making short-term high-dimensional data predictable. Proc Natl Acad Sci U S A, 115（43）: E9994-E10002.

Marbach D, Costello J C, Küffner R, et al. 2012. Wisdom of crowds for robust gene network inference. Nat Methods, 9（8）: 796-804.

Marbach D, Lamparter D, Quon G, et al. 2016. Tissue-specific regulatory circuits reveal variable modular perturbations across complex diseases. Nat Methods, 13（4）: 366-370.

Marchi S, Giorgi C, Galluzzi L, et al. 2020. Ca^{2+} fluxes and cancer. Mol Cell, 78（6）: 1055-1069.

Mardinoglu A, Nielsen J. 2012. Systems medicine and metabolic modelling. Journal of Internal Medicine, 271（2）: 142-154.

Margolin A A, Nemenman I, Basso K, et al. 2006. ARACNE: an algorithm for the reconstruction of gene regulatory networks in a mammalian cellular context. BMC Bioinformatics, 7（Suppl 1）: S7.

Marioni J C, Mason C E, Mane S M, et al. 2008. RNA-seq: an assessment of technical reproducibility and comparison with gene expression arrays. Genome Res, 18（9）: 1509-1517.

Matsumoto H, Kiryu H, Furusawa C, et al. 2017. SCODE: an efficient regulatory network inference algorithm from single-cell RNA-Seq during differentiation. Bioinformatics, 33（15）: 2314-2321.

Miao L, Yin R X, Zhang Q H, et al. 2019. A novel circRNA-miRNA-mRNA network identifies circ-YOD1 as a biomarker for coronary artery disease. Sci Rep, 9（1）: 18314.

Moore, B. 1981. Principal component analysis in linear systems: controllability, observability, and model reduction. IEEE Transactions on Automatic Control, 26（1）: 17-32.

Nakanishi S, Kageyama R, Watanabe D. 2009. Systems Biology: the Challenge of Complexity. Tokyo, New York: Springer.

Nakhai H, Siveke J T, Klein B, et al. 2008. Conditional ablation of Notch signaling in pancreatic development. Development, 135（16）: 2757-2765.

Nangreave J, Han D, Liu Y, et al. 2010. DNA origami: a history and current perspective. Curr Opin Chem Biol, 14（5）: 608-615.

Ong C T, Corces V G. 2014. CTCF: an architectural protein bridging genome topology and function. Nat Rev Genet, 15（4）: 234-246.

Organick L, Ang S D, Chen Y J, et al. 2018. Random access in large-scale DNA data storage. Nat Biotechnol, 36（3）: 242-248.

Orlando D A, Lin C Y, Bernard A, et al. 2008. Global control of cell-cycle transcription by coupled CDK and network oscillators. Nature, 453（7197）: 944-947.

Orom U A, Derrien T, Beringer M, et al. 2010. Long noncoding RNAs with enhancer-like function in human cells. Cell, 143（1）: 46-58.

Overington J P, Al-Lazikani B, Hopkins A L. 2006. How many drug targets are there? Nature reviews Drug discovery, 5（12）: 993-996.

Panda D, Molla K A, Baig M J, et al. 2018. DNA as a digital information storage device: hope or hype? 3 Biotech, 8（5）: 239.

Petronis A. 2010. Epigenetics as a unifying principle in the aetiology of complex traits and diseases. Nature, 465（7299）: 721-727.

Plowman S J, Hancock J F. 2005. Ras signaling from plasma membrane and endomembrane microdomains. Biochimica et Biophysica Acta（BBA）-Molecular Cell Research, 1746（3）: 274-283.

Portela A, Esteller M. 2010. Epigenetic modifications and human disease. Nat Biotechnol, 28（10）: 1057.

Ravasz E, Somera A L, Mongru D A, et al. 2002. Hierarchical organization of modularity in metabolic networks. Science, 297（5586）: 1551-1555.

Reis C A, Osorio H, Silva L, et al. 2010. Alterations in glycosylation as biomarkers for cancer detection. J Clin Pathol, 63（4）: 322-329.

Rendeiro A F, Schmidl C, Strefford J C, et al. 2016. Chromatin accessibility maps of chronic lymphocytic leukaemia identify subtype-specific epigenome signatures and transcription regulatory networks. Nat Commun, 7: 11938.

Rothemund P W. 2006. Folding DNA to create nanoscale shapes and patterns. Nature, 440（7082）: 297-302.

Rubiolo M, Milone D H, Stegmayer G. 2018. Extreme learning machines for reverse engineering of gene regulatory networks from expression time series. Bioinformatics, 34（7）: 1253-1260.

Rutten M G T A, Vaandrager F W, Elemans J A A W, et al. 2018. Encoding information into polymers. Nature Reviews Chemistry, 2（11）: 365-381.

Sammon J W. 1969. A nonlinear mapping for data analysis. IEEE Transactions on Computers, C-18（5）: 401-409.

Schaff J, Loew L M. 1999. The virtual cell. Pacific Symposium on Biocomputing, 4: 228-239,

Selvaggio G, Canato S, Pawar A, et al. 2020. Hybrid epithelial-mesenchymal phenotypes are controlled by microenvironmental factors. Cancer Res, 80（11）: 2407-2420.

Shaw R J, Cantley L C. 2006. Ras, PI（3）K and mTOR signalling controls tumour cell growth. Nature, 441（7092）: 424-430.

Shen J, Zhang J, Luo X, et al. 2007. Predicting protein-protein interactions based only on sequences information. Proceedings of the National Academy of Sciences, 104（11）: 4337-4341.

Shipman S L, Nivala J, Macklis J D, et al. 2017. CRISPR-Cas encoding of a digital movie into the genomes of a population of living bacteria. Nature, 547（7663）: 345-349.

Sidiropoulos K, Viteri G, Sevilla C, et al. 2017. Reactome enhanced pathway visualization. Bioinformatics, 33（21）: 3461-3467.

Silva T C, Coetzee S G, Gull N, et al. 2019. ELMER v.2: an R/Bioconductor package to reconstruct gene regulatory networks from DNA methylation and transcriptome profiles. Bioinformatics, 35（11）: 1974-1977.

Šimić G, Babić Leko M, Wray S, et al. 2016. Tau protein hyperphosphorylation and aggregation in Alzheimer's disease and other tauopathies, and possible neuroprotective strategies. Biomolecules, 6（1）: 6.

Singh A. 2020. Tools for metabolomics. Nat Methods, 17（1）: 24.

Stowell S R, Ju T, Cummings R D. 2015. Protein glycosylation in cancer. Annual Review of Pathology: Mechanisms of Disease, 10: 473-510.

Tang F, Barbacioru C, Wang Y, et al. 2009. mRNA-Seq whole-transcriptome analysis of a single cell. Nat Methods, 6（5）: 377-382.

Thakur A K, Movileanu L. 2019. Real-time measurement of protein-protein interactions at single-molecule resolution using a biological nanopore. Nat Biotechnol, 37（1）: 96-101.

Thiel D, Conrad N D, Ntini E, et al. 2019. Identifying lncRNA-mediated regulatory modules via ChIA-PET network analysis. BMC Bioinformatics, 20（1）: 292.

Tinberg C E, Khare S D, Dou J, et al. 2013. Computational design of ligand-binding proteins with high affinity and selectivity. Nature, 501（7466）: 212-216.

Tomita M, Hashimoto K, Takahashi K, et al. 2000. The E-CELL project: towards integrative simulation of cellular processes. Paper presented at: International Conference on Computational Molecular Biology.

Tomita M, Hashimoto K, Takahashi K, et al. 1999. E-CELL: software environment for whole-cell simulation. Bioinformatics, 15（1）: 72-84.

Vabulas R M, Ahmad-Nejad P, Ghose S, et al. 2002. HSP70 as endogenous stimulus of the Toll/interleukin-1 receptor signal pathway. Journal of Biological Chemistry, 277（17）: 15107-15112.

Vanunu O, Magger O, Ruppin E, et al. 2010. Associating genes and protein complexes with disease via network propagation. PLoS Comput Biol, 6（1）: e1000641.

Venter J C, Adams M D, Myers E W, et al. 2001. The sequence of the human genome. Science, 291（5507）: 1304-1351.

Venter J C, Gibson D. 2010. How We Created the First Synthetic Cell. Wall Street Journal Eastern Edition.

Wang L, Gai S. 2014. The next generation mass storage devices - Physical principles and current status. Contemporary Physics, 55（2）: 75-93.

Wang X, Lockhart S M, Rathjen T, et al. 2016. Insulin downregulates the transcriptional coregulator CITED2, an inhibitor of proangiogenic function in endothelial cells. Diabetes, 65（12）: 3680.

Wang X, Wei X, Thijssen B, et al. 2012. Three-dimensional reconstruction of protein networks provides insight into human genetic disease. Nat Biotechnol, 30（2）: 159-164.

Weigt M, White R A, Szurmant H, et al. 2009. Identification of direct residue contacts in protein-protein interaction by message passing. Proceedings of the National Academy of Sciences, 106（1）: 67-72.

Wu X, Jiang R, Zhang M Q, et al. 2008. Network - based global inference of human disease genes. Mol Syst Biol, 4（1）: 189.

Xie L, Li J, Xie L, et al. 2009. Drug discovery using chemical systems biology: Identification of the protein-ligand binding network to explain the side effects of CETP inhibitors. PLoS Comput Biol, 5（5）: e1000387.

Xu Z, Horwich A L, Sigler P B. 1997. The crystal structure of the asymmetric GroEL–GroES–（ADP）7 chaperonin complex. Nature, 388（6644）: 741-750.

Yang B, Li M, Tang W, et al. 2018. Dynamic network biomarker indicates pulmonary metastasis at the tipping point of hepatocellular carcinoma. Nature Communications, 9（1）: 1-14.

Yu X, Li G, Chen L. 2014. Prediction and early diagnosis of complex diseases by edge-network. Bioinformatics, 30（6）: 852-859.

Yuan L, Huang D S. 2019. A Network-guided association mapping approach from DNA methylation to disease. Sci Rep, 9（1）: 5601.

Zhang L, Yu G, Xia D, et al. 2019. Protein-protein interactions prediction based on ensemble deep neural networks. Neurocomputing, 324: 10-19.

Zhang Q C, Petrey D, Deng L, et al. 2012. Structure-based prediction of protein-protein interactions on a genome-wide scale. Nature, 490（7421）: 556-560.

Zhang W, Zeng T, Liu X, et al. 2015. Diagnosing phenotypes of single-sample individuals by edge biomarkers. J Mol Cell Biol, 7（3）: 231-241.

Zhao J, Zhou Y, Zhang X, et al. 2016. Part mutual information for quantifying direct associations in networks. Proc Natl Acad Sci U S A, 113（18）: 5130-5135.

Zheng G, Xu Y, Zhang X, et al. 2016. CMIP: a software package capable of reconstructing genome-wide regulatory networks using gene expression data. BMC Bioinformatics, 17（17）: 535.

Zheng M, Barrera L O, Ren B, et al. 2007. ChIP - chip: data, model, and analysis. Biometrics, 63（3）: 787-796.

Zhu C, Liang Q L, Wang Y M, et al. 2010. Integrated development of metabonomics and its new progress. Chinese Journal of Analytical Chemistry, 38（7）: 1060-1068.

Zhu Q, Tekpli X, Troyanskaya O G, et al. 2020. Subtype-specific transcriptional regulators in breast tumors subjected to genetic and epigenetic alterations. Bioinformatics, 36（4）: 994-999.

第十一章

细胞的人工改造

第一节　概　　述

合成生物学为各种功能细胞的改造或功能优化提供了一种强大的工具。合成生物学是基于工程学的思维和遗传学方法，通过理性的设计改造基因线路，赋予细胞或生物体新的功能。通过认识和优化天然生命体中的多种调控机制，改造并拓展一系列生物元件和模块，让具有活性的细胞来完成设计人员设想的各种任务。近年来，随着高通量测序技术和基因编辑技术的进步，很多物种的全基因组已经被解析，生命系统通过数十亿年进化而来的各种功能系统有望被我们全面加以改造利用，制备出性能更优化的功能性细胞。以哺乳动物细胞的合成生物学改造为基础的细胞治疗技术突飞猛进，在生物医学技术领域中展现出了巨大的潜力。

其中，干细胞由于具有自我复制能力，并且能够分化成多种功能性细胞而备受关注。根据分化潜能，可分为全能干细胞、多能干细胞和专能干细胞。2006 年，山中伸弥发现了诱导多能干细胞（induced pluripotent stem cell，iPSC）。iPSC 排除了伦理学争论，可以从患者自身的细胞产生。但在将 iPSC

衍生的细胞应用于细胞治疗之前，仍然需要解决许多挑战。这些挑战包括检测和去除未完全诱导分化的细胞，解决产生的细胞中的基因组和表观遗传学的改变以及克服移植时可能出现的致瘤性等。

在各种功能性细胞中，免疫细胞是一类在癌症、传染病和自身免疫病等各种疾病中具有关键的调控作用的细胞。免疫治疗在许多此前无药可医的晚期转移癌症患者中取得了持续性的治疗效果，从而彻底改变了癌症的传统治疗模式（June et al.，2018；Wolchok et al.，2017）。免疫细胞疗法，由于其靶向性强、可操作空间大等特点而备受瞩目，是当前科研和临床实验的前沿。通过对免疫细胞尤其是T细胞的改造让其表达嵌合抗原受体（CAR）获得CAR-T细胞，在白血病、淋巴瘤等血液系统肿瘤中取得了显著的临床成效，是细胞治疗领域的革命性突破。CAR-T细胞产品Kymriah和Yescarta于2017年获批上市，让免疫细胞治疗获得了极大的关注，同时被寄予了彻底治愈癌症的厚望。2019年初美国美国食品药品监督管理局（FDA）宣布了细胞治疗发展计划，预测截至2025年这一领域每年将有10～20款新药获得批准。

虽然以CAR-T细胞为代表的细胞治疗技术在癌症等多种疾病的精准诊疗上取得了前所未有的成就，但是我们对细胞内部各类复杂功能的理解和运用还非常有限，基于合成生物学的细胞疗法仍有巨大的开发空间。例如，细胞内部信号的调控网络涉及染色质结构调控、转录调控、转录后调控以及翻译后调控等诸多层次，各调控层次间的时空关联对于蛋白质在细胞内发挥正确功能起到关键性作用，其中各种生物过程的改造和开发对于构建更加复杂、更加精准的合成基因线路和网络具有重大意义。功能细胞的改造和功能优化取决于我们对细胞内部功能调节基因及调控环路的系统性认识，如某些调节基因的表达往往决定了细胞接收环境信号或识别抗原后反应的类型和强度。尽管，有些决定细胞功能和命运的基因已经被确定，但我们对于在复杂体内环境中细胞的功能和命运的调控网络仍缺乏系统和全面的认识。因此，需要进一步通过多组学分析和筛选来鉴别在细胞多种状态和功能中具有关键作用的基因环路，进而根据功能化或治疗需求合理利用合成生物学手段对细胞进行改造，进而开发出更切实可用的新型细胞疗法。

第二节　关键科学问题

一、解析制约细胞疗法的关键因素，优化细胞功能

细胞疗法的临床应用面临的挑战首先涉及对复杂体内微环境以及细胞命运可塑性的影响。例如，免疫细胞疗法的响应性往往受肿瘤微环境中众多动态因素如肿瘤与免疫细胞之间、免疫亚群之间复杂相互作用的影响。此外，瘤内积累的趋化因子、细胞因子和代谢产物对细胞疗法有重要的调节作用。大规模功能筛选为系统性解析免疫细胞应答中的关键基因及其功能提供了一种有效手段。目前，我国国内外多个课题组针对多种免疫细胞及其免疫调节通路进行了大规模筛选，发现了许多潜在的免疫细胞改造的靶点（Simeonov and Marson，2019）。但是，目前大多数筛选仍停留在临床相关度较低的模型上，不能良好地模拟人类肿瘤在遗传和免疫上的复杂性。

设计新型的疾病模型筛选策略，在模拟人类肿瘤、自身免疫病等疾病关键特征的环境下针对关键科学问题进行筛选是下一步要开展研究的主要方向。体内的功能细胞如干细胞、T 细胞和天然免疫细胞都存在多种亚型和非常强的可塑性，如何通过改造实现对这些细胞命运的靶向调控和塑造，摆脱不利环境的“说服”是解决细胞疗法持久性的关键之一。在免疫细胞疗法方面，要找出可以帮助免疫细胞克服免疫抑制微环境的靶基因或通路，推进免疫细胞疗法的应用。此外，对某些细胞治疗的获得性耐药尤其是随着时间的推移而获得的耐药，也是免疫细胞改造需要解决的问题。因此，需要从临床需要和免疫机制出发设计更加细致可靠的筛选和改造策略，通过对功能性基因和通路的改造，发展出更具临床应用价值的人工免疫细胞疗法。

二、制备可“现货”（off-the-shelf）的通用型免疫细胞

目前市场上的CAR-T或TCR-T疗法属于自体细胞治疗，需要对每个癌症病人自身的T细胞进行改造、扩增后重新注回病人体内。这种个性化细胞治疗存在着一些明显的局限性，例如患者自体T细胞分离制备需要较长的时间和昂贵的成本，以及部分患者前期治疗导致的T细胞受损而无法用于细胞治疗等。为解决这些问题，多家研究机构和制药公司正在推动下一代细胞治疗技术开发，即通用型（同种异体）T细胞疗法。通用型T细胞设计就是基于基因编辑技术或其他技术使同种异体方式产生的T细胞过继输注给患者，启动特定的肿瘤免疫杀伤功能，同时避免供体T细胞与患者发生移植物抗宿主病（graft versus host disease，GvHD，又称作免疫排斥）。将T细胞疗法拓展为通用型T细胞疗法，使其具备off-the-shelf供应链，具有能规模化生产、剂量恒定、疗效优越等特点是大势所趋。破坏同种异体T细胞的*TCR*基因和*HLA I*类基因来降低或消除GvHD，是目前广泛应用的制备通用CAR-T细胞的策略。然而在实际应用中，仍存在一些关键性问题，如残留TCR引起的GvHD反应、基因编辑的效率和脱靶风险，以及细胞毒性问题等仍是通用型CAR-T面临的巨大挑战。通用型T细胞的巨大治疗潜力和吸引力足以成为推动其发展的强大动力，不断突破和创新是开发新型通用细胞疗法的康庄大道。

三、开发具有高度敏感的环境信号响应型“智能细胞疗法”

智能细胞疗法可以通过识别环境或外部各类输入信号，可控地诱导细胞合成和释放各类功能性蛋白或药物，达到治疗的目的。目前已经存在多种相对原始的可用于精准调控哺乳动物细胞基因表达的新技术，即智能细胞疗法。在植入患者体内后，智能细胞犹如一个定制化的生物处理器，根据外界信号如pH、细胞因子、气味甚至是手机信号，调节各种功能性蛋白的合成。如何通过人工改造，在治疗性的功能化细胞中导入基因线路或实现某些治疗功能的时空调控，从而在功能性细胞内构建受各种环境因素控制的基因环路，赋予细胞接受特定信号后快速合成目标蛋白质或效应因子的能力。智能细胞将

有效克服此前细胞治疗中最关键的环境响应性不足的问题，推动多种基于基因组学、生物信息学和合成生物学的基因工程细胞疗法，实现真正的精准医疗。智能细胞可以通过感受环境信号或人为提供的信号来控制多种功能性基因和蛋白质的表达，从而为癌症、糖尿病等疾病治疗提供更精准可靠的新疗法。

第三节　重要研究方向的发展现状与趋势分析

一、T 细胞改造

CAR-T 细胞疗法是癌症治疗的一大进步，它主要通对患者 / 供者 T 细胞的体外基因改造和扩增，使其表达能够识别癌细胞特定抗原的嵌合抗原受体（CAR），再回输到患者体内来发挥抗癌作用。CAR 通过识别癌细胞表面的靶点，来引导 T 细胞摧毁癌细胞。2017 年，FDA 批准了两种 CAR-T 细胞疗法，诺华公司的 CAR-T 产品 Kymriah 用于治疗急性淋巴细胞白血病，Gilead Science/Kite Pharma 公司的 CAR-T 产品 Yescarta 用于治疗复发 / 难治性大 B 细胞淋巴瘤。处于各研发阶段的 CAR-T 疗法更是方兴未艾。例如，Juno Therapetics、Bluebird Bio、Cellectis、上海恒润达生生物科技股份有限公司、科济生物医药（上海）有限公司、博生吉安科细胞技术有限公司等均有多个处于临床阶段的 CAR-T 项目。截至 2020 年 6 月 30 日，中国注册 CAR-T 相关临床试验有 357 个，美国注册的有 256 个，其他国家 58 个，中美两国主导行业发展的格局已经显现。

TCR-T 疗法是细胞治疗领域的另一核心疗法。许多癌症靶点尤其是基因突变产生的靶点蛋白质位于癌细胞内，因此 CAR-T 细胞不能有效识别这些靶点，限制了其在实体瘤治疗中的应用。因此，许多生物医药公司将目光转向 TCR-T 细胞免疫疗法，因为 TCR-T 细胞可以通过扫描细胞表面人类白细胞抗原（human leucocyte antigen，HLA）来识别隐藏在癌细胞内的目标，在捕杀实体瘤方面有独特的优势。目前TCR-T的全球领先者是在纳斯达克上市的英国公司Adaptimmune Therapeutics公司，其拥有的MAGE-A4、AFP、MAGE-A10三个靶点的TCR-T（获FDA批准）处于临床试验Ⅱ期或者Ⅰ期。我国的广东香雪精准医疗技术有限公司布局第三代TCR-T产品，研发出的TAEST16001于2019年3月获得国家药品监督管理局药品审评中心（Center for Drug Evaluation, CDE）批准进行新药临床试验，是我国首个获批的TCR-T细胞治疗药物。

在CAR-T和TCR-T等细胞疗法的基础上，CRISPR-Cas9基因编辑技术为增强T细胞的天然抗癌能力进一步提供了强有力的工具。2020年2月*Science*报道了Carl H. June领导的团队利用CRISPR多重改造T细胞治疗癌症的临床试验结果。该团队通过CRISPR技术敲除T细胞内的*TRAC*和*TRBC*来避免天然TCR与基因工程插入受体产生错误配对和竞争，同时敲除*PDCD1*限制T细胞衰竭，在患者体内显示出长期安全的疗效（Stadtmauer et al.，2020）。

我国科学家在T细胞改造尤其是CAR-T开发方面的工作卓有成效，与国际研究并驾齐驱，取得了一系列国际先进成果。例如，四川大学华西医院卢铀团队于2016年10月完成了世界首例CRISPR基因编辑T细胞治疗癌症的Ⅰ期临床试验（https://www.clinicaltrials.gov/ NCT02793856）。他们利用*PD1*基因敲除的T细胞来治疗非小细胞肺癌，发现*PD1*敲除T细胞在临床上安全可行，将患者无进展中位生存期从7.7周延长至42.6周，相关成果于2020年4月发表在*Nature Medicine*上（Lu et al.，2020）。中国科学院分子细胞科学卓越创新中心许琛琦团队在*Cell*上报道了一种发新的CAR-T细胞疗法，他们将CD3e细胞质域整合到目前临床使用的CAR分子中，降低了细胞因子分泌导致的细胞因子风暴（CRS），促进了CAR-T细胞在体内的生长、存活和持续性（Wu et al.，2020）。北京大学邓宏魁、与中国人民解放军总医院陈虎与北京佑安医院吴昊合作，于2019年9月在《新英格兰医学杂志》（*The New England Journal of Medicine*）上报道了CRISPR敲除*CCR5*后的造血干细胞/祖细胞（HSPC）治疗HIV-1感染的急性淋巴细胞白血病患者的临床试验，发现急性淋巴细胞白血病完全缓解而且*CCR5*敲除供体细胞可在患者体内持续19个月以上而无基因编辑相关的不良事件（Xu et al.，2019）。北京大学肿瘤医院朱军团队与南加州大学 Si-Yi Chen 团队合作在 *Nature Medicine* 上发表论文，报道了他们开发的全新 CAR-T 疗法，可以在不损害疗效的基础上将毒副作用降到极低（Ying et al.，2019）。清华大学医学院基础医学院林欣团队将 T 细胞受体与抗体复合物结构结合开发的 STAR-T 细胞疗法，较传统产品治疗效果更优、副作用更小且易于开发设计多靶点，此外还有更长的 T 细胞体内存活时间、低 T 细胞耗竭等优势，目前正在临床试验阶段。

二、高通量筛选寻找 T 细胞改造靶点

尽管，CAR-T 等免疫细胞疗法已经取得了一系列临床成功，但是大多数实体瘤患者仍不能从中获益（Sharma et al.，2015）。免疫细胞疗法在乳腺癌和结肠直肠癌等实体瘤治疗中仍面临肿瘤浸润不足、持久性缺乏和杀伤能力差等诸多挑战，尚未取得标志性进展。如何进一步优化免疫细胞的功能，进而制备出更高效的免疫细胞疗法使更多的患者获益是目前研究的焦点和热点之一。大规模筛选为系统性地寻找可用于细胞功能优化的靶基因或基因线路提供了一种有效手段。早期的大规模筛选主要基于 RNA 干扰技术。例如，哈佛大学丹娜法伯癌症研究所的 Kai W. Wucherpfennig 团队利用 RNAi 文库在小鼠体内对 $CD8^+$T 细胞进行大规模基因沉默筛选，发现 *Ppp2r2d* 等基因的沉默能够增强 T 细胞肿瘤浸润和增殖（Zhou et al.，2014）。然而，RNAi 技术存在沉默不充分和严重的脱靶效应等缺陷。CRISPR 技术的兴起开启了功能遗传学研究的新纪元，在筛选基因靶标和细胞基因工程改造等方面显示出巨大的潜力（Doench，2018；Doudna and Charpentier，2014）。例如，利用 CRISPR 技术在 T 细胞中敲除免疫检查点分子（Ren et al.，2017；Schumann et al.，2015）或导入重组的基因组序列（Roth et al.，2018），显著提高了 T 细胞治疗的效果，在肿瘤患者中显示出持久的疗效和安全性（Lu et al.，2020；Stadtmauer et al.，2020）。CRISPR 高通量筛选为在肿瘤或免疫细胞中系统性地寻找免疫治疗新靶标提供了有效手段。利用 CRISPR 高通量筛选技术对肿瘤细胞或 $CD8^+$T 细胞进行的全基因组敲除筛选（Manguso et al.，2017；Pan et al.，2018；Patel et al.，2017；Wang et al.，2020），鉴定出了一系列与肿瘤免疫治疗效果相关的新靶标或信号通路，特别是对 T 细胞耗竭和杀伤等免疫功能具有关键调节作

用的分子（Dong et al.，2019；Shifrut et al.，2018；Wei et al.，2019；Ye et al.，2019）。加州大学旧金山分校的 Alexander Marson 团队进一步根据肿瘤微环境特征和 T 细胞生物学，提出一种新型的嵌合基因筛选系统，可以将某些免疫抑制性或凋亡信号转化为 T 细胞的刺激信号，在筛选 T 细胞改造策略中具有巨大的潜力（Roth et al.，2020）。

$CD4^+$T 细胞在调节自身免疫稳态和肿瘤免疫中的重要性被越来越多地认识到。英国剑桥大学 Sarah A. Teichmann 团队将 CRISPR 筛选技术与 RNA-seq，ATAC-seq 及 ChIP-seq 联合使用，系统剖析了控制 $CD4^+$T 细胞激活和分化的调控网络，特别是控制 $CD4^+$ 辅助型 T 细胞 2（Th2）分化和激活的基因图谱（Henriksson et al.，2019）。调节性 T 细胞（Treg）是非常重要的一类 $CD4^+$ 辅助性 T 细胞，在维持炎症和肿瘤免疫的稳态中具有重要的作用。系统性地了解 Treg 细胞功能调控网络对于发展 Treg 疗法治疗自身免疫性疾病和癌症至关重要。Alexander Marson 和 Deyu Fang 等多个团队利用 CRISPR 高通量筛选技术对调节 *Foxp3* 表达的转录调节因子和基因调控网络进行解剖，揭示了许多此前未知的对 Treg 功能具有调节作用的基因（Cortez et al.，2020；Loo et al.，2020；Schumann et al.，2020）。

我国科学家在利用高通量筛选技术寻找 T 细胞改造靶点方面的工作也卓有成效。西湖大学谢琦团队、美国希望之城国家医疗中心 Christine Brown 团队和加州大学圣迭戈分校 Jeremy Rich 团队利用全基因组水平的 CRISPR 筛选对 CAR-T 细胞和胶质瘤干细胞进行高通量筛选，发现了 TLE4、IKZF2 等多个能显著提高 CAR-T 细胞抗肿瘤效力的新靶点（Wang et al.，2020）。上海科技大学王皞鹏、加州大学旧金山分校 Arthur Weiss 和中山大学魏来通过 CRISPR 全基因组遗传筛选系统性地研究了 T 细胞激活的分子机制，绘制了人类 T 细胞功能的调控图谱，并指出 FAM49B 是一个能增强 T 细胞抗癌能力的靶点（Shang et al.，2018）。但目前大部分研究仍集中在 $CD8^+$T 细胞的筛选和改造方面。$CD4^+$T 细胞作为辅助类 T 细胞在持久性抗肿瘤免疫中具有核心地位。但是 $CD4^+$T 细胞由于存在 Th1、Th2、Th17、Treg 等诸多亚型和可塑性，增加了大规模筛选和功能研究的难度（Zhu et al.，2010）。解析控制 $CD4^+$T 细胞分化尤其是维持其抗肿瘤免疫的关键基因，制备基因工程改造的 $CD4^+$T 细胞也许是解决 T 细胞治疗持久性的关键。总之，以上研究表明，大规模筛选发现的靶基因能够让抗肿瘤 T 细胞疗法更强大更持久，为发展下一代更有效的基因工程 T 细胞提供了依据（Guo et al.，2020）。通过修改靶基因来微调 T 细胞的免疫抑制性或促炎性功能，是开发可用于癌症和自身免疫的新型细胞疗法的关键。

三、天然免疫细胞改造和筛选

为了开发可以使更多的人受益的免疫疗法，研究人员一直在寻找更多的免疫细胞类型来对抗癌症。天然免疫细胞作为免疫系统的主要参与者，在多种疾病的形成和控制中发挥着关键作用。自然杀伤细胞（NK 细胞）是一类关键的先天免疫细胞，也是感染性疾病病原和癌细胞的主要杀手之一。NK 细胞作为改造的效应细胞具有很多优良特性。例如，许多临床试验表明异体 NK 细胞治疗不会引起 GvHD 反应，而且 NK 细胞不分泌诱导细胞因子释放综合征（cytokine release syndrome，CRS）的 IL6 等炎症因子。在 NK 细胞中导入 CAR 制备的 CAR-NK 细胞，除了具有 CAR 介导的靶向杀伤，还能发挥 NK 细胞本身抗肿瘤的特性，识别并杀伤 CAR 靶标下调或缺失的肿瘤细胞，提高免疫治疗效果。

此外，髓样细胞尤其是树突状细胞和巨噬细胞是免疫系统的关键参与者，在维持免疫监视、组织稳态中发挥着重要的调节作用（Lapenna et al.，2018；Worbs et al.，2017）。树突状细胞是专职的抗原呈递细胞，通过将病原或肿瘤抗原呈递给 T 细胞来激活抗原特异性的获得性免疫。在脂多糖刺激下对树突状细胞进行的 CRISPR 全基因组敲除筛选，系统性地分析了病原感染过程中 TLR4 激活通路，发现了一些此前未知的调节因子（Parnas et al.，2015）。此外，针对树突状细胞交叉呈递的功能性 CRISPR 筛选发现 WDFY4 蛋白是树突状细胞交叉表达相关抗原所必需的，在抗病毒和抗肿瘤免疫中发挥关键作用（Theisen et al.，2018）。此外，巨噬细胞是免疫反应不可或缺的一部分，对免疫系统调控有至关重要的作用。巨噬细胞不仅是专业的抗原呈递细胞，也通过噬细胞化和清除细胞碎片积极参与免疫反应，因此也被作为免疫治疗靶标而受到广泛关注（de Nardo and Ruffell，2019）。宾夕法尼亚大学的 Michael Klichinsky 与 Saar Gill 团队将嵌合抗原受体（CAR）转入巨噬细胞制备出 CAR-Ms。CAR-Ms 在体内外实验中显示出抗原特异性吞噬作用和肿瘤清除并可诱导促炎性肿瘤微环境，有效地促进抗肿瘤反应（Klichinsky et al.，2020）。随后，利用对 $CD14^+$ 单核细胞的改造和筛选系统（Hiatt et al.，2021），可以系统地分析多种基因与巨噬细胞或树突状细胞分化以及激活之间的关系。先天免疫细胞尤其髓系细胞作为炎性微环境主要组成部分，对免疫治疗效果具有关键的调节作用。针对各种先天免疫细胞的功能进行靶向改造，是开辟新型免疫细胞疗法的新工具和新方向。

四、iPSC 来源免疫细胞改造

由人体外周血单核细胞经过重编程产生的 iPSC 具有分化成为多种体细胞的潜能。由于其具有易于获得、容易体外扩增，以及便于基因编辑等优点，iPSC 受到广泛的关注，iPSC 分化制备免疫细胞已经成为其重要应用之一。纪念斯隆－凯特琳癌症中心的 Michel Sadelain 团队利用来自患者的干细胞 iPSC 重编程生成了一些 T 细胞，将 iPSC 与 CAR 技术集合，生成了大量能够寻找和破坏肿瘤的 CAR-T 细胞（Themeli et al.，2013）。由 iPSC 衍生制备的 iNK 细胞识别血液系统肿瘤和实体瘤后不仅能够产生炎症因子和强烈的细胞毒作用杀伤肿瘤，还会招募 T 细胞促进抗肿瘤免疫。iNK 细胞源于可再生的 iPSC，且能够一次性生产出大量细胞，代表了一种用于免疫疗法的可“现货”免疫细胞疗法（Cichocki et al.，2020）。在 NK 细胞中敲除 *CISH* 基因，能够消除 IL-15 信号传导通路的负向调节作用，增强 NK 细胞的活化和功能，提高其能量利用率，改善体内治疗效果（Zhu et al.，2020）。加州大学圣地亚哥分校 Dan Kaufman 团队进一步将 CAR 分子导入人 iPSC 后，将它们诱导分化为 CAR-NK 细胞，在人卵巢癌模型中显示出良好的治疗效果，且比 CAR-T 疗法更加安全（Li et al.，2018）。

我国科学家在利用 iPSC 制备治疗性免疫细胞方面也取得了一些进展。浙江大学医学院张进团队利用 iPSC 诱导获得的巨噬细胞（iMac）来制备 CAR-iMac，不仅具有很高的产率和纯度，还具有巨噬细胞的基因表达谱及成熟巨噬细胞的吞噬、极化等功能（Zhang et al.，2020）。CAR-iMac 细胞展现出抗原依赖性的吞噬和杀伤肿瘤细胞的功能，以及抗原依赖性的分泌促炎、抑肿瘤细胞因子和向 M1 型巨噬细胞极化的功能，在小鼠的血液肿瘤和实体瘤中展现出良好的肿瘤杀伤功能。

五、通用型免疫细胞

通用型免疫细胞疗法是指能够实现同种异体、可“现货”、可规模生产的免疫细胞疗法，旨在突破传统模式中昂贵且耗时的自体免疫细胞分离、改造和扩增等方面的限制。目前处于研发或临床前阶段的通用型免疫细胞包括通用型 CAR-T、TCR-T 及 CAR-NK 疗法。宾夕法尼亚大学 Daniel J. Powell Jr 团队基于生物素 - 亲和素开发的 BBIR CAR 系统，主要由 T 细胞表面的亲和素及标记抗体的生物素组成。该系统通过亲和素和生物素特异性结合实现 T 细胞的肿瘤抗原靶向激活，而且添加相应的生物素抗体能够让 T 细胞识别多种肿瘤相关抗原，说明 BBIR 系统可拓展常规 CAR-T 细胞疗法。随后 Wilson Wong 与 Jim Collins 团队进一步拓展开发了一种通用、可分离、可编程式的 SUPER CAR 系统，由带有亮氨酸适配器的 T 细胞通用受体（zipCAR）以及带有亮氨酸适配器的靶向肿瘤抗原的 scFV 配体（zipFV）组成。该 SUPER CAR 系统能够感知并逻辑响应多种抗原信号，能够对 T 细胞激活反应的强度进行调节从而减轻副作用，能够调控不同类型免疫细胞的信号通路以实现对不同类型 T 细胞的控制。随着基因编辑技术的发展，基于基因编辑的通用型 T 细胞成为目前研究的热点。例如，法国 Cellectis 公司率先利用 Talent 基因编辑技术开发了同种异体 CAR-T，用于治疗急性淋巴性白血病（ALL）、急性髓系白血病（acute myeloid leukemia，AML）、多发性骨髓瘤（multiple myeloma，MM）等。2018 年美国 Fate Therapeutics 公司公布了一款来自 iPSC 的“可现货”、TCR-less 的 CD19 CAR-T 细胞产品，临床前研究结果显示它具有良好的特异性、功能性及安全性。该公司研发的 iPSC 来源的通用型 CAR-NK 目前也处于临床试验阶段。Emmanuelle Charpentier 创办的 CRISPR Therapeutics 公司也专注于开发通用型 CAR-T 疗法，他们通过 CRISPR 敲除 T 细胞受体来制备同种异体产品，消除 MHC Ⅰ类抗原以改善 CAR-T 细胞的耐久性。最新研究数据证明，多个基因编辑的同种异体 CAR-T 细胞具有高编辑率、表达一致性、强选择性及强杀伤力等特点。美国 Poseida Therapeutics 公司利用其专有的非病毒 piggyBac（PB）DNA 转座技术，与高保真 CRISPR 基因编辑系统结合在记忆性干细胞样 T 细胞亚群中产生 CAR-T 细胞，开发了一种可“现货”的同种异体 CAR-T 细胞产品。该产品克服了自体 CAR-T 细胞面临的诸如制造时间、重现性、一致性和成本等重大挑战，以及由病毒载体产生的 CAR-T 细胞倾向于由分化的 T 细胞亚群组成及较差的体内耐受性相关等这些限制。此外，比利时 Celyad Oncology SA 公司利用 TCR 抑制分子降低 GvHD，开发了全球第一个非基因编辑的同种异体 CAR-T 产品，该产品于 2018 年 7 月被美国 FDA 批准进行研究性新药临床试验。

在通用型 TCR-T 细胞疗法方面，美国 TCR2 Therapeutics 公司专注开发新型通用型 T 细胞疗法，该公司利用其独有的 TRuC 平台对天然 TCR 进行改造，TRuC-T 细胞无需 HLA 匹配，就能让 T 细胞识别肿瘤细胞特有抗原，从而激活 T 细胞并杀死癌细胞。荷兰 Gadeta BV 公司与美国 Gilead Science/Kite Pharma 公司合作开发基于 $\gamma\delta$ T 细胞受体的新型通用型同种异体 T 细胞疗法，结合了 $\alpha\beta$ T 细胞的高增殖和记忆能力及 $\gamma\delta$ 受体的抗癌特异性和活性，目前正在多发性骨髓瘤治疗的 I 期临床试验。挪威 Zelluna Immunotherapy 公司，开发一种基于 NK 细胞的 TCR-NK 细胞疗法，认为这些细胞非常适合应用于通用型疗法，因为它们攻击患者健康组织的风险比 T 细胞低，而且杀死癌细胞的速度更快。

我国在通用型免疫细胞疗法开发方面也取得了一系列成果，临床研究也正在迎头赶上。亘喜生物科技（上海）有限公司在 2020 年公布了利用其独有的 TruUCAR 专利技术开发的通用型 CAR-T 疗法治疗复发性或难治性急性 T 淋巴细胞白血病的 I 期临床试验结果，显示出良好的安全性和治疗效果。此外，恒瑞源正（上海）生物科技有限公司、北京可瑞生物科技有限公司、优瑞科生物技术公司、深圳因诺免疫有限公司等多家公司也正在致力于通用型 CAR-T 或 TCR-T 疗法的开发。

六、智能细胞疗法

通过合成生物学改造赋予某些功能性底盘细胞尤其是免疫细胞新的功能也是目前研究的热点。加州大学旧金山分校 Wendell Lim 团队构建出能够精确靶向癌症病灶，并执行一系列可定制反应的治疗性免疫细胞，他们利用 synNotch 重编程 T 细胞使其能通过靶向运载药物、调控 T 细胞状态等多种方式杀死癌细胞（Roybal et al.，2016）。随后，他们和普林斯顿大学 Olga G.Troyanskaya 团队合作利用机器学习分析了癌症和正常细胞中数千种蛋白质的大量数据库，通过对数百万种可能的蛋白质组合梳理，开发了一种既能杀死癌细胞又能使正常组织毫发无损的智能细胞疗法（Dannenfelser et al.，2020）。他们同期还在 *Science* 上报道了使用 synNotch 模块将多个抗原受体串在一起，设计出可以同时识别癌细胞表面或内部表达的三个抗原的智能 T 细胞（Williams et al.，2020）。随后，他们设计了两步正反馈基因环路触发 CAR 的机制来改造人类 T 细胞，允许 T 细胞根据 S 型抗原密度阈值来区分目标，即只杀死大量表达癌症抗原的癌细胞（Hernandez-Lopez et al.，2021）。以上研究表明，通过免疫细胞的工程设计和改造，可以实现细胞治疗靶向癌细胞的精确控制。基于 CRISPR 干扰、CRISPR 激活或其他基因工程技术来研发能够在患者体内自发做出各种抗癌决策的编程控制的治疗性智能细胞仍有广阔的发展空间。

我国科学家在通过合成生物学改造制备治疗性“智能细胞”方面的工作也取得了一系列进展。华东师范大学叶海峰团队结合光遗传学和细胞生物学开发的用于糖尿病的远程诊疗系统，将无线血糖仪所测出的血糖数值转化成远红光信号，通过调控人工设计的远红光响应启动子激活胰岛素基因的转录来缓解 1 型糖尿病（Shao et al.，2017）。此外，他们还设计开发了一种治疗基因回路，使齐墩果酸（oleanolic acid）和胰腺素类肽 1 两种药物能够协同作用，改善肝脏和胰腺功能，协同促进了它们的靶向疗效（Xue et al.，2017）。

七、干细胞疗法在神经系统疾病中的应用

迄今为止，世界卫生组织国际临床试验注册平台已注册了超过 3000 项涉及使用成体干细胞的临床试验，其中Ⅲ期临床试验及以上 262 个。适应证涉及神经性疾病、心血管疾病、糖尿病、血液病和癌症等多个不同领域。截至目前，CDE 已受理 14 项干细胞治疗临床试验的申请，10 项已通过临床试验默示许可。全球已上市 16 种干细胞治疗产品。

神经系统疾病包括一系列不同的中枢和外周神经系统紊乱，它们的治疗选择范围有限，与其他治疗领域相比，改进治疗的药物批准率仍然较低。干细胞疗法给神经性疾病治疗提供了希望。

干细胞在神经性疾病治疗上有多种治疗潜力，可以通过以下多种途径进行神经损伤修复。①移植的干细胞可以迁移到损伤的神经部位，通过细胞替代作用更换机体已经死亡或受损伤的神经细胞，修复受损神经网络。②移植的干细胞可以分泌大量神经细胞活性生长因子和营养因子，激活神经细胞，促进新细胞的再生和重建；分泌血管生成因子，促进病变部位血管生成。③可进行免疫调节，干细胞可以调节免疫细胞到达病理部位的数量并且分泌不同水平的细胞因子相互影响，起到抗炎的保护作用。

目前在神经性疾病中，干细胞疗法主要用于帕金森病（PD）、肌萎缩侧索硬化（amyotrophic lateral sclerosis，ALS）和脑卒中等的治疗。目前认为iPSC为PD最有前景的干细胞选择。在2018年8月，日本京都大学团队被批准了首个使用iPSC治疗帕金森病的临床试验，该试验招募了7名患有中度帕金森病的患者，使用同种异体的iPSC产生多巴胺能祖细胞，然后通过特殊装置将其手术移植到患者的大脑中，同时给予免疫抑制剂药物以避免免疫排斥反应，试验初步验证了该治疗的安全性。中国首款iPSC来源细胞疗法在2022年4月获批临床试验："异体内皮祖细胞（EPCs）注射液"，拟用于治疗大动脉粥样硬化型急性缺血性卒中。我国开展的其他试验大多使用脐带间充质干细胞和自体骨髓干细胞。ALS是一种进行性神经系统疾病，可攻击大脑皮质、脑干和脊髓神经细胞。治疗ALS的第一个尝试是通过在小鼠模型中移植间充质干细胞（MSC），该试验结果证明，干细胞治疗ALS是有希望的，将干细胞注射到小鼠脊髓中可以延迟ALS的发生，提高生存率。Neuronata-R于2014年7月获得韩国食品与药品安全部批准上市，并于2015年2月投放市场，但至今未开展Ⅲ期临床试验。目前在进行的临床试验几乎都为基于安全性的研究，尚未有证明有效性的临床最终试验结果。值得注意的一点是，尽管临床前研究报告说，来自未患病个体的细胞要优于ALS患者的细胞，但大多数临床试验都采用了自体移植，这可能解释了ALS缺乏优秀有效性数据的问题。卒中患者的包括中枢神经元、星形胶质细胞、少突胶质细胞和其他细胞在内的多种细胞在卒中发作后突然死亡，并且随后可发生继发性损伤导致更多的细胞变性。因此卒中的恢复与PD的不同之处在于需要再生多种类型的细胞，细胞间的相互作用被视作再生的关键，在现有情况下高质量且持续的细胞再生非常困难，因此阻止神经元－胶质细胞－血管单位死亡的目标对于设计干细胞治疗的再生医学至关重要。MSC已经显示出其在再生神经元－胶质细胞－血管单位方面再生的希望。MSC治疗卒中的优点是：①可从骨髓中快速分离；②在培养物中高效扩增；③在培养物中易于维护；④即使在卒中的急性期也适合自体移植；⑤神经营养作用。MSC已成为许多卒中临床研究的重点，在血液病安全性方面的过往试验以及大量临床前卒中研究证明了其安全性和有效性。2005年，全球首次对缺血性卒中患者进行了静脉内自体MSC的Ⅰ期研究。2014年，发表了第一项对缺血性卒中患者进行异体MSC静脉内给药的Ⅱ期临床试验情况，在这项研究中，报告了缺血性卒中患者静脉注射异体MSC的治疗效果。在一项使用动脉内同种异体MSC给药的研究中，40%接受干细胞治疗的卒中患者在发病后3～7天内表现出良好的临床效果。目前日本也正在进行一些干细胞治疗卒中的临床试验。

第四节　我国发展战略和重点研究方向

一、发展总体思路与发展目标

1. 发展总体思路

建立并发展寻找免疫细胞等功能性细胞中关键调节因子的新技术和新方法，尤其注重拓展学科交叉和多种新技术融合的应用；在免疫细胞的改造方面既要注重具有研究基础和理论支撑的研究团队深入挖掘突破瓶颈，也要鼓励运用新方法、新思路、新概念来开展原始创新的团队开展更多探索性的研究。重点关注和培育一批具有临床应用前景和自主知识产权的新型“干细胞或免疫细胞改造技术和产品”推向社会或临床应用。当然同时也应该加强和完善相关技术平台和数据库的建设，鼓励合作和共享。

2. 发展目标

积极探索新型的免疫细胞改造技术，为突破免疫细胞疗法目前所面临的瓶颈及解决相关科学问题做出贡献。在可用于改造的功能性细胞来源、可用于改造的分子靶点筛选以及改造技术方面取得突破性成果，进一步提升我国在细胞治疗和基因治疗领域的国际影响力，在某些研究领域与国际先进水平并行甚至超越。

二、优先发展领域和重要研究方向

1. 高通量筛选可用于免疫细胞功能优化的靶基因

建立和发展新型的能够克服免疫抑制微环境的免疫细胞疗法，重点在于寻找癌症和相关病灶关键的细胞类型，以及对细胞功能状态具有关键调控作用的基因。根据肿瘤微环境中的免疫细胞状态、免疫细胞间通信，建立筛选体系寻找能够帮助 T 细胞克服免疫抑制肿瘤微环境的靶基因。针对 $CD4^+$T 细胞的功能可塑性，探索并建立筛选体系解析控制 $CD4^+$ T 分化和抗肿瘤功能的关键基因，制备基因工程改造的 $CD4^+$ T 细胞也许是解决 T 细胞治疗持久性的关键。

对髓样细胞的改造和调控是未来免疫细胞治疗的重要部分。需要筛选并鉴定影响髓样细胞向树突状细胞分化、成熟和迁移的关键基因，找出对树突状细胞疫苗效力有关键调控作用的因素。巨噬细胞在肿瘤微环境中的高度可塑性是其应用的一个关键问题。我们需要建立合理的筛选体系寻找在巨噬细胞可塑性调节中有关键作用的靶标，进一步推动基因工程改造巨噬细胞疗法的应用。

2. 高通量技术联合应用寻找可用于改善功能细胞疗法的新方向

各种功能性细胞在炎性环境中呈现出多种表型和高度的可塑性。因此，将来的功能性细胞改造需要获取各种细胞更详细的数据，通过大规模测序和筛选深入剖析它们在各种炎性环境中的功能状态以及调节其状态和治疗功能的关键信号通路。通过流式分析、单细胞测序和大数据分析为系统性解析疾病发展和治疗的不同阶段，各种功能性细胞的状态、亚群以及特异性转录图谱提供了有效手段，然后根据细胞不同功能阶段的差异性转录图谱有针对性地设计大规模筛选文库，更加精准地分析对细胞亚群的分化、功能和状态起关键调节作用的基因。深度剖析各种免疫细胞在炎性疾病和癌症进展过程中的可塑性和分化发展轨迹，并找出其可塑性和功能调节中的一系列调控回路，最终通过对相关基因和通路的改造来促进新型细胞疗法的发展。总之，通过单细胞测序、功能基因组学、表观遗传学、高通量筛选和合成生物学等多学科技术的综合运用来实现对各种细胞的调控网络和功能状态进行深度解析，然后根据疾病需求有针对性地设计出功能优化改造模块，实现对免疫细胞各种状态和功能的靶向调控，开发精准免疫细胞疗法。

3. 利用 iPSC 制备新型“免疫细胞疗法”

iPSC 和造血干细胞为体外制备大量免疫细胞提供了良好途径（Themeli et al.，2013）。发展和建立更加安全可靠的通过 iPSC 制备各种功能性细胞尤其是功能性免疫细胞的技术和方法。通过结合 iPSC 分化过程中的细胞谱系示踪和转录组分析，来指导如何更加高效地利用 iPSC 制备可用于免疫细胞疗法的功能性免疫细胞尤其是 T 细胞和 NK 细胞。进一步发展和建立在 iPSC 制备功能性免疫细胞路线上的 CRISPR 筛选体系，研究对 iPSC 向功能性免疫细胞分化及相应功能调控中有核心作用的靶基因。研究如何通过靶基因通路的改造实现 iPSC 产生的免疫细胞功能的精确修饰和调控，为开发个性化免疫细胞疗法提供了新思路。鉴于免疫细胞本身的特性也会对疗效有影响。例如，T 细胞的分化过程为从初始 T 细胞（naive T cell, Tn）逐步分化成 T 记忆干细胞（memory stem T cell，Tscm）和中央记忆 T 细胞（central memory T cell，Tcm），然后分化成生存时间较短的效应记忆 T 细胞（effector memory T cell，Tem）和效应 T 细胞（effector T cell，Te）。这就意味着选择分化程度较低的 Tn、Tscm 或 Tcm 作为基因工程的原材料可能产生疗效更高的治疗性 T 细胞。利用分化诱导体系和基因编辑，诱导 iPSC 获取干性更强的记忆性 T 细胞，来制备更强疗效的免疫细胞疗法。

4. 开发基于“智能细胞”的新型免疫细胞疗法

利用高通量筛选技术和合成生物学技术，探索如何通过人工改造，在功能细胞尤其是治疗性免疫细胞中导入基因线路来实现治疗功能的时空调控；开发出类似于分子计算机的功能性细胞，智能地感知微环境各种信号，如抗原表达、趋化因子浓度等，然后整合这些信息来做出决定触发不同信号环路，或快速合成杀伤性效应蛋白以及相关细胞因子，或分泌趋化信号募集更多的“智能细胞”到微环境，达到智能反应和杀伤。结合单细胞测序、表观遗传学和系统生物学手段研究炎症环境或肿瘤微环境中多种免疫细胞的功能状态、转录组、表观遗传方面的变化规律，结合高通量筛选找出各种功能状态的“开关分子”，设计开发出可以根据环境或外部信号智能做出决定，重塑炎症微环境的智能细胞疗法。例如，将能够重塑自身和其他免疫细胞状态的基因线路置于“智能细胞”内部，收集到肿瘤微环境中的免疫抑制性细胞募集信号后可以分泌新型调控蛋白，抑制甚至逆转某些细胞抑制性功能。

5. 增强干细胞疗法的有效性和安全性

干细胞治疗被认为是治疗神经性疾病的最有希望的治疗方法，但是必须要优化干细胞的治疗潜力，包括增强干细胞的分化、迁移或神经网络的形成，这些都是中枢神经系统再生的关键指标。电刺激成为一种更重要方式，试验证明电刺激可触发脑卒中大鼠脑内移植的间充质基质细胞通过趋化因子 SDF-1α 进行信号传递，增强海马体中的神经细胞再生能力，此外，电刺激可改善突触的形成，从而促进干细胞疗法的治疗效果，并提示组合疗法的潜力。

自体患者来源的细胞可以通过伦理审查；此外，对成体细胞进行基因改造，进行干细胞扩增有可能会导致细胞繁殖不受控制，潜在风险的可控性成为重要挑战。但是不对患者来源的细胞进行编辑，又很可能会限制其生存能力和治疗潜力。使用健康的供体进行同种异体移植可以避免上述问题，但是这些细胞具有免疫排斥的风险。因此干细胞治疗需要消除不确定的细胞分化及繁殖因素，除去有致瘤风险的细胞，并纯化分化的细胞。

作为自体干细胞的 iPSC 已成为一种有吸引力的移植细胞来源。将 iPSC 过渡到临床应用，减少将这些细胞分化为所需的细胞的时间，以及建立、维护和使用用于治疗目的的 iPSC 培养相关的成本是关键。当使用 iPSC 分化的神经元、少突胶质细胞或星形胶质细胞作为中枢神经系统疾病的细胞产物时，增殖的差异、分化的非同质性和不同的遗传背景都是技术方面的障碍。

总之，各种高通量测序技术、表观遗传组学、功能基因组学和高通量筛选技术，加速了人们对于各种功能性细胞内部基因调控环路和关键调节基因的理解和认识，而合成生物学的发展使人们可以根据个体需求，非常精准地进一步优化甚至设计出智能的细胞疗法，来靶向到病灶部位精确治疗各种疾病。目前细胞疗法尤其是免疫细胞疗法飞速发展，在各种疾病的治疗上显示出广阔的应用前景，然而细胞疗法在靶向性、持久性、精确打击性以及通用性等方面仍面临各种挑战。接下来，需要根据疾病的治疗应答机制和临床需求，更加有针对性地提出科学问题并设计更加精确的筛选体系或靶向改造环路，开发出切实可用的细胞治疗产品。此外，还需要建立从体外到类器官再到体内，一系列能够高度还原人类疾病部位的疾病模型评价体系和筛选体系，为系统性寻找、改造和评价细胞治疗效果提供可靠的临床前评价体系。

本章参考文献

Cichocki F, Bjordahl R, Gaidarova S, et al. 2020. iPSC-derived NK cells maintain high cytotoxicity and enhance in vivo tumor control in concert with T cells and anti-PD-1 therapy. Sci Transl Med, 12（568）: eaaz5618.

Cortez J T, Montauti E, Shifrut E, et al. 2020. CRISPR screen in regulatory T cells reveals modulators of Foxp3. Nature, 582（7812）: 416-420.

Dannenfelser R, Allen G M, VanderSluis B, et al. 2020. discriminatory power of combinatorial antigen recognition in cancer T Cell Therapies. Cell Syst, 11（3）: 215-228 e215.

de Nardo D G, Ruffell B. 2019. Macrophages as regulators of tumour immunity and immunotherapy. Nat Rev Immunol, 19（6）: 369-382.

Doench J G. 2018. Am I ready for CRISPR? A user's guide to genetic screens. Nat Rev Genet, 19（2）: 67-80.

Dong M B, Wang G, Chow R D, et al. 2019. Systematic immunotherapy target discovery using genome-scale *in vivo* CRISPR screens in CD8 T cells. Cell, 178（5）: 1189-1204 e1123.

Doudna J A, Charpentier E. 2014. Genome editing. The new frontier of genome engineering with

CRISPR-Cas9. Science, 346（6213）: 1258096.

Guo J, Xu C. 2020. Screening for the next-generation T Cell Therapies. Cancer Cell, 37（5）: 627-629.

Henriksson J, Chen X, Gomes T, et al. 2019. Genome-wide CRISPR screens in T helper cells reveal pervasive crosstalk between activation and differentiation. Cell, 176（4）: 882-896 e818.

Hernandez-Lopez R A, Yu W, Cabral K A, et al. 2021. T cell circuits that sense antigen density with an ultrasensitive threshold. Science, 371（6534）: 1166-1171.

Hiatt J, Cavero D A, McGregor M J, et al. 2020. Efficient generation of isogenic primary human myeloid cells using CRISPR-Cas9 ribonucleoproteins. BioRxiv, 35（6）:109105.

June C H, O'Connor R S, Kawalekar O U, et al. 2018. CAR T cell immunotherapy for human cancer. Science, 359（6382）: 1361-1365.

Klichinsky M, Ruella M, Shestova O, et al. 2020. Human chimeric antigen receptor macrophages for cancer immunotherapy. Nat Biotechnol, 38（8）: 947-953.

Lapenna A, De Palma M, Lewis C E. 2018. Perivascular macrophages in health and disease. Nat Rev Immunol, 18（11）: 689-702.

Li Y, Hermanson D L, Moriarity B S, et al. 2018. Human iPSC-derived natural killer cells engineered with chimeric antigen receptors enhance anti-tumor activity. Cell Stem Cell, 23（2）: 181-192 e185.

Loo C S, Gatchalian J, Liang Y, et al. 2020. A Genome-wide CRISPR screen reveals a role for the non-canonical nucleosome-remodeling BAF complex in Foxp3 expression and regulatory T cell function. Immunity, 53（1）: 143-157 e148.

Lu Y, Xue J, Deng T, et al. 2020. Safety and feasibility of CRISPR-edited T cells in patients with refractory non-small-cell lung cancer. Nat Med, 26（5）: 732-740.

Manguso R T, Pope H W, Zimmer M D, et al. 2017. *In vivo* CRISPR screening identifies Ptpn2 as a cancer immunotherapy target. Nature, 547（7664）: 413-418.

Pan D, Kobayashi A, Jiang P, et al. 2018. A major chromatin regulator determines resistance of tumor cells to T cell-mediated killing. Science, 359（6377）: 770-775.

Parnas O, Jovanovic M, Eisenhaure T M, et al. 2015. A genome-wide CRISPR screen in primary immune cells to dissect regulatory networks. Cell, 162（3）: 675-686.

Patel S J, Sanjana N E, Kishton R J, et al. 2017. Identification of essential genes for cancer immunotherapy. Nature, 548（7669）: 537-542.

Ren J, Liu X, Fang C, et al. 2017. Multiplex genome editing to generate universal CAR T cells resistant to PD1 inhibition. Clin Cancer Res, 23（9）: 2255-2266.

Roth T L, Li P J, Blaeschke F, et al. 2020. Pooled knockin targeting for genome engineering of cellular immunotherapies. Cell, 181（3）: 728-744 e721.

Roth T L, Puig-Saus C, Yu R, et al. 2018. Reprogramming human T cell function and specificity with non-viral genome targeting. Nature, 559（7714）: 405-409.

Roybal K T, Williams J Z, Morsut L, et al. 2016. Engineering T cells with customized therapeutic response programs using synthetic Notch receptors. Cell, 167（2）: 419-432 e416.

Schumann K, Lin S, Boyer E, et al. 2015. Generation of knock-in primary human T cells using Cas9 ribonucleoproteins. Proc Natl Acad Sci U S A, 112（33）: 10437-10442.

Schumann K, Raju S S, Lauber M, et al. 2020. Functional CRISPR dissection of gene networks controlling human regulatory T cell identity. Nat Immunol, 21（11）: 1456-1466.

Shang W, Jiang Y, Boettcher M, et al. 2018. Genome-wide CRISPR screen identifies FAM49B as a key regulator of actin dynamics and T cell activation. Proc Natl Acad Sci U S A, 115（17）: E4051-E4060.

Shao J, Xue S, Yu G, et al. 2017. Smartphone-controlled optogenetically engineered cells enable semiautomatic glucose homeostasis in diabetic mice. Sci Transl Med, 9（387）:eaal2298.

Sharma P, Allison J P. 2015. The future of immune checkpoint therapy. Science, 348（6230）: 56-61.

Shifrut E, Carnevale J, Tobin V, et al. 2018. Genome-wide CRISPR screens in primary human T cells reveal key regulators of immune function. Cell, 175（7）: 1958-1971 e1915.

Simeonov D R, Marson A. 2019. CRISPR-based tools in immunity. Annu Rev Immunol, 37: 571-597.

Stadtmauer E A, Fraietta J A, Davis M M, et al. 2020. CRISPR-engineered T cells in patients with refractory cancer. Science, 367（6481）: eaba7365

Theisen D J, Davidson J T t, Briseno C G, et al. 2018. WDFY4 is required for cross-presentation in response to viral and tumor antigens. Science, 362（6415）: 694-699.

Themeli M, Kloss C C, Ciriello G, et al. 2013. Generation of tumor-targeted human T lymphocytes from induced pluripotent stem cells for cancer therapy. Nat Biotechnol, 31（10）: 928-933.

Wang D, Prager B C, Gimple R C, et al. 2020. CRISPR screening of CAR T cells and cancer stem cells reveals critical dependencies for cell-based therapies. Cancer Discov, 11（5）: 1192-

1211.

Wang G, Chow R D, Zhu L, et al. 2020. CRISPR-GEMM pooled mutagenic screening identifies KMT2D as a major modulator of immune checkpoint blockade. Cancer Discov, 10（12）: 1912-1933.

Wei J, Long L, Zheng W, et al. 2019. Targeting REGNASE-1 programs long-lived effector T cells for cancer therapy. Nature, 576（7787）: 471-476.

Williams J Z, Allen G M, Shah D, et al. 2020. Precise T cell recognition programs designed by transcriptionally linking multiple receptors. Science, 370（6520）: 1099-1104.

Wolchok J D, Chiarion-Sileni V, Gonzalez R, et al. 2017. Overall survival with combined nivolumab and ipilimumab in advanced melanoma. N Engl J Med, 377（14）: 1345-1356.

Worbs T, Hammerschmidt S I, Forster R. 2017. Dendritic cell migration in health and disease. Nat Rev Immunol, 17（1）: 30-48.

Wu W, Zhou Q, Masubuchi T, et al. 2020. Multiple signaling roles of CD3epsilon and its application in CAR-T cell therapy. Cell, 182（4）: 855-871 e823.

Xu L, Wang J, Liu Y, et al. 2019. CRISPR-edited stem cells in a patient with HIV and acute lymphocytic leukemia. N Engl J Med, 381（13）: 1240-1247.

Xue S, Yin J, Shao J, et al. 2017. A synthetic-biology-inspired therapeutic strategy for targeting and treating hepatogenous diabetes. Mol Ther, 25（2）: 443-455.

Ye L, Park J J, Dong M B, et al. 2019. *In vivo* CRISPR screening in CD8 T cells with AAV-Sleeping Beauty hybrid vectors identifies membrane targets for improving immunotherapy for glioblastoma. Nat Biotechnol, 37（11）: 1302-1313.

Ying Z, Huang X F, Xiang X, et al. 2019. A safe and potent anti-CD19 CAR T cell therapy. Nat Med, 25（6）: 947-953.

Zhang L, Tian L, Dai X, et al. 2020. Pluripotent stem cell-derived CAR-macrophage cells with antigen-dependent anti-cancer cell functions. J Hematol Oncol, 13（1）: 153.

Zhou P, Shaffer D R, Alvarez A D A, et al. 2014. *In vivo* discovery of immunotherapy targets in the tumour microenvironment. Nature, 506（7486）: 52-57.

Zhu H, Blum R H, Bernareggi D, et al. 2020. Metabolic reprograming via deletion of CISH in human iPSC-derived NK cells promotes *in vivo* persistence and enhances anti-tumor activity. Cell Stem Cell, 27（2）: 224-237 e226.

Zhu J F, Yamane H, Paul W E. 2010. Differentiation of effector CD4 T cell populations*. Annu Rev Immunol, 28: 445-489.

关键词索引

A

B

C

D

E

F

G

H

J

K

L

M

N

P

Q

R

S

T

W

X

Y

Z

其他